M. A. Krasnosel'skiĭ A. V. Pokrovskiĭ

Systems with Hysteresis

Translated from the Russian by
Marek Niezgódka

With 81 Figures

Springer-Verlag
Berlin Heidelberg New York
London Paris Tokyo

Mark A. Krasnosel'skiĭ
Aleksei V. Pokrovskiĭ
Institute of Control Sciences
ul. Profsojuznaja 65
117806 Moscow GSP-7, USSR

Marek Niezgódka
Systems Research Institute, Polish Academy of Sciences
Newelska 6
01-447 Warsaw, Poland

Title of the Russian original edition:
Sistemy s gisteresisom
Publisher Nauka, Moscow 1983

Mathematics Subject Classification (1980): 47H15, 34A60, 49E10, 49E15

ISBN-13: 978-3-642-64782-6 e-ISBN-13: 978-3-642-61302-9
DOI: 10.1007/978-3-642-61302-9

Library of Congress Cataloging-in-Publication Data
Krasnosel'skiĭ, M. A. (Mark Aleksandrovich), 1920- . [Sistemy s gisterezisom. English]
Systems with hysteresis / Mark A. Krasnosel'skiĭ, Aleksei
V. Pokrovskiĭ ; translated from the Russian by Marek Niezgódka. p. cm.
Translation of: Sistemy s gisterezisom.
Bibliography: p.
Includes index.
ISBN-13: 978-3-642-64782-6
1. Hysteresis. I. Pokrovskiĭ, A. V. (Aleksei Vadimovich) II. Title.
QC754.2.H9K7313 1989 538'.3-dc19 88-39156 CIP

Softcover reprint of the hardcover 1st edition 1989

Printing: Weihert-Druck GmbH, Darmstadt
Binding: Schäffer GmbH, Grünstadt
2141/3140-543210 - Printed on acid-free paper

Preface

Forwards, once more - told me my instinct
And led me wisely, as an old scholar,
Through a virginal, nonpassable reed
Of warmed trees, lilacs and sadness horrors...
B. Pasternak

... The most important is that everything develops systematically. Everything should comprehend a system.
- A systematic system, - said senior scribe Vanek with a sceptic smile.
- Yes... a systematized systematic system...
J. Hašek

Nonlinearities of hysteresis type are common for various branches of science and technology, including physics, mechanics, biology and civil engineering, in particular. Effects like magnetic hysteresis, dielectric hysteresis or plastic hysteresis are all well-known phenomena. This book discloses general methods which are applicable to describing and studying a large class of systems with hysteresis nonlinearities. New mathematical techniques based on resolving the systems into elementary hysteresis carriers - hysterons - have been developed for this purpose. The hysterons are treated as transducers with well-defined state space, input-output relations, and input-state relations.

The way we interpret hysteresis phenomena in this book is similar to the standard approach within continuum mechanics (related to the sixth Hilbert problem), developed by Coleman, Noll, Truesdell et al. Models of hysteresis phenomena, dating back to Maxwell, Boltzmann, Madelung, Prandtl, Masing, Volterra, von Mises, Saint-Venant, Tresca, Ishlinskii and others, are also made use of in this book. Techniques of the linear systems theory, recently developed by Zadeh, Desoer, Arbib, Falb, Kalman and others, play an important role as well.

In the study of any system, two essentially different situations have to be distinguished.

In the first case, at given external actions (inputs) the problem consists in determining the response (output) of a system. For this purpose, an algorithm should be set up which permits a numerical or analytic construction of the input-output and input-state relations (for example, the correspondences between velocity and position in mechanics, external field - magnetic induction in magnetism theory, deformation - stress in plasticity theory, etc.). In this construction, it is often sufficient to admit only inputs with the simplest structure: piecewise linear, piecewise smooth, splines, etc.

The second case arises when the system under consideration is merely an element of a larger structure and therefore it cannot be treated in isolation. Complex systems of such an origin are typical for control engineering, mechanics and physics. If an element with hysteresis is included in a complex system which is subject to external influences (controls) and inevitable noises (non-controllable), then the actual inputs of this element are unknown (and their nature may be rather complicated). Any description of hysteresis effects in such a case must thus allow for inputs of a general form (for example, arbitrary continuous functions) and must be applicable to an analysis of the dynamics of the whole system. For this purpose, it must be possible to consider the appropriate input-output and input-state relations as operators with as many "good" properties as possible on various classes of the inputs. The classic problem of trajectory determination can serve as a trivial example of such a situation: to perform necessary computations, it is sufficient to use integral sums or mechanical quadratures. In order to derive and study equations that describe a certain process, an integral notion is needed together with information concerning some its properties (mean value theorem, integration by parts, linearity, etc.). Another example of the same situation is given by problems where, instead of difference operations, differentiation must be performed.

Throughout the book, the *hysteron* - a deterministic, static transducer with scalar input and scalar output, controllable and stable with respect to small-amplitude noises at the input (vibro-correct) - will be considered as an elementary carrier of hysteresis effects. A special construction based on passing to some limit will be used to describe the dynamics of the hysteron. First, piecewise monotone inputs are assumed and analysis techniques standard for nonlinearities such as those in models of a play (backlash) or stop are used. Then, the construction is extended onto arbitrary continuous inputs (similarly as at passing from finite integral sums to an appropriate integral).

A characterization of the hysteron is given and its properties are studied in Part 1 . Plays and stops, extended to all continuous inputs, are considered as simple examples of hysterons. Part 2 contains formulations and proofs of several identification theorems.

If an elasto-plastic element (fibre) has time-independent mechanical properties, it can often be treated as a hysteron with variable deformation and variable stress as the input and output, respectively. If, due to ageing, temperature changes, etc., the properties are variable in time, it proves convenient to introduce a variable hysteron as a suitable model. A general theory of variable hysterons is developed in Part 3 . That theory is based on an analysis of so-called vibro-correct differential equations.

Hysterons with vector inputs and outputs are considered in Part 4 (for the models due to von Mises, Saint-Venant, Tresca and others).

In Part 5, a study of transducers with discontinuous input-output relations is developed. Static elements described by discontinuous superposition operators (of special importance in the theory of systems with variable structure), as well as ideal and non-ideal relays (on--off elements) are studied there in detail. Special attention is given to the role of small noises at the input; taking them into account simplifies an analysis of the dynamics of transducers.

In Part 6, the notion of hysteron is modified so as to cover phenomena of self-magnetization type. The first of those modifications refers to the classical Madelung's model. The others employ notions of stochastic integrals. All the constructions are based on the use of vibro-correct differential equations. Relationships between solutions of the vibro-correct equations and solutions of some stochastic equations (understood in the Ito or Stratonovich sense) are established.

In Part 7, a study of complex non-deterministic hysteresis nonlinearities is developed. By extending the constructions originally due to Masing, Ishlinskii, Preisach, Giltay and others, some nonlinear systems composed of hysterons and relays are considered.

The techniques developed in this book turn out useful at studying closed-loop systems which contain hysteresis elements. The tools we use include functional analysis and topology methods applied to problems such as stability analysis (including absolute stability questions), an analysis of forced periodic oscillations and self-oscillations, a study of singular perturbations, construction of averaged equations, the selection of special functioning conditions and their study, numerical analysis, etc. Concerning all of the problems mentioned above, numerous results have been obtained by many authors. This, however, would have to be the subject of another book !

My personal interest in hysteresis phenomena arose in connection with the theory of nonlinear oscillations. A small seminar had been organized at the Voronezh University; besides myself, there were also two young mathematicians, B.M. Darinskii and P.P. Zabreiko, as well as three students, I.V. Emelin, E.A. Lifshitz and A.V. Pokrovskii, who took part in the seminar. With extraordinary patience, B.M. Darinskii acquainted us with various hysteresis models in plasticity theory.

More than ten years ago I began working at the Institute of Control Problems (Institute of Automatic Control and Remote Mechanics) in Moscow. It was at this time that I started dealing with various problems of automatic control and general systems theory. It turned out that, despite their importance in many concrete problems, systems with hysteresis represented an area within the general systems theory which had hardly been researched. From then on, hysteresis itself, physical phenomena with hysteresis effects and mathematical techniques applicable to an analysis of systems with hysteresis nonlinearities have remained of interest to me as well as to several students and young colleagues of mine.

General program frames for a study of hysteresis nonlinearities from the viewpoint of the systems theory have been clear to me for a long time. A realization of that program required overcoming diverse difficulties. Unexpected and unusual associations of hysteresis with stochastic processes were discovered, a special class of so-called vibro-correct differential equations were introduced and studied, a number of non-classical geometric problems had to be solved, properties of some new types of discontinuous operators required a study, etc. The first formulations of the fundamental theorems were clumsy and cumbersome; their first proofs took up dozens of pages; numerous heuristically obvious facts, confirmed by numerical experiments, waited long for rigorous mathematical foundations. So far, basic points of the general program have been completed, although many questions still remain open.

The very first results were obtained by the participants of the above-mentioned Voronezh seminar. Further research was, in a diverse form, developed by T.S. Gil'man, N.I. Grachev, A.F. Kleptsyn, V.S. Kozyakin, A.A. Kravchenko, N.P. Panskikh, V.B. Proval'skii, A.Yu. Veretennikov and A.A. Vladimirov. During all these years, my co-author, A.V. Pokrovskii, has played a decisive role in this work; within a short time he worked his way up from a young assistant to a collaborator of equal standing. He has obtained a number of the fundamental results and recently took over, to a large degree, supervision of colleagues of the next generation.

We have written this book together. A first version of Chapter 19 was worked on and prepared by A.A. Vladimirov. Chapter 39 was prepared especially for the English translation of the book by A.A. Kravchenko.

Taking advantage of this opportunity, I would like to express my gratitude to N.N. Bogolyubov, A. Yu. Ishlinskii, A.I. Lur'e, Yu.A. Mitropol'skii, M.A. Rozenblat and Ya.Z. Tsypkin for sometimes short, sometimes long (but always interesting and profitable) discussions concerning various aspects of hysteresis phenomena.

M.A. Krasnosel'skii

Contents

PART 1 STATIC HYSTERON 1

1. Short-memory transducer 1
   1.1 Transducer 1
   1.2 States of transducer 2
   1.3 Some properties of transducers 3
   1.4 Admissible inputs 4
   1.5 Vibro-correctness 6

2. Generalized play 6
   2.1 Ordinary play 6
   2.2 Generalized play with piecewise monotone inputs 8
   2.3 Estimates 10
   2.4 Generalized play with continuous inputs 14
   2.5 Dependence of outputs upon initial states 15
   2.6 Correctness of the definition of the play 16
   2.7 Monotonicity 17
   2.8 Periodic inputs 19
   2.9 Inputs defined on the whole real axis 20

3. Hysteron 22
   3.1 Stop 22
   3.2 Determining systems of curves 24
   3.3 Piecewise monotone inputs 28
   3.4 Passage to arbitrary continuous inputs 28

4. Canonical representation of hysteron and proof of Theorem 3.2 30
   4.1 Canonical hysteron 30

4.2 Canonical representation theorem 32
4.3 Proof of Theorem 3.2 35
4.4 Properties of hysteron 35
4.5 Rectification of hysteron 39

5. Distances 40
5.1 Definition of distance 40
5.2 Estimates on differences of output signals 43

6. Various input spaces 47
6.1 Statement of the problem 47
6.2 Spaces of continuously differentiable functions 48
6.3 Play in the space S of absolutely continuous functions 50
6.4 Hysteron in the space S 51
6.5 Hysterons in spaces H_α 53
6.6 Discontinuous inputs 54
6.7 Hysteron in the space of functions with bounded variation 56
6.8 Hysteron in Wiener spaces 57

PART 2 IDENTIFICATION THEOREM 59

7. Identification problem 59
7.1 General identification problem 59
7.2 Prehysteron 61
7.3 Basic identification theorem 63
7.4 Concluding remarks 64

8. Proof of Theorem 7.1 65
8.1 Singular points of the domain $\Omega(V)$ 65
8.2 Construction of curves $\Pi(M)$ 69
8.3 Construction of curves Φ_ℓ, Φ_r 74
8.4 Completion of the proof of Theorem 7.1 75

9. α - identifiability 78
9.1 Statement of the problem 78
9.2 Normal hysteron 79
9.3 Theorem on α- identification 80
9.4 A remark 81

10. Approximate construction of hysteron 81

10.1 Distance between hysterons 81
10.2 Bounded inputs 86
10.3 Frames of hysterons 88
10.4 Approximation by operators different from hysterons 90

PART 3 VIBRO-CORRECT DIFFERENTIAL EQUATIONS AND VARIABLE HYSTERONS 94

11. Necessary condition of vibro-correctness 94
11.1 Integrator 94
11.2 Simple examples 96
11.3 Necessary condition of vibro-correctness 97
11.4 Vibro-correctness in a point 102

12. Sufficient condition of vibro-correctness 103
12.1 Main result 103
12.2 An auxiliary equation 105
12.3 A substitution 107
12.4 Proof of Theorem 12.1 109
12.5 Lemma on differential inequalities 110
12.6 Vibro-correctness on smooth inputs 111

13. Vibro-solutions 113
13.1 Definition 113
13.2 Global vibro-correctness 115
13.3 Inputs on finite time interval 117
13.4 Inputs on infinite time interval 119

14. Equations with constraints 122
14.1 Equations with discontinuous right-hand sides 122
14.2 Arbitrary continuous constraints 125
14.3 Vibro-correct equations with constraints 128
14.4 Properties of vibro-solutions to equations with constraints 132
14.5 Vibro-solutions of parametrized equations 134

15. Variable hysteron 138
15.1 Description of hysteron by differential equations 138
15.2 Variable hysteron 139
15.3 Variable hysteron governed by differential equations 143
15.4 Infinitesimal hysteron 145
15.5 A special class of transducers 147

PART 4 MULTIDIMENSIONAL HYSTERONS 151

16. Multidimensional play and stop defined on smooth inputs 151
16.1 A simple example 151
16.2 A general notion 154
16.3 Correctness of the definitions of play and stop 158
16.4 Properties of play and stop 160
16.5 On the classical solutions of equations with discontinuous right-hand sides 160

17. Strictly convex characteristics 162
17.1 Vibro-correctness modulus 162
17.2 Hölder condition 164
17.3 Passage to continuous inputs 166
17.4 Strong convergence 167
17.5 Perturbation of characteristics 167
17.6 Vibro-correctness modulus and differential inclusions 170
17.7 Lower bound for vibro-correctness moduli 175

18. Polyhedral characteristics 177
18.1 Basic theorems 177
18.2 Estimates of the Lipschitz constant 179
18.3 Proofs of Lemma 18.1 and Theorem 18.4 181
18.4 Proof of Theorem 18.3 184
18.5 Remarks 186

19. Arbitrary convex characteristics 187
19.1 Vibro-correctness of play and stop 187
19.2 Estimate for the variation of output 189
19.3 Proof of Theorem 19.1 191

20. Inputs with summable derivatives 193
20.1 Statement of the problem 193
20.2 Lipschitz condition 194
20.3 Remarks 196

21. Vibro-correct equations with vector input 196
21.1 Statement of the problem 196
21.2 Frobenius condition 198
21.3 Necessary condition of vibro-correctness 200
21.4 Sufficient condition of vibro-correctness 204
21.5 Remarks 205

22. Equations with vector inputs and smooth constraints 206
22.1 Constraints 206
22.2 Planar motion 208
22.3 Other descriptions 209

PART 5 DISCONTINUOUS NONLINEARITIES 212

23. Static elements 212
23.1 Continuous characteristics 212
23.2 Elements with discontinuous characteristics 214
23.3 Estimates of outputs 216
23.4 Proper characteristics 219
23.5 Continuity on a fixed input 220
23.6 Additional remarks 221

24. Elements with monotone characteristics 223
24.1 Cones 223
24.2 Special classes of cones 224
24.3 Monotone characteristics 226
24.4 Proof of Theorem 24.1 228
24.5 Proof of Theorem 24.2 230
24.6 Remarks 231

25. Elements with multi-valued characteristics 233
25.1 Selection problem 233
25.2 General theorems on selectors 234
25.3 Monotone selectors 238
25.4 Measurable selectors 240
25.5 Input-output relations 241

26. Closures of static element 243
26.1 Closure of transducer 243
26.2 Characteristic of the closure 244
26.3 Closure modulo a negligence class 247
26.4 Comments 248

27. Weak closures and convexification procedure 249
27.1 Weak closures 249
27.2 Convexification 251
27.3 Weak closures and convexification of static element 252
27.4 Proof of Theorem 27.1 253
27.5 Proof of Theorem 27.2 257

27.6 Convexification of static element modulo negligence class 259
27.7 Examples of open nonlinear systems composed of static elements 260

28. Relay 262
28.1 Ideal relay 262
28.2 Non-ideal relay 263
28.3 Periodic inputs 266
28.4 Closure of relay 268
28.5 Convexification of relay 269
28.6 Relay and "slow" controls 271
28.7 Discontinuous inputs 272

PART 6 SELF-MAGNETIZATION PHENOMENON 274

29. Madelung's hysterons 274
29.1 Non-correct prehysteron 274
29.2 Periodic inputs 277
29.3 Madelung's prehysteron 277
29.4 Properties of Madelung's prehysteron 279
29.5 Madelung's hysteron 281
29.6 Discontinuous inputs with bounded variation 284

30. Proofs of Theorems 29.1 and 29.2 285
30.1 Passage to classical solutions 285
30.2 Lemma on differential inequalities 288
30.3 Proof of Theorem 29.1 289
30.4 Proof of Theorem 29.2 293

31. Response to small perturbations of the input 294
31.1 General scheme 294
31.2 Intensities 296
31.3 Construction of κ-outputs to Madelung's hysteron 299
31.4 Construction of κ-vibrosolutions to differential equations 304
31.5 Construction of κ-outputs for hysterons 311

32. Closure modulo sets of Wiener measure zero 311
32.1 A general scheme 311
32.2 Main theorem 312

32.3 Passage to integral equations 314
32.4 Equations with constraints 317
32.5 Implications for stochastic equations 317

PART 7 COMPLEX HYSTERESIS NONLINEARITIES 320

33. Parallel connections and bundles of hysterons 320
33.1 Complex nonlinearities 320
33.2 Parallel connections 321
33.3 Completely controllable restrictions 323
33.4 Periodic inputs 326
33.5 An important example 330
33.6 Remarks 331

34. Sequential connections of hysterons 332
34.1 Sequential connections and cascades 332
34.2 Sequential connections of plays and stops 335
34.3 Compensators 338
34.4 Complex connections 341

35. Ishlinskii's material 342
35.1 Continual systems of hysterons 342
35.2 Ishlinskii's transducer 343
35.3 Loading and unloading functions 346
35.4 Normal states of Ishlinskii's transducer 351
35.5 Periodic inputs 354
35.6 Davidenkov's model 357
35.7 Controllable restrictions of Ishlinskii's bundles 358

36. Properties of Ishlinskii's transducer 359
36.1 Continuity of Ishlinskii's operator 359
36.2 Correctness with respect to weight functions 363
36.3 Unilateral estimates 365

37. Finite systems of relays 367
37.1 Block-diagrams with relays 367
37.2 Parallel connections and bundles of relays 367
37.3 Independent perturbations of inputs 369
37.4 General perturbation of input 370

38. Continual systems of relays 371
38.1 Bundles of relays and CRS-transducers 371

38.2 Monotonicity of CRS-transducers 373
38.3 Demagnetization function 373
38.4 Periodic inputs 374
38.5 Evaluation of outputs 375
38.6 Vibro-correctness 377
38.7 Controllable restrictions 381

39. Rheological models 384
39.1 Construction of the model 384
39.2 Graphs 386
39.3 Transducer M 389
39.4 Properties of the transducer M 391
39.5 Transducer W 393
39.6 Remarks 395

Bibliographic comments 397

References 400

Subject index 407

Part 1. Static hysteron

"Must a name mean something?" Alice asked doubtfully.
"Of course it must," Humpty Dumpty said with a short laugh: "my name means the shape I am - and a good handsome shape it is, too. With a name like yours, you might be any shape, almost."
L. Carroll

1. Short-memory transducer

1.1. Transducer

Nonlinearities of hysteresis type represent effects typical for transducers. For a transducer W (see Fig. 1.1), a variable *input* u(t) and variable *output* x(t) are to be distinguished. Often the names *input signal* and *output signal* are used too. The functions u(t), x(t) may be scalar- or vector-valued; they may assume values in sets of any arbitrary nature as well.

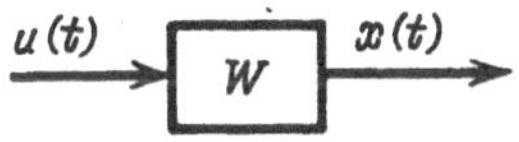

Fig. 1.1

Now we have to specify the class of all admissible inputs u(t) and the relation (single- or multivalued) between the input u(t) and the corresponding output x(t). We shall express this relation in the form

$$x(t) = W\,u(t) \quad ; \tag{1.1}$$

this equality will be given a precise interpretation in the sequel.

As a simplest example, consider linear transducers for which the relation (1.1) assumes either of the forms

$$W\,u(t) = u'(t) \quad , \qquad W\,u(t) = x(t_0) + \int_{t_0}^{t} u(s)\, ds \tag{1.2}$$

where respectively differentiable and integrable functions are to be taken as the admissible inputs.

In turn, as a very simple example of a nonlinear transducer one can take the *superposition operator*

$$W\,u(t) = f[t,u(t)] \quad , \tag{1.3}$$

defined for a function $f(t,u)$ of two variables (without any restriction on the inputs). These examples have nothing to do with hysteresis.

A plastic body is characterized by the deformation $u(t)$ and the stress $x(t)$. In this case the variable deformation can be considered as an input and the variable stress as an output of the transducer W. In the magnetism theory, a magnetic element can be regarded as a transducer, provided a variable external field is taken as an input and the magnetic induction of the element as an output. Both transducers show strong hysteresis effects.

1.2. States of transducer

Transducers with hysteresis nonlinearities may attain any *state* in a certain set. The state is variable and depends either on the changes of the input or some other external factors. In this connection, the knowledge of the values $u(t)$ of the input at $t \geq t_0$ is not sufficient for determining the output $x(t)$ at $t \geq t_0$.

If the state of a transducer can be uniquely determined for any fixed time moment $t = t_0$ by the values $u_0 = u(t_0)$, $x_0 = x(t_0)$ of the input and the corresponding output, the transducer will be called *deterministic* (or *short-memory transducer*). Thus, the state of any short-memory transducer can be characterized by the pair u_0, x_0 or, equivalently, by the point $\{u_0, x_0\}$ in the plane. The corresponding domain (set) of feasible states can be defined in a natural way.

Alternatively, a short-memory transducer can be characterized by the equality

$$x(t) = W[t_0,u_0,x_0]u(t) \tag{1.4}$$

with $W[t_0,u_0,x_0]$ considered as a single-valued operator which depends on the initial values t_0, u_0, x_0 and is defined on a certain class of inputs $u(t)$.

Examples (1.2) and (1.3) were concerned with short-memory transducers. Many other examples of short-memory transducers can be given for dependences between the deformation and stress in problems of plasticity theory and dependences between the external field and magnetic induction in problems of magnetism.

1.3. Some properties of transducers

Denote by $u(t)$, $v(t)$ $(t \geq t_0)$ two admissible input signals for the transducer W with a given state $x(t_0)$ at $t = t_0$. If the equality $u(t) = v(t)$ $(t_0 \leq t \leq t_1)$ always implies that the output signals $Wu(t)$, $Wv(t)$ are identical for $t_0 \leq t \leq t_1$, then the transducer W is called *physically realizable*. This means that the transducer W is physically realizable if its state $x(t_1)$ is independent of all input values at $t > t_1$. For the physically realizable transducers, operators (1.4) are referred to as *Volterra operators*.

Throughout this book only physically realizable transducers will be considered.

Let $u(t)$ $(t \geq t_0)$ be an admissible input signal for a certain transducer W which assumes a given state $\{u_0,x_0\}$ at the initial time moment $t = t_0$. If the transducer W is deterministic, then for $t_0 \leq t_1 \leq t$ the following equality is true: *)

$$W[t_0,u_0,x_0]u(t) = W[t_1,u(t_1), W[t_0,u_0,x_0]u(t_1)]u(t) \ . \tag{1.5}$$

---

*) From now on $W[t_0,u_0,x_0]u(t_1)$ will denote the value of the function $x(t) = W[t_0,u_0,x_0]u(t)$ at $t = t_1$.

This equality is called the *semigroup property* of the transducer or the *semigroup identity*.

Directly by the notion of an *admissible* input signal $u(t)$ $(t \geq t_0)$ (at the state $\{u_0,x_0\}$ of the deterministic transducer W) it follows that

$$u_0 = u(t_0) \quad , \quad x_0 = W[t_0,u_0,x_0]u(t_0) \quad . \tag{1.6}$$

Let the domain of the feasible states for a deterministic transducer W be independent of t. Assume that the admissibility of an input $u(t)$ $(t \geq t_0)$ at the state $\{u_0,x_0\}$ implies that the input $v(t) = u(t+t_0-t_1)$ $(t \geq t_1)$ is admissible at the same state and the equality

$$W[t_0,u_0,x_0]u(t) = W[t_1,u_0,x_0]v(t-t_0+t_1) \quad (t \geq t_0) \tag{1.7}$$

holds. Then the transducer W is called *autonomous*. In other words, a transducer is autonomous if its properties are time-invariant. A transducer is called *static* if it is autonomous and if the change of the input $u(t)$ $(t \geq t_0)$ for $v(t) = u[\alpha t + (1-\alpha)t_0]$ $(t \geq t_0)$, where $\alpha > 0$, results in the change of $x(t)$ $(t \geq t_0)$ for $y(t) = x[\alpha t + (1-\alpha)t_0]$ $(t \geq t_0)$ at the output. In particular, the transducer (1.3) will be static in the case of any function $f(t,u)$ independent of t.

Often, one uses a notion of the *controllable transducer*. We shall introduce this notion for the class of static deterministic transducers. A static deterministic transducer W is *controllable* if for any pair of the feasible states $\{u_0,x_0\}$, u_1,x_1 there exists an input $u(t)$ $(t_0 \leq t \leq t_1)$ which is admissible at the initial state $\{u_0,x_0\}$, such that $u(t_0) = u_0$, $u(t_1) = u_1$, $W[t_0,u_0,x_0]u(t_1) = x_1$.

1.4. Admissible inputs

The transducers with hysteresis that are studied in this book represent an idealization of real physical systems. In real-life situations any complete characterization of all admissible inputs for those transducers is hardly possible.

To illustrate the arising difficulties, let us refer to the following two classical problems.

To begin with, let us consider the problem of determining the area of a curvilinear trapezoid bounded by the graph of a function $u(t)$. If the function $u(t)$ is piecewise constant, one can easily compute such an area. In general, such a curvilinear trapezoid is at first approximated by figures composed of rectangles whose areas (integral sums) after taking a limit determine the area of our interest. This may be considered as a prototype of an integration procedure. Without integration or any similar operation based on passing to some limits, it would be impossible to determine the area of the curvilinear trapezoid bounded by the graph of a sufficiently general "input signal" $u(t)$.

In the second problem, the current velocity of a point which has passed the distance $u(t)$ within time interval $[0,t]$ is to be determined. For $u(t) = at$, this velocity can be found by taking the difference increments. In other cases one has to use a special operation of finding the derivative $x(t) = u'(t)$, again based on passing to a limit.

In the first problem, once the integration operation has been introduced, the set of "admissible inputs" $u(t)$ contains not only all continuous functions but also all functions that are integrable in Riemann's or any more general sense. In the second problem, only the differentiable functions $u(t)$ are admissible.

It is by no means clear a priori for any concrete transducer with hysteresis, how to choose the relevant classes of admissible inputs $u(t)$. Usually we shall proceed as follows.

At first, relations which determine the outputs corresponding to monotone continuous inputs are to be specified. Next, the outputs corresponding to piecewise monotone continuous inputs should be determined by means of the semigroup identity (1.5) . For more general classes of inputs $u(t)$, sequences $u_n(t)$ of piecewise monotone continuous inputs are to be constructed which converge to $u(t)$ in a suitable sense, such that the corresponding outputs $x_n(t)$ are convergent (perhaps in a different sense) to a certain function $x(t)$. The function $x(t)$ will be then the output corresponding to the input $u(t)$.

Sometimes it may be quite difficult to implement the above scheme. Once the relevant implementation is available, one should still verify correctness of the performed limit constructions, analyze properties of the resulting operators, etc.

1.5. Vibro-correctness

The scheme of constructing hysteresis nonlinearities by using some suitable limit passages, we have just described, may seem doubtful .

To improve it, some additional physical properties of transducers should be taken into account. In particular, subjected to small noises of some special types imposed upon the input, the transducer should react correctly (continuously in a certain sense). This property will be further referred to as a *vibro-correctness* .

Let us give an example. Let a deterministic static transducer W represent a model of a certain physical system. Assume that some simple rules defining the values of operator (1.4) that correspond to piecewise monotone continuous inputs are available. Admit that all continuous inputs are physically admissible, though the above relations are not true in the general situation. At the same time, suppose that the physical system under consideration is vibro-correct with respect to noises of small amplitude at the input.

The above behaviour can be equivalently characterized in terms of the operator $W[t_0,u_0,x_0]$ in (1.4) . The vibro-correctness is equivalent to the norm continuity of $W[t_0,u_0,x_0]$ defined on a set of continuous inputs. If this is true, then the values of $W[t_0,u_0,x_0]$, corresponding to any continuous input, can be determined by passing to the limit in the appropriate set of piecewise monotone functions.

Therefore, if a transducer is vibro-correct, the limit construction should completely specify the values of operator (1.4). If, however, such a procedure is not available, one only can claim that the employed mathematical formalism which should determine the outputs corresponding to any piecewise monotone input is useless since it has contributed to the loss of vibro-correctness.

2. Generalized play

2.1. Ordinary play

Let us consider a physical system (P,S) which comprises a cylinder of the length S and a piston P . Both elements which are placed as shown in Figure 2.1 can move along y-axis . The position of piston is characterized by the coordinate u of point A , while the position of cylinder is defined by the coordinate x of point B . The state of system (P,S) is then characterized by the pair $\{u,x\}$.

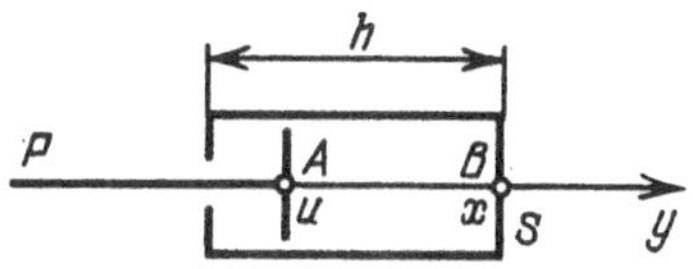

Fig. 2.1

We shall treat the piston P and the cylinder S as a driving and driven element of the system (P,S) , respectively. The system (P,S) can be then considered as a transducer L_0 with the input signal $u(t)$ - the variable position of the piston, and the output signal $x(t)$ - the variable location of the cylinder. Such system is referred to as a *play* . The set of feasible states of the play takes form of the strip $u \leq x \leq u + h \quad (-\infty < u < \infty)$.

As a natural hypothesis one assumes that the location of the piston can vary according to any continuous rule $u(t) \quad (t \geq t_0)$, and the position of the cylinder corresponding to any arbitrary initial state $\{u(t_0),x(t_0)\}$ can be prescribed deterministically. In other words, the variable location $x(t)$ of the cylinder is determined by the equality

$$x(t) = L_0[t_0,x(t_0)]u(t) \quad (t \geq t_0) , \tag{2.1}$$

with an operator $L_0[t_0,x_0]$ defined for arbitrary $x_0 = x(t_0)$ on the set of all continuous inputs $u(t) \quad (t \geq t_0)$, such that $x_0 - h \leq u(t_0) \leq x_0$.

The output (2.1) corresponding to a monotone input $u(t) \quad (t \geq t_0)$ is defined by

$$x(t) = \begin{cases} x(t_0) , & \text{for all } t \text{ such that } x(t_0) - h \leq u(t) \leq x(t_0) \\ u(t) , & \text{for all } t \text{ such that } u(t) \geq x(t_0) \\ u(t)+h , & \text{for all } t \text{ such that } u(t) \leq x(t_0)-h . \end{cases} \tag{2.2}$$

Expressing this in a different way, to construct the output $x(t)$, one should take the initial state $M_0 = \{u(t_0),x(t_0)\}$ and draw (see Figure 2.2) the horizontal segment with end-points N_1, N_2 on the lines $x = u$, $x = u + h$, then notice that the variable state $\{u(t),x(t)\}$ at $t \geq t_0$ is a point of the broken line, thickened in Fig. 2.2.

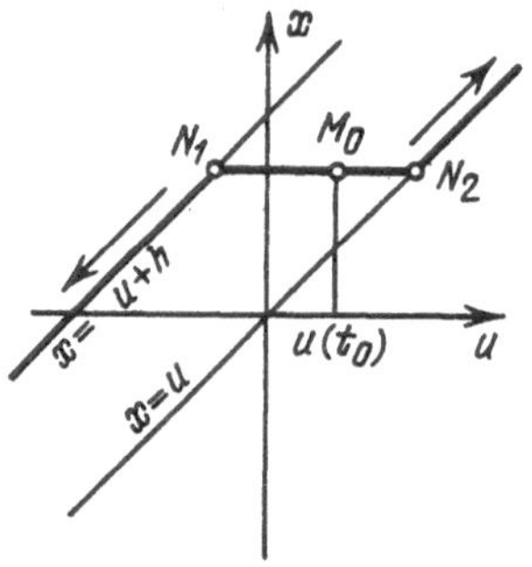

Fig. 2.2

As it follows from the semigroup identity (1.5), the above characterization is applicable to piecewise monotone inputs. For such inputs, (1.5) assumes the form

$$L_o[t_1,L_o[t_o,x_o]u(t_1)]u(t) = L_o[t_o,x_o]u(t) \quad (t_o \leq t_1 \leq t) . \tag{2.3}$$

For arbitrary continuous inputs, the construction we have just developed does not define the operator (2.1).

2.2. Generalized play with piecewise monotone inputs

The ordinary play was completely characterized by the motion of its variable state $\{u(t),x(t)\}$ along horizontal segments with end-points on two prescribed lines. By taking some more general curves Γ_ℓ and Γ_r instead of those lines, we come to the notion of a generalized play.

Consider two continuous non-decreasing functions $\Gamma_\ell(u)$ and $\Gamma_r(u)$, defined on the intervals $(-\infty,a_\ell)$ and (b_r,∞), respectively. Denote their graphs by Γ_ℓ and Γ_r (see Figure 2.3). Assume that the functions $\Gamma_\ell(u)$ and $\Gamma_r(u)$ have the same range D (the set of all values). The set D has then the form of an interval, either finite or infinite (closed, open or unilaterally open). We shall say that functions $\Gamma_\ell(u)$ and $\Gamma_r(u)$ form a *regular pair*, if the curve Γ_ℓ is located left of Γ_r, i.e., $\Gamma_\ell(u) \geq \Gamma_r(u)$ for $a_\ell < u < b_r$.

Now, for any regular pair $\Gamma_\ell(u)$, $\Gamma_r(u)$ we define a transducer $L(\Gamma_\ell,\Gamma_r)$, further referred to as a *generalized play*. For this transducer, the set of admissible states is defined as the union Ω =

$= \Omega(\Gamma_\ell,\Gamma_r)$ of all horizontal segments with end-points belonging to Γ_ℓ and Γ_r, respectively. The generalized plays may be taken as members of the class of deterministic static transducers. For their characterization one therefore has to define the operators (dependent on initial state as a parameter)

$$x(t) = L[t_0,x_0;\Gamma_\ell,\Gamma_r]u(t) \quad (t \geq t_0) \quad , \tag{2.4}$$

which transform the continuous inputs $u(t)$ $(\{u(t_0),x_0\} \subset \Omega(\Gamma_\ell,\Gamma_0))$ into the output signals $x(t)$. To begin with (as in the preceding chapter), let us only take piecewise monotone functions $u(t)$ into account.

For a monotone function $u(t)$ we shall define the values of operator (2.4) by

$$x(t) = \begin{cases} \max\{x_0,\Gamma_r[u(t)]\} & \text{, if } u(t) \text{ is non-decreasing,} \\ \min\{x_0,\Gamma_\ell[u(t)]\} & \text{, if } u(t) \text{ is non-increasing.} \end{cases} \tag{2.5}$$

This means that the output $x(t)$, corresponding to the initial state $M_0 = \{u(t_0),x_0\}$ and to the monotone input $u(t)$, is defined so that the variable state $\{u(t),x(t)\}$ belong to the curve represented by the thickened line in Fig. 2.3.

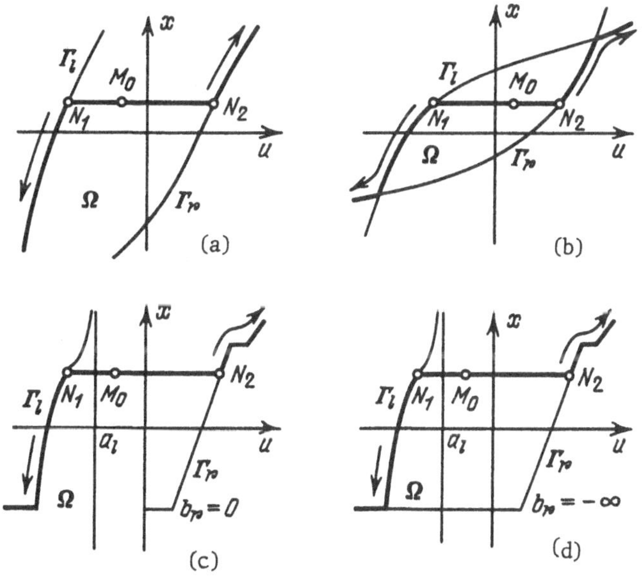

Fig. 2.3

For piecewise monotone inputs, in turn, operator (2.4) can be defined by using the semigroup identity (2.3). To this purpose, the domain of the input $u(t)$ is to be divided into intervals $[t_0,t_1]$, $[t_1,t_2]$, $\dots$, $[t_{i-1},t_i]$, $\dots$ of its monotonicity. On the interval $[t_o,t_1]$, we define the output $x(t)$ by relation (2.5). Next, we take the pair $\{u(t_1),x(t_1)\}$ as a new initial state and, again by using the identity (2.5), we define the output on $[t_1,t_2]$. After a finite number of steps, the outputs $x(t)$ will be then defined on the interval $[t_{i-1},t_i]$. Clearly, this procedure would produce the same output signal also for other decompositions of the domain of the input $u(t)$ into intervals of its monotonicity.

Let us conclude this section with a few observations. The functions $\Gamma_\ell(u)$ and $\Gamma_r(u)$ may be constant on some segments (see Figures 2.3 c, d). These functions may assume the same values at various points or on various intervals (see Figure 2.3,b). The same generalized play may be defined by various pairs of functions $\Gamma_\ell(u)$ and $\Gamma_r(u)$ (see Figures 2.3 c,d). If $\Gamma_\ell(u) \equiv \Gamma_r(u)$, the corresponding generalized play degenerates to standard superposition operator $L[t_o,x_o]u(t) = \Gamma_\ell[u(t)]$.

2.3. Estimates

We now define the operator (2.4) for all continuous inputs (of course, the initial state should be in $\Omega(\Gamma_\ell,\Gamma_r)$). For this, we need some estimates for the operator (2.4) with piecewise monotone inputs. The following obvious estimates hold.

Lemma 2.1. Let the input $u(t)$ $(t_o \le t \le T)$ be piecewise monotone and

$$\{u(t_o),x_o\} \in \Omega(\Gamma_\ell,\Gamma_r) \quad , \quad a \le u(t) \le b \quad (t_o \le t \le T) \tag{2.6}$$

Then

$$\min\{x_o,\Gamma_\ell(a)\} \le L[t_o,x_o;\Gamma_\ell,\Gamma_r]u(t) \le \max\{x_o,\Gamma_r(b)\} \quad (t_o \le t \le T) \tag{2.7}$$

The subsequent lemmas comprehend some estimates on the increments of the operator (2.4).

In the sequel we shall use the notation

$$\|z(t)\|_{t_0,t_1} = \sup_{t_0 \le t \le t_1} |z(t)| \tag{2.8}$$

Let $I_\ell = I_\ell(x_0;a,b)$ be a projection of the intersection of the curve Γ_ℓ and the rectangle $\{\{(u,x)\}: a \le x \le b ; \min\{x_0,\Gamma_\ell(a)\} \le x \le \max\{x_0,\Gamma_r(b)\}\}$ onto the abscissa, and $I_r = I_r(x_0;a,b)$ be the corresponding projection of the intersection of the curve Γ_r with the same rectangle. In Fig. 2.4 the projected parts of the curves are thickened.

<u>Lemma 2.2.</u> Assume piecewise monotone inputs $u(t)$ and $v(t)$ satisfy the conditions

$$\{u(t_0),x_0\}, \{v(t_0),x_0\} \in \Omega(\Gamma_\ell,\Gamma_r) \; ; \quad a \le u(t), v(t) \le b \quad (t_0 \le t \le T). \tag{2.9}$$

Let the functions $\Gamma_\ell(u)$ and $\Gamma_r(u)$ fulfil a Lipschitz condition with constant λ on the intervals $I_\ell(x_0;a,b)$ and $I_r(x_0;a,b)$, respectively. Then

$$|L[t_0,x_0;\Gamma_\ell,\Gamma_r]u(t) - L[t_0,x_0;\Gamma_\ell,\Gamma_r]v(t)| \le$$
$$\le \lambda\|u(s) - v(s)\|_{t_0,t} \quad (t_0 \le t \le T) \quad . \tag{2.10}$$

<u>Proof.</u> First assume that the functions $\Gamma_\ell(u)$ and $\Gamma_r(u)$ are Lipschitz continuous with constant λ on the whole real axis.

Were the assertion of the lemma incorrect, there would exist $\tau_0, \tau_1 \in [t_0,T]$ such that

$$|x(\tau_0) - y(\tau_0)| = \lambda\|u(s) - v(s)\|_{t_0,\tau_0} \tag{2.11}$$

and

$$|x(t) - y(t)| > \lambda\|u(s) - v(s)\|_{t_0,t} \quad (\tau_0 < t \le \tau_1) \quad , \tag{2.12}$$

where $x(t) = L[t_0,x_0;\Gamma_\ell,\Gamma_r]u(t)$, $y(t) = L[t_0,x_0;\Gamma_\ell,\Gamma_r]v(t)$. With no loss of generality one can assume that the functions $u(t)$ and $v(t)$ are monotone on $[\tau_0,\tau_1]$. For definiteness, let $x(t) > y(t)$ for $\tau_0 < t < \tau_1$.

Consider only the case of a non-decreasing input $u(t)$ on $[\tau_0,\tau_1]$ (an analogous treatment would also apply to the case of a non-increasing input). At first, let $x(\tau_0) = \Gamma_r[u(\tau_0)]$. Then $x(t) = \Gamma_r[u(t)]$ $(\tau_0 \leq t \leq \tau_1)$ and, by the obvious inequality $y(t) \geq \Gamma_r[v(t)]$ $(\tau_0 \leq t \leq \tau_1)$, the estimate

$$x(t) - y(t) \leq \Gamma_r[u(t)] - \Gamma_r[v(t)] \quad (\tau_0 \leq t \leq \tau_1) \quad , \tag{2.13}$$

follows, implying that

$$|x(t) - y(t)| \leq \lambda|u(t) - v(t)| \quad (\tau_0 \leq t \leq \tau_1) \quad . \tag{2.14}$$

The relation (2.14) contradicts (2.12), thus the inequality $x(\tau_0) > \Gamma_r[u(\tau_0)]$ is true . Consequently, for the values t which are close to τ_0 and larger, the output signal $x(t)$ assumes the constant value $x(\tau_0)$. By (2.11) and (2.12), it follows then that the function $y(t)$ is non-increasing, and for the same range of time instants t it assumes values less than $y(\tau_0)$. This means that the function $v(t)$ is non-increasing and $y(t) = \Gamma_\ell[v(t)]$ $(\tau_0 \leq t \leq \tau_1)$. Hence, in view of the obvious inequality $x(t) \leq \Gamma_\ell[u(t)]$ $(\tau_0 \leq t \leq \tau_1)$, the estimate (2.13) follows, yielding also (2.14) which contradicts (2.12). This completes the proof for the case of functions which are globally Lipschitz continuous (on the whole real line) .

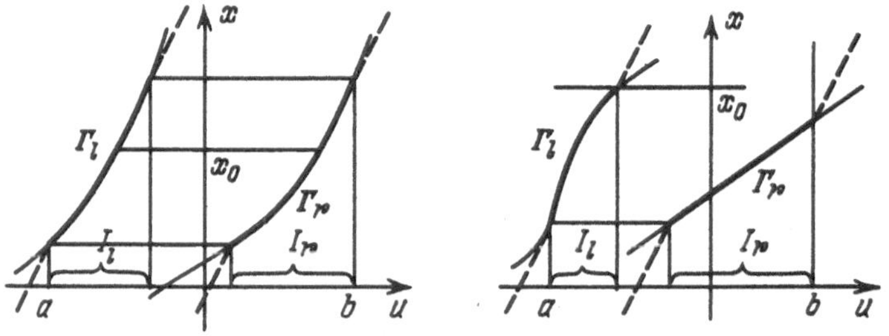

Fig. 2.4

Let us return to the general situation. Given the functions $\Gamma_\ell(u)$ and $\Gamma_r(u)$, construct a new regular pair of functions $\Gamma_\ell^*(u)$ and $\Gamma_r^*(u)$, appropriately equal to $\Gamma_\ell(u)$ and $\Gamma_r(u)$ on the intervals $I_\ell(x_o;a,b)$ and $I_r(x_o;a,b)$, and linear with the slope λ outside.

The curves Γ_ℓ^* and Γ_r^* in Fig. 2.4 consist of the thickened parts of the curves Γ_ℓ, Γ_r and half-lines pointed out by the dashed lines. The functions Γ_ℓ^* and Γ_r^* satisfy the Lipschitz condition with the constant λ on the whole real line. Hence,

$$|L[t_o,x_o;\Gamma_\ell^*,\Gamma_r^*]u(t) - L[t_o,x_o;\Gamma_\ell^*,\Gamma_r^*]v(t)| \leq$$

$$\leq \lambda \; \|u(s) - v(s)\|_{t_o,t} \quad (t_o \leq t \leq T) \quad , \tag{2.15}$$

and it remains to remark that, due to Lemma 2.1, the output signals of the transducers $L(\Gamma_\ell^*,\Gamma_r^*)$ and $L(\Gamma_\ell,\Gamma_r)$, corresponding to the inputs $u(t)$ and $v(t)$, are equal.

Let $\mu(r)$ $(r \geq 0)$ be a continuous, increasing function such that $\mu(0) = 0$. The function $\Gamma(u)$ is said to be *continuous, with the modulus of continuity* $\mu(r)$ on an interval $[u_1,u_2]$, if

$$|\Gamma(u)-\Gamma(v)| \leq \mu(|u-v|) \quad (u_1 \leq u,\ v \leq u_2) \quad . \tag{2.16}$$

If $\mu(r) = \lambda r$, then (2.16) reduces to Lipschitz condition. In the case of $\mu(r) = \lambda r^\alpha$ $(0 < \alpha < 1)$, it has the form of a Hölder condition. Any function $\Gamma(u)$ which is continuous on the closed interval $[u_1,u_2]$, has the modulus of continuity

$$\mu_o(r) = \max_{\substack{|u-v|\leq r;\\ u_1\leq u,v\leq u_2}} |\Gamma(u) - \Gamma(v)| \quad . \tag{2.17}$$

<u>Lemma 2.3.</u> Let piecewise monotone inputs $u(t)$ and $v(t)$ satisfy conditions (2.9). Assume that the functions $\Gamma_\ell(u)$ and $\Gamma_r(u)$ are continuous with the modulus of continuity $\mu(r)$ on the intervals $I_\ell(x_o;a,b)$ and $I_r(x_o;a,b)$, respectively. Then

$$|L[t_0,x_0;\Gamma_\ell,\Gamma_r]u(t) - L[t_0,x_0;\Gamma_\ell,\Gamma_r]v(t)| \leq$$
$$\leq \mu\left[\|u(s) - v(s)\|_{t_0,t}\right] \quad (t_0 \leq t \leq T) \quad . \tag{2.18}$$

This lemma is an extension of Lemma 2.2 and its proof proceeds similarly. ∎

2.4. Generalized play with continuous inputs

Let $u_*(t)$ $(t \geq t_0)$ be a continuous input and $\{u_*(t_0),x_0\} \in \Omega(\Gamma_\ell,\Gamma_r)$. Consider any arbitrary sequence $u_n(t)$ $(t \geq t_0)$ of piecewise monotone continuous inputs which converge to $u_*(t)$ uniformly on any finite interval $[t_0,T]$ and satisfy the condition $\{u_n(t_0),x_0\} \in \Omega(\Gamma_\ell,\Gamma_r)$. Let

$$x_n(t) = L[t_0,x_0;\Gamma_\ell,\Gamma_r]u_n(t) \quad (t \geq t_0;\ n = 1,\ 2,\ldots) \quad . \tag{2.19}$$

The values of functions $u_n(t)$ $(t_0 \leq t \leq T)$ are contained in an interval $[a,b]$. Hence, due to Lemma 2.3, the estimates

$$\|x_n(t) - x_m(t)\|_{t_0,T} \leq \mu\left[\|u_n(s) - u_m(s)\|_{t_0,T}\right] \tag{2.20}$$

hold, with $\mu(r)$ being the modulus of continuity of the functions $\Gamma_\ell(u)$ and $\Gamma_r(u)$ on intervals $I_\ell(x_0;a,b)$ and $I_r(x_0;a,b)$, respectively. Therefore , (2.19) is a Cauchy sequence and it converges uniformly to some function $x_*(t)$ $(t_0 \leq t \leq T)$.

Let us define

$$L[t_0,x_0;\Gamma_\ell,\Gamma_r]u_*(t) = x_*(t) \quad (t \geq t_0) \quad . \tag{2.21}$$

The above definition provides a characterization of the generalized play for all continuous inputs. The function $x_*(t)$ does not depend on the choice of a sequence $u_n(t)$ convergent to $u_*(t)$.

The generalized play defined by (2.21) for all continuous inputs, is the only vibro-correct transducer which coincides with the operator (2.4) in the case of piecewise monotone continuous inputs. By virtue of definition of the operator $L[t_0,x_0;\Gamma_\ell,\Gamma_r]$, the following statement holds.

Theorem 2.1. The generalized play $L(\Gamma_\ell,\Gamma_r)$, defined for all continuous inputs, is a deterministic, static, controllable and vibro-correct transducer. ∎

Being deterministic means here, in particular, that the semigroup identity is true for the operator (2.21) over all continuous inputs. Controllability occurs already within the class of piecewise monotone inputs; moreover, the transfer from one feasible state to another can always be performed by taking a piecewise monotone input with at most two monotonicity intervals. Vibro-correctness means there that any operator (2.21) is continuous on its domain in the metric of the space $C(t_0,T)$ (for t_0, $T \in (-\infty,\infty)$).

The assertions of Lemmas 2.1 - 2.3 are preserved after passing from the piecewise monotone to all continuous inputs. For the proof of these properties one has to verify the appropriate assertions in the case of piecewise monotone inputs and next take some limit.

2.5. Dependence of outputs upon initial states

In the definition of vibro-correctness, output signals were considered for various inputs but their initial values were fixed. We now compare the outputs which correspond to different initial values x_0 .

Let ξ, η and a, b be fixed reals, with $\xi \le \eta$ and $a < b$. Denote by $I_\ell(\xi,\eta;a,b)$ a projection of intersection of the curve Γ_ℓ and the rectangle

$$\{\{u,x\}:\ a \le u \le b;\ \min\{\xi,\Gamma_\ell(a)\} \le a \le \max\{\eta,\Gamma_r(b)\}\} \quad , \tag{2.24}$$

onto the abscissa, and by $I_r(\xi,\eta;a,b)$ the corresponding projection of the intersection of Γ_r with the rectangle (2.24). Let $\mu(r;\xi,\eta;a,b)$ be the modulus of continuity of the functions $\Gamma_\ell(u)$ and $\Gamma_r(u)$ on $I_\ell(\xi,\eta;a,b)$ and $I_r(\xi,\eta;a,b)$, respectively.

Theorem 2.2. Let $\{u(t_0),x_0\}$ and $\{v(t_0),y_0\}$ be some points in the domain $\Omega(\Gamma_\ell,\Gamma_r)$ of the feasible states for the generalized play $L(\Gamma_\ell,\Gamma_r)$. Let $u(t)$, $v(t)$ $(t_0 \le t \le T)$ be continuous inputs such that $a \le u(t)$, $y(t) \le b$. Then

$$|L[t_o,x_o;\Gamma_\ell,\Gamma_r]u(t) - L[t_o,y_o;\Gamma_\ell,\Gamma_r]v(t)| \le$$
$$\le \max\{|x_o - y_o|,\ \mu[\|u(s) - v(s)\|_{t_o,T};\ \xi,\eta,a,b]\} \quad (t_o \le t \le T) \qquad (2.25)$$

The proof is left to the reader. ∎

If, in particular, the functions $\Gamma_\ell(u)$ and $\Gamma_r(u)$ are defined on the whole real axis and satisfy there the Lipschitz condition with constant λ, then the estimate (2.25) admits particularly simple form

$$|L[t_o,x_o;\Gamma_\ell,\Gamma_r]u(t) - L[t_o,y_o;\Gamma_\ell,\Gamma_r]v(t)| \le$$
$$\le \max\{|x_o - y_o|,\ \lambda\|u(s) - v(s)\|_{t_o,T}\} \quad (t_o \le t \le T) \ . \qquad (2.26)$$

By (2.25), it follows in particular that for a fixed input $u(t)$ the *difference* $|L[t_o,x_o;\Gamma_\ell,\Gamma_r]u(t) - L[t_o,y_o;\Gamma_\ell,\Gamma_r]u(t)|$ *is a non-increasing function*. This property follows from elementary geometric considerations.

2.6. Correctness of the definition of the play

Let us consider two generalized plays $L(\Gamma_\ell,\Gamma_r)$ and $L(\Gamma_\ell^*,\Gamma_r^*)$. Construct the corresponding intervals $I_\ell(x_o;a,b)$, $I_r(x_o;a,b)$, $I_\ell^*(x_o^*;a,b)$, $I_r^*(x_o^*;a,b)$. Define

$$\rho(\Gamma_\ell,\Gamma_\ell^*;x_o,x_o^*;a,b) = \max_{u\in I_\ell(x_o;a,b)\cap I_\ell^*(x_o^*;a,b)} \{|\Gamma_\ell(u) - \Gamma_\ell^*(u)|\}$$

and

$$\rho(\Gamma_r,\Gamma_r^*;x_o,x_o^*;a,b) = \max_{u\in I_r(x_o;a,b)\cap I_r^*(x_o^*;a,b)} \{|\Gamma_r(u) - \Gamma_r^*(u)|\} \ .$$

If either $I_\ell \cap I_\ell^*$ or $I_r \cap I_r^*$ is empty, then the appropriate expression on the right-hand side will be considered as equal to zero.

Without overcoming any difficulties (although in a quite tedious way) one can prove that *for any continuous input* $u(t)$ $(t_o \le t \le T)$ *relations* $a \le u(t) \le b$, $\{u(t_o),x_o\} \in \Omega(\Gamma_\ell,\Gamma_r)$, $\{u(t_o),x_o^*\} \in \Omega(\Gamma_\ell^*,\Gamma_r^*)$ *imply the estimate*

$$\|L[t_o,x_o;\Gamma_\ell,\Gamma_r]u(t) - L[t_o,x_o^*;\Gamma_\ell^*,\Gamma_r^*]u(t)\|_{t_o,T} \le$$
$$\le \max\{|x_o - x_o^*|,\ \rho(\Gamma_\ell,\Gamma_\ell^*;x_o,x_o^*;a,b),\ \rho(\Gamma_r,\Gamma_r^*;x_o,x_o^*;a,b)\} \quad . \qquad (2.27)$$

The estimate (2.27) reflects the correctness of a generalized play with respect to perturbations of its defining curves Γ_ℓ and Γ_r .

2.7. Monotonicity

Monotonicity is an important feature of various physical transducers - a reasonable "increase" of the input signal brings on an "increase" of the output. This property can be revealed at the very beginning of an analysis of real transducers. Preservation of the monotonicity property (or its loss) upon passing to a mathematical model is an important characteristic of the model.

Monotonicity of a model is also important for a mathematical convenience of its use. Monotone operators (similarly as monotone functions) have a number of specific properties which significantly simplify a study of closed-loop systems that contain monotone transducers as single elements.

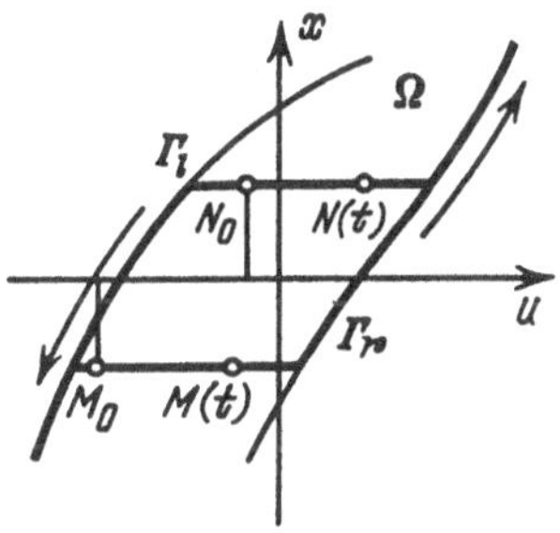

Fig. 2.5

Theorem 2.3. Each generalized play is monotone in the following sense:
the relations

$$x_o \le y_o;\ u(t) \le v(t) \quad (t_o \le t \le T);\ \{u(t_o),x_o\},\ \{v(t_o),y_o\} \in \Omega(\Gamma_\ell,\Gamma_r) \qquad (2.28)$$

imply the inequality

$$L[t_o,x_o;\Gamma_\ell,\Gamma_r]u(t) \le L[t_o,y_o;\Gamma_\ell,\Gamma_r]v(t) \quad (t_o \le t \le T) \quad . \qquad (2.29)$$

Proof. Let the points t_o, t_1, ..., $t_n = T$ divide the interval $[t_o,T]$ into n equal parts. Denote by $u_n(t)$, $v_n(t)$ continuous functions, linear on each interval $[t_{i-1},t_i]$ and assuming the values $u(t_i)$, $v(t_i)$ in the points t_i. By the construction, $u_n(t) \le v_n(t)$ $(t_o \le t \le T)$ and the sequences $u_n(t)$, $v_n(t)$ converge to $u(t)$, $v(t)$ uniformly over $[t_o,T]$. In this connection it is enough to show that the inequality (2.29) holds for piecewise monotone and then conclude its validity for monotone inputs.

Let the inputs $u(t)$ and $v(t)$ which satisfy conditions (2.28), be monotone. Then the state variables

$$M(t) = \{u(t),\, x(t)\}, \quad N(t) = \{v(t),\, y(t)\},$$

where

$$x(t) = L[t_o,x_o;\Gamma_\ell,\Gamma_r]u(t), \quad y(t) = L[t_o,y_o;\Gamma_\ell,\Gamma_r]v(t) \quad ,$$

move along the curves indicated in Fig. 2.5 by the thickened lines. The point $M(t)$ belongs there to the lower curve, whereas $N(t)$ - to the upper one. The inequality $u(t) \le v(t)$ means that the point $M(t)$ cannot be located to the right of $N(t)$. Hence, for every $t \ge t_o$, the point $M(t)$ is not placed over $N(t)$, i.e., the inequality (2.29) is satisfied.

∎

Now let us note that *the generalized play* $L = L(\Gamma_\ell,\Gamma_r)$ *is monotone with respect to* Γ_ℓ *and* Γ_r *in the following sense: if* $\Gamma_\ell(u) \le \Gamma_\ell^*(u)$ *and* $\Gamma_r(u) \le \Gamma_r^*(u)$ *(for all* u *such that these inequalities make sense); if, moreover,* $\{u(t_o),x_o\} \in \Omega(\Gamma_\ell,\Gamma_r)$, $\{u^*(t_o)x_o^*\} \in \Omega(\Gamma_\ell^*,\Gamma_r^*)$ *and, finally,* $u(t) \le u^*(t)$ $(t \ge t_o)$, $x_o \le x_o^*$, *then*

$$L[t_o,x_o;\Gamma_\ell,\Gamma_r]u(t) \le L[t_o,x_o^*;\Gamma_\ell^*,\Gamma_r^*]u^*(t) \quad (t \ge t_o) \quad . \qquad (2.30)$$

2.8. Periodic inputs

Let $u(t)$ $(t \geq t_0)$ be an input signal such that $u(t_0) \leq u(t) \leq u(t_1)$ $(t_0 \leq t \leq t_1)$. Then $L[t_0,x_0;\Gamma_\ell,\Gamma_r]u(t_1) = y(t_1)$ where $y(t)$ is the output which corresponds to the linear input signal $v(t)$ such that $v(t) = u(t)$ at $t = t_0$ and $t = t_1$. An analogous effect occurs if $u(t_0) \geq u(t) \geq u(t_1)$ $(t_0 \leq t \leq t_1)$.

Consider the output

$$x(t) = L[t_0,x_0;\Gamma_\ell,\Gamma_r]u(t) \quad (t \geq t_0) , \tag{2.31}$$

corresponding to an input signal $u(t)$ $(t \geq t_0)$ which is T-periodic, i.e., $u(t+T) = u(t)$ $(t \geq t_0)$. Denote respectively by t_m and t_M the very left minimum and maximum points of the input $u(t)$. We can confine ourselves to only those inputs whose minimal and maximal values are different; for definiteness we shall assume $t_0 \leq t_m \leq t_M \leq t_0 + T$.

According to the remark at the beginning of this chapter, in order to determine the values of the output at

$$t_m,\ t_M,\ t_m+T,\ t_M+T,\ t_m+2T,\dots , \tag{2.32}$$

one can treat t_m as an initial time moment and replace $u(t)$ by the piecewise linear input $v(t)$ that coincides with $u(t)$ in the points (2.32). The output $y(t)$ which corresponds to the input $v(t)$ will be T-periodic at $t \geq t_M$ (this property is geometrically clear). Consequently, also the output $x(t)$ will be T-periodic at $t \geq t_M$.

The least $\delta > 0$ such that the identity

$$x(t+T) = x(t) \quad (t \geq t_0 + \delta) \tag{2.33}$$

holds will be called a *periodicity stabilization time* of the output. If, for any periodic input, this time does not exceed the value of one period then the transducer is named *monocyclic*. As we have shown, *every generalized play is monocyclic.*

Let us assume that the variable state $\{u(t),x(t)\}$ of a certain transducer W, corresponding to a periodic input $u(t)$, for $t > \delta$ (see (2.33)) describes a closed curve, further referred to as a *hysteresis loop*.

If the hysteresis loops are oriented clockwise, we shall say that the transducer has a *negative spin*. If the variable state always moves counter-clockwise along the hysteresis loop, then W is said to have a *positive spin*. There are also transducers which may have either of the spins in dependence on the particular periodic input. As it can be easily seen, *every generalized play has positive spin*.

The hysteresis loop of a generalized play may in some cases degenerate to a horizontal segment. If, however, such a degeneracy does not occur, then the points of the hysteresis loop which have the minimal and maximal abscissae belong appropriately to Γ_ℓ and Γ_r. Therefore, this hysteresis loop can be characterized solely by the input signal. The form of such a loop does not depend on an initial state.

A continuous function $u(t)$ $(t \geq t_0)$ is called *asymptotically periodic (asymptotically almost periodic)* if there is a T-periodic (respectively, almost periodic) function $v(t)$ $(-\infty < t < \infty)$ such that

$$\lim_{t \to \infty} |u(t) - v(t)| = 0 .$$

The output $L[t_0,x_0;\Gamma_\ell,\Gamma_r]u(t)$ corresponding to an asymptotically periodic or asymptotically almost periodic input $u(t)$ is appropriately asymptotically periodic or asymptotically almost periodic. The first of those properties follows directly by vibro-correctness of the generalized play. The proof of the second property utilizes the equivalence of the asymptotic almost periodicity of the function $u(t)$ $(t \geq t_0)$ and the compactness of every sequence of functions $v_n(t) = u(t+h_n)$ $(t \geq t_0)$ with $h_n \to \infty$ in the sense of the uniform convergence over $[t_0,\infty)$.

2.9. Inputs defined on the whole real axis

Up to now we have considered only the input signals $u(t)$ defined on intervals having the form $[t_0,t_1]$ or $[t_0,\infty)$. Sometimes it makes sense to admit also inputs defined on the whole real axis $(-\infty,\infty)$. Certainly, one needs there some additional notions.

We shall say that the output $x(t)$ $(-\infty < t < \infty)$ corresponds to the input $u(t)$ $(-\infty < t < \infty)$ if, for every t_*, the equality

$$x(t) = L[t_*,x(t_*);\Gamma_\ell,\Gamma_r]u(t) \quad (t \geq t_*) \tag{2.34}$$

holds. Unfortunately, in the case of an arbitrary generalized play it may happen that there is no output corresponding to any input defined over $(-\infty,\infty)$. In turn, if such an output exists, it may be non-unique.

The simplest situation refers to $\Gamma_\ell(u)$ and $\Gamma_r(u)$ defined for all $u \in (-\infty,\infty)$. In this case, for any continuous input $u(t)$ $(-\infty < t < \infty)$ there exists at least one output $x(t)$ defined on $(-\infty,\infty)$. This output may be constructed in the following way. First, for all $k, \ell = 0,1,2,\ldots$ let us construct the closed segments

$$\Lambda_{k,\ell} = \{x: x = L[-k-\ell,x_0;\Gamma_\ell,\Gamma_r]u(-k),\ \Gamma_r[u(-k-\ell)] \leq x_0 \leq$$

$$\leq \Gamma_\ell[u(-k-\ell)]\}$$

and intersections

$$\Lambda_k^* = \Lambda_{k,0} \cap \Lambda_{k,1} \cap \ldots \cap \Lambda_{k,\ell} \cap \ldots \quad (k = 0,\ 1,\ 2,\ \ldots)$$

(which are non-empty, since $\Lambda_{k,0} \supset \Lambda_{k,1} \supset \Lambda_{k,2} \supset \ldots$) ; then choose any point x_0 from the set Λ_0^* , and by induction determine points $x_k \in \Lambda_k^*$ such that

$$x_{k-1} = L[-k,x_k;\Gamma_\ell,\Gamma_r]u(-k+1) \quad ;$$

subsequently, for each k assume

$$x(t) = L[-k,x_k;\Gamma_\ell,\Gamma_r]u(t) \qquad (t \geq -k)$$

(by virtue of the semigroup property, the last formula uniquely determines the output values at all t).

In a somewhat more complicated case, one of the functions $\Gamma_\ell(u)$, $\Gamma_r(u)$ (or both of them) is defined only for some u . To analyze the possibilities which may arise there, let us introduce the notations

$$\alpha = \varliminf_{t\to-\infty} u(t) \quad , \qquad \beta = \varlimsup_{t\to-\infty} u(t) \quad .$$

If the values α and β are finite, then for the input $u(t)$ $(-\infty < t < \infty)$ there exists always an output $x(t)$ defined on $(-\infty,\infty)$;

this output is unique if and only if the values $\Gamma_\ell(\alpha)$ and $\Gamma_r(\beta)$ are well-defined and $\Gamma_\ell(\alpha) \leq \Gamma_r(\beta)$. If $\alpha = -\infty$ and there exists a sequence $t_n \to -\infty$, such that $u(t_n) > b_r$, then there is a unique output defined on $(-\infty,\infty)$, which corresponds to $u(t)$; if $\alpha = -\infty$ and $u(t) \leq b_r$ for all negative t with sufficiently large absolute value, then there is no output defined on $(-\infty,\infty)$ which would correspond to the input $u(t)$. At last, if $\alpha = \infty$, then the output defined on $(-\infty,\infty)$ does exist (and is unique) if and only if the function $\Gamma_\ell(u)$ is defined for all $u \in (-\infty,\infty)$.

If the input $u(t)$ $(-\infty < t < \infty)$ is bounded, then every corresponding output $x(t)$, defined on $(-\infty,\infty)$, also remains bounded. The periodicity (almost periodicity) of the input $u(t)$ implies periodicity (almost periodicity) of the output $x(t)$, with the same period. Furthermore, if the input $u(t)$ is defined only for $t \geq t_0$, and the equality

$$\lim_{t\to\infty} |u(t) - v(t)| = 0$$

holds with an input $v(t)$ defined for all $t \in (-\infty,\infty)$ and periodic (almost periodic), then there exists a periodic (almost periodic) output $y(t)$ defined on $(-\infty,\infty)$, such that

$$\lim_{t\to\infty} |L[t_0,x(t_0);\Gamma_\ell,\Gamma_r]u(t) - y(t)| = 0 .$$

3. Hysteron

3.1. Stop

In various applications, transducers can be faced which, although different from generalized plays, have still an intrinsic structure reminiscent of that for a play.

Let us consider a transducer W with scalar-valued inputs $u(t)$ and outputs $x(t)$, whose states can be defined as the input-output pairs $\{u,x\}$. Let the set of all feasible states of the transducer W take the form of a strip $\Omega = \Omega(W)$ located between two horizontal straight lines Φ_ℓ and Φ_r (Fig. 3.1). If the input $u(t)$ $(t \geq t_0)$ is continuous and monotone, then we define the output

$$x(t) = W[t_o,x_o]u(t) \quad (t \geq t_o) \tag{3.1}$$

so that the variable state $\{u(t),x(t)\}$ is a point of the broken line indicated in Fig. 3.1 by thickening; this line comprises two horizontal half-lines and a segment with slope 1, which contains the initial state $M_o = \{u(t_o),x_o\}$ and has the ends on the straight lines Φ_ℓ, Φ_r. Hence, the output $x(t)$ corresponding to a monotone input $u(t)$ is given by

$$x(t) = \begin{cases} \min\{h,u(t)-u(t_o)+x(t_o)\}, & \text{if } u(t) \text{ is non-decreasing} \\ \max\{-h,u(t)-u(t_o)+x(t_o)\}, & \text{if } u(t) \text{ is non-increasing} \end{cases} \tag{3.2}$$

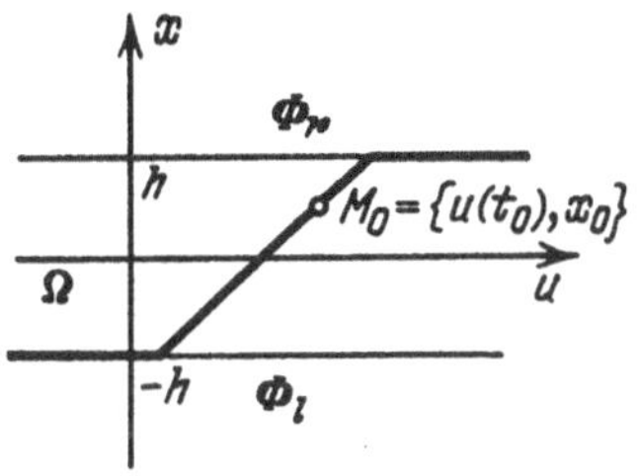

Fig. 3.1

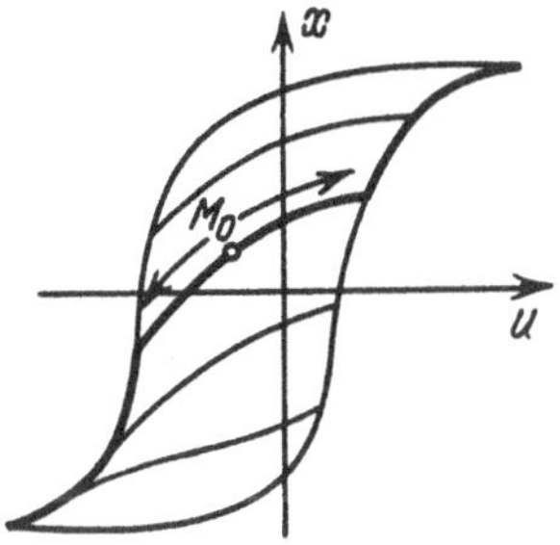

Fig. 3.2

To define the output (3.1) in the case of piecewise-monotone continuous inputs, we shall use *the semigroup property*

$$W[t_o,x_o]u(t) = W[t_1,W[t_o,x_o]u(t_1)]u(t) \quad (t_o \leq t_1 \leq t) . \tag{3.3}$$

The transducer which we have just described is referred to as a *stop* .

In most of the typical models of elasto-plastic fibres, the states are completely characterized by values of the deformation u and the stress x. Such fibres can be treated as transducers with variable deformation as an input and with variable stress as the suitable output.

In particular, for Prandtl's model, the dependence of the stress on deformation is similar to the input-output relation of a stop. The only difference is in the slope E which now may be different from 1 (for small deformations, the fibre is assumed elastic, with elasticity modulus E) .

Similar constructions are applicable also for determining the magnetic induction $x(t)$ of a magnet in a variable field $u(t)$. The whole system is considered as a certain transducer W. One assumes that the variable state $M(t) = \{u(t),x(t)\}$, corresponding to the periodic input $u(t) = A \sin t$ with large amplitude, describes a *hysteresis loop* (see Fig. 3.2).

For more complex inputs and for initial states M_0 located in the interior of the loop, the determination of the output turns out more complicated. In several cases one postulates that the interior of the loop is filled up by graphs of some continuous functions, and that the output corresponding to a monotone input is defined so that the variable states $\{u(t),x(t)\}$ belong to the curve indicated in Fig. 3.2 by the thickened line which includes the initial state M_0. Further, in order to cover also the case of piecewise monotone inputs, one exploits the semigroup identity.

In the examples under consideration, two specific curves Φ_ℓ and Φ_r were selected in the domain Ω of all feasible states for the transducer. The remaining part of the domain Ω was filled by graphs of some continuous functions. By using those curves and graphs, we were able to determine the outputs that corresponded to piecewise monotone continuous inputs.

3.2. Determining systems of curves

In this section a broad class of deterministic vibro-correct transducers including those considered in Section 3.1 will be given a more detailed characterization. Such transducers will be referred to as *hysterons*.

In order to describe a concrete hysteron W, one has to determine the domain $\Omega(W)$ of its feasible states and the operator (3.1) which maps the input signals $u(t)$ into the outputs $x(t)$. Similarly to the construction of the play, the first step will consist in determining outputs that correspond to monotone and piecewise monotone continuous inputs. Next, again by the same procedure as for the play, we shall pass to arbitrary continuous inputs by a limit construction.

The domain $\Omega(W)$ must satisfy the following hypotheses (which are to be regarded as axioms):

(a1) The intersection $K(u_0)$ *of the domain* $\Omega(W)$ *with any vertical straight line* $u = u_0$ *is a non-empty interval. This interval can be*

closed, open or semi-open; it can be finite or infinite, it can also consist of a single point.

(a2) Two curves $\Phi_\ell(u)$ *and* $\Phi_r(u)$ *, being graphs of continuous functions* $\Phi_\ell(u)$ $(-\infty < u < a_\ell)$ *and* $\Phi_r(u)$ $(b_r < u < \infty)$ *are selected in* $\Omega(W)$ *. The points* $\{u_0,\Phi_\ell(u_0)\}$ *and* $\{u_0,\Phi_r(u_0)\}$ *define two ends of the interval* $K(u_0)$ *for all* $u = u_0$ *such that both functions* $\Phi_\ell(u)$ *and* $\Phi_r(u)$ *are defined. At those* $u = u_0$ *for which only one of the functions* $\Phi_\ell(u)$ *and* $\Phi_r(u)$ *is defined, the appropriate point* $\{u_0,\Phi_\ell(u_0)\}$ *or* $\{u_0,\Phi_r(u_0)\}$ *is one of the ends of the interval* $K(u_0)$ *; the intervals* $K(u_0)$ *consisting of one point are subsets of* Φ_ℓ *and* Φ_r , at the same time.

(a3) A part $\Omega_0(W)$ *of the domain* $\Omega(W)$ *, containing all the points which belong neither to* Φ_ℓ *nor to* Φ_r *is selected; this part is stratified into a system of non-intersecting graphs of some continuous functions defined on finite closed intervals. The left end-points of all graphs belong to* Φ_ℓ *(but not to* Φ_r *); accordingly, the right end-points belong to* Φ_r *(but not to* Φ_ℓ *). The remaining points of the graphs belong neither to* Φ_ℓ *nor to* Φ_r .

The graph which contains $M \in \Omega_0(W)$ will be denoted by $\Gamma(M)$; the corresponding function by $\Pi(u;M)$; the interval where $\Pi(u;M)$ is defined - by $[u_\ell(M),u_r(M)]$. For the same graph different notations will be used, because it contains various points M (the same concerns the appropriate functions and intervals).

(a4) If the points $M, N \in \Omega(W)$ *belong simultaneously to different graphs, then*

$$[u_\ell(M) - u_\ell(N)][u_r(M) - u_r(N)] > 0 \quad . \tag{3.4}$$

Hypothesis (a4) can be given a simple geometric interpretation: one of the two non-intersecting graphs $\Pi(M)$ and $\Pi(N)$ is located to the "left" of the other.

Let us denote respectively by A_ℓ and A_r the projections of the sets containing left and right end-points of the graphs $\Pi(M)$ onto the abscissa. By A_0 we shall denote the projection of the intersection $\Phi_0 = \Phi_\ell \cap \Phi_r$ (clearly, this set may be empty) onto the same axis.

(a5) If $u < a_\ell$ *and* $u \notin A_\ell \cup A_o$ *, then there exists a finite number*

$$\gamma_\ell(u) = \min\{v: v > u, v \in A_\ell \cup A_o\} \tag{3.5}$$

and $\gamma_\ell(u) \in A_\ell \cup A_o$ *. Analogously, if* $u > b_r$ *and* $u \notin A_r \cup A_o$ *, then there exists a finite number*

$$\gamma_r(u) = \max\{v: v < u; v \in A_r \cup A_o\} \tag{3.6}$$

and $\gamma_r(u) \in A_r \cup A_o$ *.*

Hypotheses (a1) ÷ (a5) are not independent. One can show that (a5) follows from (a1) ÷ (a4) .

In order to get used to the hypotheses (a1) ÷ (a5), let us have a look into their consequences in the situations portrayed in Figures 2.2, 2.3, 2.5, 3.1, 3.2. Other situations are presented in Fig. 3.3.

In Fig. 3.3 a, the curve Φ_ℓ is indicated by the thickened line, Φ_r - by the double line; the interval $K(u_1)$ degenerates to a single point; the intervals $K(u_2)$ and $K(u_3)$ are closed, their lower end-points are located on Φ_ℓ and Φ_r, respectively; the interval $K(u_4)$ is unbounded from above. In Fig. 3.3b , two of the graphs that reflect the stratified structure of $\Omega_o(W)$ are shown. The thickened lines in Fig. 3.3 e, d refer to the parts of the curves Φ_ℓ and Φ_r which include neither the left and right end-points of the curves $\Pi(M)$ nor the points of $\Phi_o = \Phi_\ell \cap \Phi_r$. The curves shown in Fig. 3.3 e do not satisfy (a4) .

In turn, the family of curves portrayed in Fig. 3.3 f do not have the properties (a2) and (a5) (in the last example, one can easily re-define the curve Φ_r so that it will fulfil the hypotheses (a1) - (a5)).

To each point $M \in \Omega(W)$ let us assign a curve $T(M)$, contained in $\Omega(W)$, which is the graph of a continuous function $T(u;M)$ $(-\infty < u < \infty)$. The function $T(u;M)$ will be defined as follows.

If $M = \{u_o,x_o\} \in \Omega_o(W)$, then let

$$T(u;M) = \begin{cases} \Phi_\ell(u) , & \text{for } u \le u_\ell(M) , \\ \Pi(u;M), & \text{for } u_\ell(M) \le u \le u_r(M) , \\ \Phi_r(u) , & \text{for } u \ge u_r(M) . \end{cases}$$

If $M = \{u_o,x_o\} \in \Phi_\ell \cap \Phi_r$, then

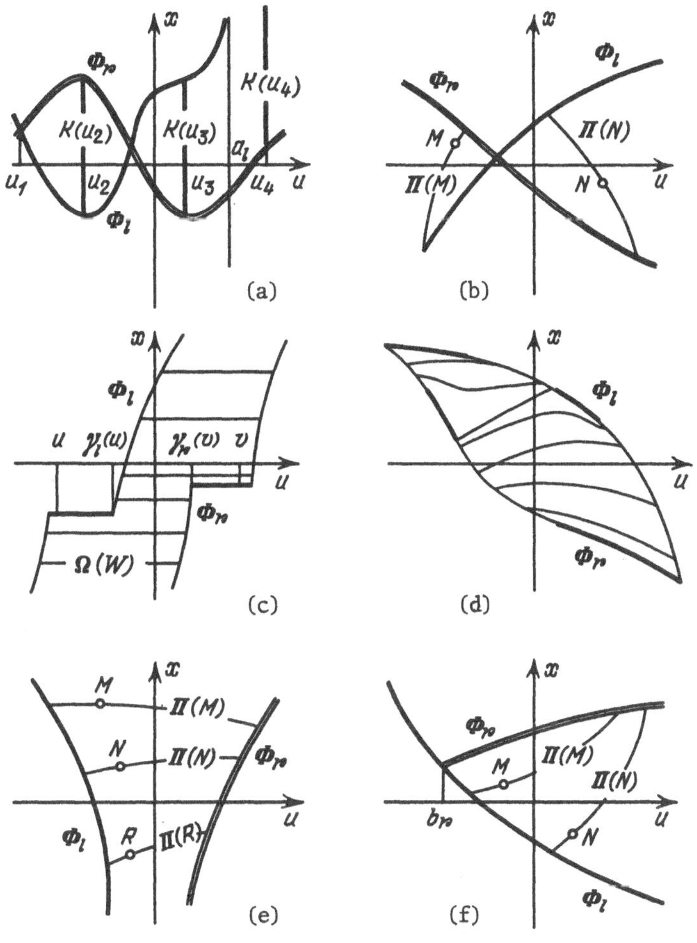

Fig. 3.3

$$T(u;M) = \begin{cases} \Phi_\ell(u)\,, & \text{for} \quad u \le u_0\,, \\ \Phi_r(u)\,, & \text{for} \quad u \ge u_0\,. \end{cases}$$

If $M = \{u_0, x_0\} \in \Phi_\ell$ and $u_0 \notin A_\ell \cup A_0$, then construct the point $M_1 = \{\gamma_\ell(u_0), \Phi_\ell[\gamma_\ell(u_0)]\}$ and take $T(u;M) = T(u;M_1)$. Finally, if

$M = \{u_0, x_0\} \in \Phi_r$ and $u_0 \notin A_r \cup A_0$, then choose the point $M_1 = \{\gamma_r(u_0), \Phi_r[\gamma_r(u_0)]\}$ and define $T(u;M) = T(u;M_1)$.

The family of curves $T(M)$ $(M \in \Omega(W))$ will be called *a defining system of curves* or *a defining system of curves for the hysteron* . The curves $T(M)$ will be called *defining curves*.

By the above description of the defining curves, for any pair M , $N \in \Omega(W)$, either $T(u;M) \equiv T(u;N)$ or the difference $T(u;M) - T(u;N)$ has fixed sign.

3.3. Piecewise monotone inputs

Now we are going to give a description of the operator $W[t_0,x_0]$. Let $M_0 = \{u(t_0),x_0\} \in \Omega(W)$ and the input $u(t)$ $(t \geq t_0)$ be monotone. Then the corresponding output $x(t) = W[t_0,x_0]u(t)$ will be defined by the equality

$$W[t_0,x_0]u(t) = T[u(t);M_0] \quad (t \geq t_0) \quad . \tag{3.7}$$

If the input $u(t)$ is only piecewise monotone, then we shall divide its domain into monotonicity intervals, and, in order to define the output $x(t) = W[t_0,x_0]u(t)$, we shall use the semigroup property (3.3).

Let $M = \{t_0,x_0\}$, $N = \{t_1,x_1\} \in T(M)$ and $u_0 < u_1$; then, by definition, $T(u;N) = T(u;M)$ for $u \geq u_1$. Analogously, it follows from $u_0 > u_1$ that $T(u;N) = T(u;M)$ for $u \leq u_1$. This implies that the definition of the operator $W[t_0,x_0]$ is correct.

Theorem 3.1. A hysteron defined on piecewise-monotone inputs belongs to the class of deterministic, static and controllable transducers .

The first two assertions of the theorem are evident. The third assertion is a consequence of the property (a4) (the first two are true also without that property). ∎

3.4. Passage to arbitrary continuous inputs

In order to determine values of the operator (3.1) which describes a hysteron for any continuous input $u(t)$ $(t \geq t_0)$, we shall use a limit construction (as in the case of any generalized play) .

Let us consider a continuous input $u_*(t)$ $(t_0 \leq t \leq t_1)$ and construct a sequence of piecewise monotone continuous inputs $u_n(t)$ $(t_0 \leq t \leq t_1,\ n = 1, 2, \ldots)$ which are convergent to $u_*(t)$ and satisfy the condition $u_n(t_0) = u_*(t_0)$. Providing $\{u_*(t_0), x_0\} \in \Omega(W)$, the outputs

$$x_n(t) = W[t_0, x_0]u_n(t) \quad (t_0 \leq t \leq t_1;\ n = 1, 2, \ldots) \ . \tag{3.8}$$

are well-defined.

As it can be shown, the sequence (3.8) is uniformly convergent to a certain function $x_*(t)$ $(t_0 \leq t \leq t_1)$; we shall treat this continuous limit function as the output of the hysteron W, corresponding to the input $u_*(t)$. Correctness of the above construction is guaranteed by the following.

<u>Theorem 3.2.</u> The hysteron W , defined on piecewise monotone continuous inputs, admits a unique extension to a deterministic, static, controllable and vibro-correct transducer - a hysteron defined for any arbitrary initial state $\{u_0, x_0\} \in \Omega(W)$ on all continuous inputs $u(t)$ $(t \geq t_0)$ such that $u(t_0) = u_0$.

Vibro-correctness means here , as already mentioned, that from $\{u_0, x_0\} \in \Omega(W)$ and by the convergence of the sequence of continuous inputs $u_n(t)$ $(t \geq t_0)$ to an input $u_*(t)$ $(t \geq t_0)$, where $u_n(t_0) = u_*(t_0) = u_0$, uniformly on any finite interval $[t_0, t_1]$, it follows that the corresponding outputs

$$x_n(t) = W[t_0, x_0]u_n(t) \quad (t \geq t_0;\ n = 1, 2, \ldots) \tag{3.9}$$

are convergent to the output

$$x_*(t) = W[t_0, x_0]u_*(t) \quad (t \geq t_0) \ , \tag{3.10}$$

uniformly on finite intervals.

The generalized play considered in Chapter 2, is an example of the hysteron. By an extension of the transducer representing a stop, treated in Section 3.1, from the class of piecewise-monotone inputs onto all admissible continuous inputs, one also will get a certain hysteron (in this case we shall preserve the name "stop").

Theorem 3.2 will be proved in Chapter 4. In the subsequent sections of this part, we shall discuss various properties of the hysterons.

4. Canonical representation of hysteron and proof of Theorem 3.2

4.1. Canonical hysteron

Let W^o be a certain deterministic transducer with the domain $\Omega(W^o)$ of the feasible states and with the defining operator $x(t) = W^o[t_o,x_o]u(t)$. Let $f(u,z)$ be a continuous function defined on $\Omega(W^o)$ and strictly monotone with respect to z (in the case of some fixed values u this function can be strictly increasing, at some others - decreasing). By a *superposition of the function* $f(u,z)$ *and transducer* W^o we shall understand the transducer

$$W = f(u,W^o) \quad , \tag{4.1}$$

defined by the operator

$$W[t_o,x_o]u(t) = f\{u(t), W^o[t_o,z_o]u(t)\} \tag{4.2}$$

with

$$x_o = f[u(t_o),z_o] \quad . \tag{4.3}$$

The domain $\Omega(W)$ of all feasible states for the transducer (4.1) is defined in a natural way, as the set of points $\{u,x\}$ such that $x = f(u,z)$ and $\{u,z\} \in \Omega(W^o)$. The inputs $u(t)$ admissible for the transducer W at the state $\{u,f(u,z)\}$ will be the same as for the transducer W^o at the state $\{u,z\}$.

If the transducer W^o is deterministic, the same follows for the transducer (4.1); if W^o is static, the same holds for W; similar implications are true for controllability, etc.

For visualization, it is useful to introduce the representation of the transducer (4.1) in the form of the block-diagram shown in Fig. 4.1.

Since the function $f(u,z)$ is strictly monotone with respect to z, for any fixed u one can determine the corresponding inverse function $f_z^{-1}(u,z)$ with a natural domain. Obviously,

$$z \equiv f[u,f_z^{-1}(u,z)] \equiv f_z^{-1}[u,f(u,z)] \quad .$$

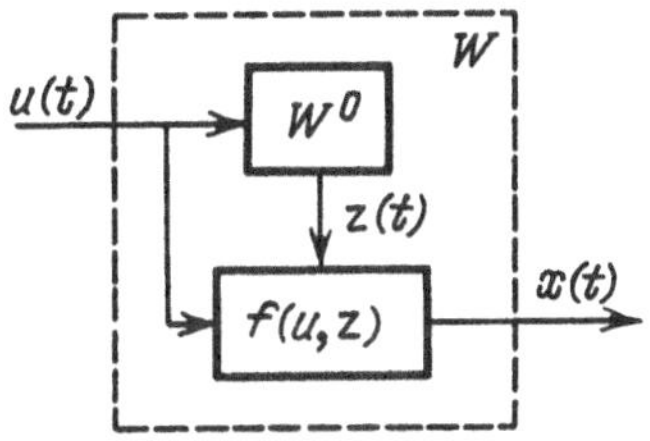

Fig. 4.1

By (4.1), it follows that

$$W^o = f_z^{-1}(u,W) \quad . \tag{4.4}$$

If W^o represents a hysteron, then also W will be a hysteron. The transfer from the curves Φ_ℓ^o and Φ_r^o (in the definition of the hysteron W^o) to the curves Φ_ℓ and Φ_r is performed via the equalities

$$\Phi_\ell(u) = f[u,\Phi_\ell^o(u)] \ , \quad \Phi_r(u) = f[u,\Phi_r^o(u] \quad . \tag{4.5}$$

If

$$M^o = \{u_o,z_o\} \in \Omega(W^o) \quad , \quad M = \{u_o,f(u_o,z_o)\} \quad ,$$

then the functions $\Pi^o(u;M^o)$ and $\Pi(u;M)$ are defined for the same values u , and

$$\Pi(u;M) = f[u,\Pi^o(u;M^o)] \qquad (u_\ell(M^o) \le u \le u_r(M^o)) \quad . \tag{4.6}$$

The most important case for our purposes concerns W^o which represents a play $L(\Gamma_\ell,\Gamma_r)$. In this case, the transducer (4.1) will be called a *canonical hysteron*. If W^o is a play , $f(u,z) = g(z)$ and $g(z)$ is strictly increasing, then the canonical hysteron (4.1) will be a play, too.

As an example of the canonical hysteron one can assume the stop considered in Section 3.1; it admits a representation in the form (4.1), provided $f(u,z) = u - z + h$ and the standard play with $\ell = 2h$ (see Fig. 2.2) is taken as W^o .

4.2. Canonical representation theorem

Let us consider a hysteron W defined on piecewise monotone continuous inputs. We now give some additional characterization of the set $\Omega(W)$ containing all feasible states of W. We shall use the notations and definitions of Chapter 3.

Let us denote by $\Delta_\ell(z)$ the set of those u $(u < a_\ell,\ u \notin A_\ell \cup A_0)$ for which the function (3.5) assumes the fixed value z. By virtue of the property (a5), the set $\Delta_\ell(z)$ is then a right-sided open interval. We are going to show that *this interval is always bounded.*

Suppose that $\Delta_\ell(z)$ is unbounded and $z \in A_\ell$. Draw the graph $\Pi(M)$ through the point $M = \{z, \Phi_\ell(z)\}$ and denote by N its right end-point (it belongs to Φ_r, see Fig. 4.2). Due to (a5), there is a point N_1 in Φ_r, which is located on the left of N and is the right end-point of the graph $\Pi(N_1)$, at the same time. By virtue of (a4), the left end-point M_1 of this graph is then located to the left of M, but this contradicts the left-sided unboundedness of the interval $\Delta_\ell(z)$.

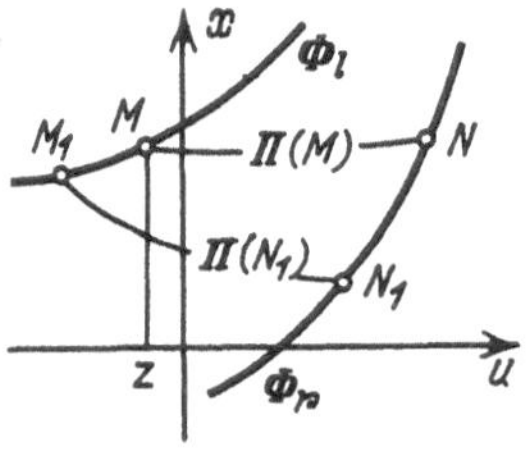

Fig. 4.2

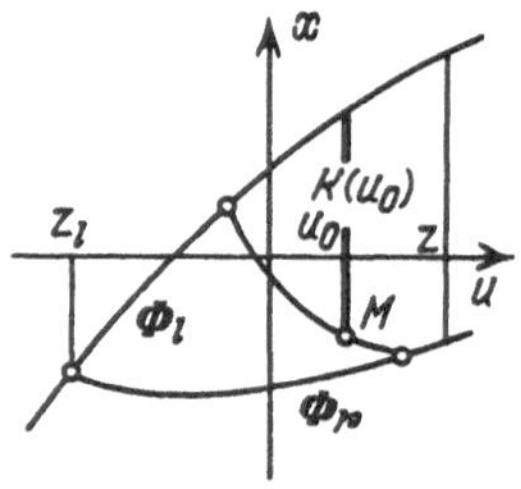

Fig. 4.3

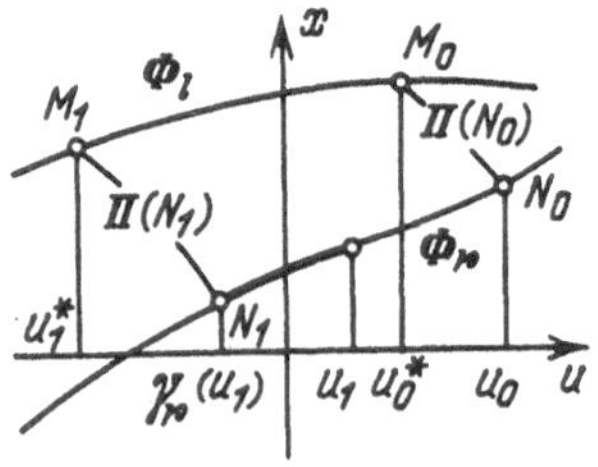

Fig. 4.4

In turn, suppose that the interval $\Delta_\ell(z)$ is unbounded and $z \in A_0$. Then $b_r < z$ and, by virtue of the unboundedness of $\Delta_\ell(z)$, the functions $\Phi_\ell(u)$ and $\Phi_r(u)$ do not coincide for any input $u \in (b_r, z)$.

Thus, every vertical interval $K(u)$ $(b_r < u < z)$ contains interior points which belong to the graphs $\Pi(M)$ whose left end-points are located in Φ_ℓ to the left of the point $\{z,\Phi_\ell(z)\}$. This, in turn, contradicts the unboundedness of $\Delta_\ell(z)$.

Therefore, the interval $\Delta_\ell(z)$ is bounded. Denote its left end-point by z_ℓ . We are going to prove that $z_\ell \in \Delta_\ell(z)$. To this end, it suffices to show that $z_\ell \notin A_o \cup A_\ell$.

Indeed, if $z_\ell \in A_o$ then the functions $\Phi_\ell(u)$ and $\Phi_r(u)$ assume different values on (z_ℓ,z) , hence the vertical segments $K(u)$ $(z_\ell < u < z)$ contain interior points. Choose one of those points and draw the graph $\Pi(M)$ through it. By (a3), the part of the graph, located to the left of the point M (Fig. 4.3), cannot have common points with the curve Φ_r . Thus, the left end-point of the graph ought to be located in Φ_ℓ on the right of the point $\{z_\ell,\Phi_\ell(z_\ell)\}$. We have come to contradiction. We would come to analogous contradiction also in the case of $z_\ell \in A_\ell$.

In an analogous way one can introduce the intervals $\Delta_r(z)$ as the sets of those u $(u > b_r,\ u \notin A_r \cup A_o)$ for which the function (3.6) assumes the value z . Any interval $\Delta_r(z)$ is then bounded, open to the left and contains its right end-point z_r .

In the sequel we shall also consider the corresponding closed intervals $\bar\Delta_\ell(z) = [z_\ell,z]$ and $\bar\Delta_r(z) = [z,z_r]$.

Lemma 4.1. Let a finite or countable family of non-intersecting finite intervals Δ_i be given on $(-\infty,a)$. Then one can define on $(-\infty,a)$ a continuous non-decreasing function $\Gamma(u)$ which is piecewise constant, assumes constant values at most on the intervals Δ_i , and for which the equalities

$$\lim_{u\to-\infty} \Gamma(u) = -\infty , \quad \lim_{u\to a} \Gamma(u) = \infty \tag{4.7}$$

are true .

The proof is left to the reader. ∎

Theorem 4.1 (on the canonical representation).
Any hysteron W , defined on piecewise monotone, continuous inputs, admits a canonical representation

$$W = f[u,L(\Gamma_\ell,\Gamma_r)] \quad . \tag{4.8}$$

Proof. All what we have to do, is to construct functions $\Gamma_\ell(u)$, $\Gamma_r(u)$ and $f(u,z)$.

By $\Gamma_\ell(u)$ $(-\infty < u < a_\ell)$ we shall denote a non-decreasing, continuous function which assumes constant values within the intervals $\bar{\Delta}_\ell(z)$, and such that the equalities (4.7) are satisfied. Such a construction of $\Gamma_\ell(u)$ can be performed by virtue of Lemma 4.1 .

Now let us construct the function $\Gamma_r(u)$. We shall assume that this function equals $\Gamma_\ell(u)$ over the whole domain A_0 . If $u_0 \in A_r$, then consider the graph $\Pi(N_r)$ where $N_r = \{u_0,\Phi_r(u_0)\}$, find the abscissa u_0^* of its left end-point $N_\ell = \{u_0^*,\Phi_\ell(u_0^*)\}$ and take $\Gamma_r(u_0) = \Gamma_\ell(u_0^*)$. If $u_0 \notin A_0 \cup A_r$ and $\gamma_r(u_0) \in A_0$, then set $\Gamma_r(u_0) = \Gamma_\ell[\gamma_r(u_0)]$. If, at last, $u_0 \notin A_0 \cup A_r$ and $\gamma_r(u_0) \in A_r$, then consider the graph $\Pi(N_r)$ where $N_r = \{\gamma_r(u_0),\Phi_r[\gamma_r(u_0)]\}$, find the abscissa u_0^* of its left end-point N_ℓ and set $\Gamma_r(u_0) = \Gamma_\ell(u_0^*)$. The second and fourth cases are depicted in Fig. 4.4.

In this way, we have defined $\Gamma_r(u)$ on (b_r,∞) . By construction, this function is non-decreasing; it is constant on the intervals $\bar{\Delta}_r(z)$. The ranges of functions $\Gamma_\ell(u)$ and $\Gamma_r(u)$ are identical. Thus, these ranges cover the whole real line. Hence it follows that the function $\Gamma_r(u)$ is continuous. Clearly, the graph of the function $\Gamma_r(u)$ is located to the right of the graph of the function $\Gamma_\ell(u)$.

Given the functions $\Gamma_\ell(u)$ and $\Gamma_r(u)$, let us define the play $L(\Gamma_\ell,\Gamma_r)$. To this end, we only need to construct the function $f(u,z)$ $(\{u,z\} \in \Omega(\Gamma_\ell,\Gamma_r))$.

On the curve $z = \Gamma_\ell(u)$, let $f[u,\Gamma_\ell(u)] = \Phi_\ell(u)$. On the curve $z = \Gamma_r(u)$, in turn, let $f[u,\Gamma_r(u)] = \Phi_r(u)$. Through every point $\{u,z\} \in \Omega(\Gamma_\ell,\Gamma_r)$ which does not belong to $\Gamma_\ell \cup \Gamma_r$, draw a horizontal segment such that only its left end-point $P_\ell = \{u_\ell,\Gamma_\ell(u_\ell)\}$ is located on Γ_ℓ ; clearly, $u_\ell \in A_\ell$ and thus the graph $\Pi(M_\ell)$ includes the point $M_\ell = \{u_\ell,\Phi_\ell(u_\ell)\}$ of the plane $\{u,x\}$; we then define $f(u,z) = \Pi(u;M\ )$.

The strong monotonicity of the function $f(u,z)$ with respect to z and its joint continuity are evident. Also the equality (4.8) can be easily verified, one only has to notice that the mapping under consideration transforms the horizontal segments located in $\Omega(\Gamma_\ell,\Gamma_r)$ into the graphs $\Pi(M)$.

∎

Since the above construction of $\Gamma_{\ell}(u)$ is non-unique, also the canonical representation of any hysteron is non-uniquely determined.

4.3. Proof of Theorem 3.2

Proof. All assertions of Theorem 3.2 follow from Theorem 4.1 and the properties of the generalized play, established in Chapter 2. ∎

As a consequence of Theorems 4.1 and 3.2 we can conclude the following.

Theorem 4.2 (on the canonical representation).
Let a hysteron W with any initial state $\{u_0,x_0\} \in \Omega(W)$ be defined for all continuous inputs $u(t)$ $(t \geq t_0)$ such that $u(t_0) = u_0$. Then this hysteron admits the representation (4.8). ∎

4.4. Properties of hysteron

The representation (4.8) and the properties of the generalized play directly imply several important properties of hysterons. We shall mention here some of them.

a. Let a hysteron W be defined by the operator

$$x(t) = W[t_0,x_0]u(t) \quad (t \geq t_0) \tag{4.9}$$

on suitable continuous inputs. In this case, the operators (4.9) are continuous with respect to initial states. Moreover, if a sequence of continuous inputs $u_n(t)$ $(t \geq t_0)$ is uniformly convergent to an input $u_*(t)$ $(t \geq t_0)$ on any finite interval, $u_n(t_0) = u_*(t_0) = u_0$, and the sequence of initial states $\{u_0,x_n\} \in \Omega(W)$ is convergent to the initial state $\{u_0,x_0\} \in \Omega(W)$, then the sequence of outputs

$$x_n(t) = W[t_0,x_n]u_n(t) \quad (t \geq t_0) \tag{4.10}$$

converges to the output

$$x_*(t) = W[t_0,x_0]u_*(t) \quad (t \geq t_0) \tag{4.11}$$

uniformly on any finite interval.

If the inputs $u_n(t)$ $(t \geq t_0)$ in the above hypotheses are uniformly bounded and converge to $u_*(t)$ uniformly on the interval $[t_0,\infty)$, then the corresponding outputs (4.10) converge to (4.11) uniformly on $[t_0,\infty)$.

b. In contrast to the generalized play, not all hysterons are monotone. To discuss possible situations, we introduce a special ordering in the domain $\Omega(W)$ which comprises all attainable states of the hysteron W.

Let the defining curves $T(M_1)$ and $T(M_2)$ coincide with the curve Φ_ℓ respectively on intervals $(-\infty,u_1]$ and $(-\infty,u_2]$, where $u_1 \leq u_2$. We shall then say that the point M_1 is *located west of* the point M_2, and write $M_1 \prec M_2$. The points N_1, N_2, N_3 in Fig. 4.5 are located west of the point M. Clearly, the relationship "located west of" is transitive and sustains a limit passage. If W is a play, the relation $M \prec N$ means that the point M is not located over the point N.

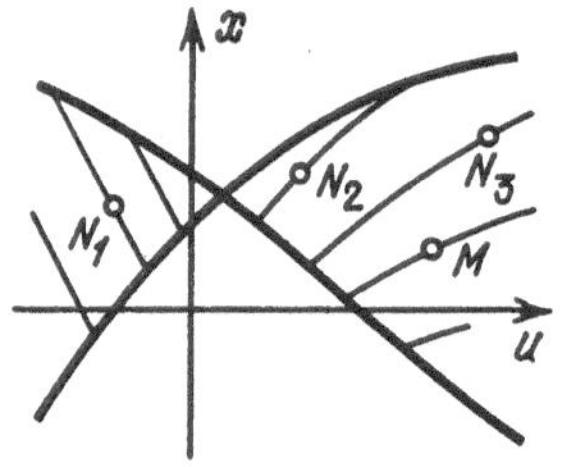

Fig. 4.5

As an analogue of Theorem 2.3, the following statement holds.

Lemma 4.2. Suppose that continuous inputs $u(t)$, $v(t)$ $(t_0 \leq t \leq t_1)$ satisfy the inequality

$$u(t) \leq v(t) \quad (t_0 \leq t \leq t_1) , \tag{4.12}$$

and the initial states of the hysteron W are related by

$$\{u(t_0),x_0\} \prec \{v(t_0),y_0\} . \tag{4.13}$$

Then

$$\{u(t), W[t_0,x_0]u(t)\} < \{v(t), W[t_0,y_0]v(t)\} \quad (t_0 \leq t \leq t_1) \quad . \qquad (4.14)$$

∎

The hysteron W is called *monotone* if all its defining curves T(M) are graphs of some monotone non-decreasing functions and the upper end-point of each interval K(u) belongs to the curve Φ_ℓ . Only the left of the hysterons depicted in Fig. 4.6 is monotone. By Lemma 4.2, we can conclude the following.

Theorem 4.3. Let the hysteron W be monotone. Then by (4.12) and (4.13) it follows that the inequality

$$W[t_0,x_0]u(t) \leq W[t_0,y_0]v(t) \quad (t_0 \leq t \leq t_1) \qquad (4.15)$$

is satisfied. ∎

All hysterons are monotone with respect to the initial value x_0 of the output (for any input u(t)).

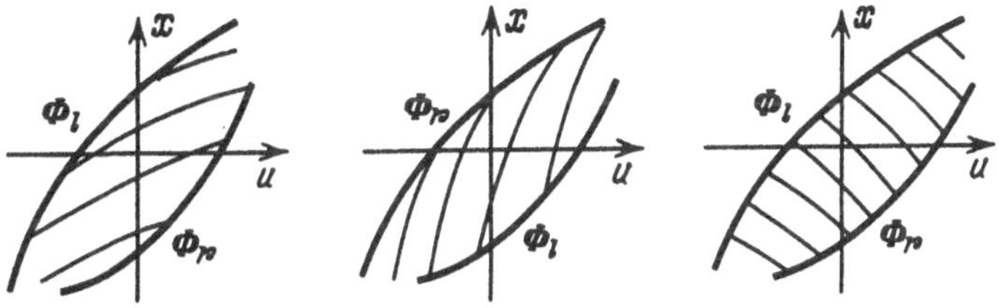

Fig. 4.6

Let us also note that the hysteron W is a generalized play if for every x_0 and for arbitrary inputs u(t), v(t) $(t \geq t_0)$, satisfying the hypothesis $\{u(t_0),x_0\}, \{v(t_0),x_0\} \in \Omega(W)$, relation (4.12) implies the inequality

$$W[t_0,x_0]u(t) \leq W[t_0,x_0]v(t) \quad (t \geq t_0) \quad .$$

c. If $M_1 < M_2$, then we shall say that the defining curve $T(M_1)$ is *situated west of the curve* $T(M_2)$, and the curve $T(M_2)$ - *east of the curve* $T(M_1)$.

Let $G \subset \Omega(W)$ be a compact set. Then among T(M) $(M \in G)$ there exist two curves $T_{west}(G)$ and $T_{east}(G)$, which are situated to the west and east of all elements of $\Omega(M)$, respectively. These curves represent the graphs of functions $T_{west}(u;G)$ and $T_{east}(u;G)$, correspondingly. Assume that the projection of a compact set G onto the axis u is contained in the interval $[a,b]$ and denote by $G(a,b)$ the set (in Fig. 4.7 traced out by the thickened line) of those states of hysteron W which are attainable from the state $\{u(t_0),x_0\} \in G$ by means of the inputs $u(t)$ with values in $[a,b]$. This definition immediately implies the following

Lemma 4.3. The set $G(a,b)$ consists of the points $M = \{u,x\} \in \Omega(W)$ which fulfil the relations

$$\{a,T_{west}(a;G)\} \prec M \prec \{b,T_{east}(b;G)\} , \quad a \leq u \leq b . \tag{4.16}$$

∎

Lemma 4.3 yields estimates of the values of the output signals.

d. In the case of a T-periodic input $u(t)$ $(t \geq t_0)$, the corresponding output (4.9) is T-periodic for $t \geq t_0 + T$, i.e., the hysteron is monocyclic. The closed curve

$$u = u(t), \ x = W[t_0,x_0]u(t) \quad (t_0 + T \leq t \leq t_0 + 2T) \tag{4.17}$$

is a *hysteresis loop* corresponding to the input $u(t)$.

Unlike for the generalized play, the spin of a hysteron is not necessarily positive. The fine lines in Fig. 4.8 indicate some defining curves, and the thickened ones - the hysteresis loops corresponding to the output signals $u(t) = \sin t$; the spin of the first loop is negative, the second loop has positive spin, and the third - undefined.

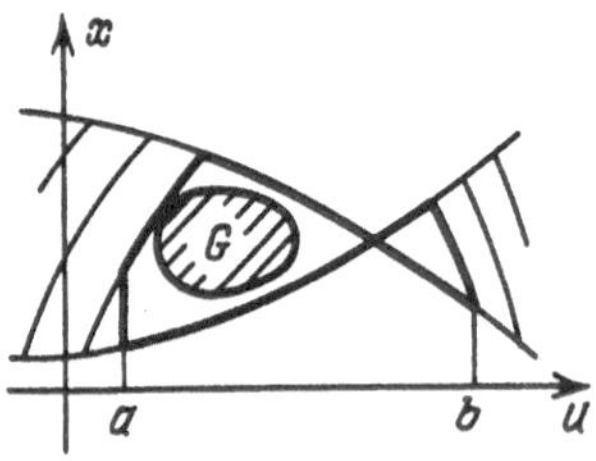

Fig. 4.7

In the case of an asymptotically periodic input $u(t)$, also the output (4.9) is asymptotically periodic; in the case of an asymptotically almost periodic input, (4.9) becomes asymptotically almost periodic.

The properties of inputs defined over the whole real axis, established in the Section 2.9 for the play, in a natural way apply also to the hysteron.

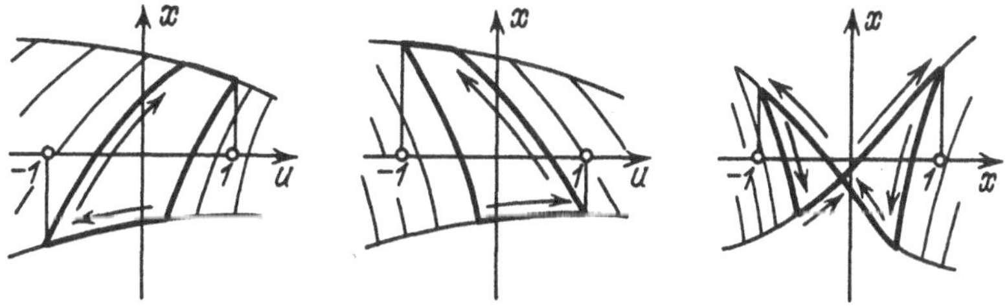

Fig. 4.8

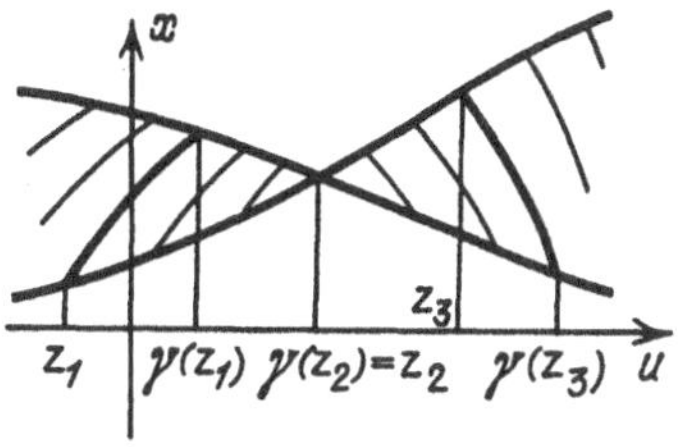

Fig. 4.9

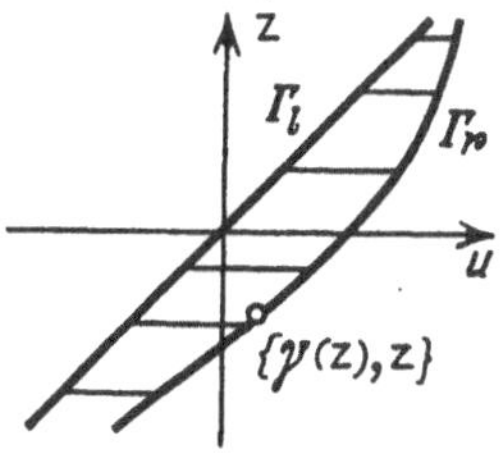

Fig. 4.10

4.5. Rectification of hysteron

The generalized play $L(\Gamma_\ell,\Gamma_r)$ will be called *left- sided ordinary* (*left-ordinary*) if $\Gamma_\ell(u) \equiv u$.

The hysteron W will be called *left-nondegenerate* if the function $\Phi_\ell(u)$ is defined for all u , and every point of the curve Φ_ℓ either belongs to Φ_0 or is the left end-point of a certain curve $\Pi(M)$.

Analogously, one can introduce the notion of a *right-nondegenerate* hysteron. It follows from the proof of Theorem 4.1 that *in the canonical representation* (4.8) *of the left-nondegenerate hysteron, the play* $W^0 = L(\Gamma_\ell,\Gamma_r)$ *can be regarded as left-ordinary.*

Let us note that the hysteron W admits the representation

$$W[t_0,x_0]u(t) = f\{u(t),L_0[t_0,z_0]g[u(t)]\} , \tag{4.18}$$

where $g(x)$ and $f(u,x)$ are continuous and strictly monotone in x . L_0 is an ordinary play if and only if W is both left- and right-

-nondegenerate and moreover the curves Φ_ℓ and Φ_r do not intersect at any point.

Let the hysteron W be left- and right-sided nondegenerate, at the same time. We then introduce $\gamma(z)$ $(-\infty < z < \infty)$ and $\Pi_0(u,z)$ $(z \leq u \leq \gamma(z),\ -\infty < z < \infty)$ according to the following rule. If the point $\{z,\Phi_\ell(z)\}$ belongs to Φ_r, set $\gamma(z) = z$ and $\Pi_0(z,z) = z$. If, in turn, the point $\{z,\Phi_\ell(z)\}$ is the left end-point of the graph $\Pi(M)$ of some function $\Pi(u;M)$ defined on the interval $z \leq u \leq u_r(M)$, then set $\gamma(z) = u_r(M)$ and $\Pi_0(u,z) = \Pi(u;M)$. The characteristic cases are shown in Fig. 4.9.

The functions $\gamma(z)$ and $\Pi_0(u,z)$ are continuous and strictly monotone with respect to z. By the definition of these functions, the hysteron W admits the following representation

$$W[t_0,\Pi_0[u(t_0),z_0]]u(t) = \Pi_0\{u(t),L[t_0,z_0;\Gamma_\ell,\Gamma_r]u(t)\}\ , \tag{4.19}$$

where the generalized play $L(\Gamma_\ell,\Gamma_r)$ is left-sided ordinary and Γ_r is the graph of function $u = \gamma(z)$ (see Fig. 4.10).

By virtue of Theorem 2.2, the representation (4.19) gives rise to estimates on increments of the output signals corresponding to variations of the initial states and input signals. In particular, if all the functions whose graphs build up a certain hysteron satisfy the general Lipschitz condition, and the Lipschitz condition is fulfilled by the function $\gamma(z)$, then also for the hysteron W the Lipschitz condition

$$|W[t_0,x_0]u(t) - W[t_0,y_0]v(t)| \leq$$
$$\leq \lambda \max\{|x_0 - y_0|,\ \|u(s) - v(s)\|_{t_0,t}\} \quad (t \geq t_0) \tag{4.20}$$

holds, where λ is a certain constant.

5. Distances

5.1. Definition of distance

Let us continue the study of the hysteron W characterized in Section 3.2 by a defining system of curves.

Denote by $D_\ell(M)$ $(M \in \Omega(W))$ the set which contains u such that

$M \in T(N_u)$, $N_u = \{u,\Phi_\ell(u)\}$; the set $D_r(M)$ is defined analogously.

If the hysteron W is nondegenerate, then each of the sets $D_\ell(M)$ and $D_r(M)$ is a singleton. Let us define

$$d_\ell(M,N) = \min_{u\in D_\ell(M),\, v\in D_\ell(N)} |u - v| \;,$$

$$d_r(M,N) = \min_{u\in D_r(M),\, v\in D_r(N)} |u - v| \;. \tag{5.1}$$

$d_\ell(M,N)$ and $d_R(M,N)$ will be appropriately called a *left* and *right distance* between the points M and N . In general, the left distance is different from the right one; however, if either of them equals zero, the same is true for the other.

The distances do not have properties of a metric. Firstly, the distance between two different points can be zero (in Fig. 5.1, the distance between M and N), secondly, the triangle inequality is not always satisfied (as in Fig. 5.1, for the points M, M_1, M_2) .

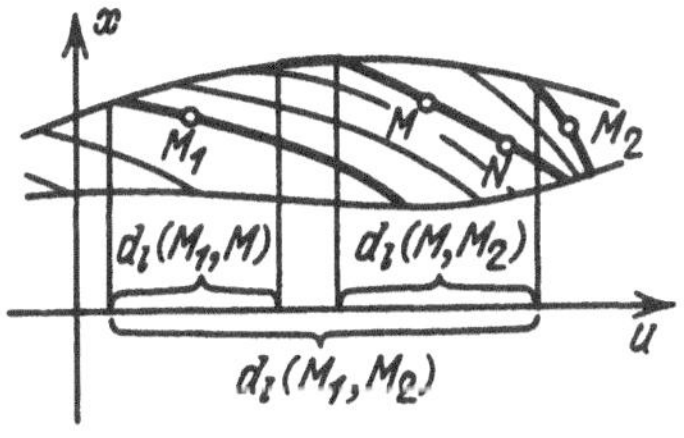

Fig. 5.1

In Section 3.2 , the functions $u_\ell(M)$ and $u_r(M)$ have been defined for all points $M \in \Omega_0(W)$. In our further considerations, we shall use extensions of those functions onto all points $M = \{u_0,x_0\} \in \Omega(W)$. We define

$$u_\ell(M) = \max\{u\colon u \le u_0,\ \{u,\Phi_\ell(u)\} \in T(M)\} \;, \tag{5.2}$$

$$u_r(M) = \min\{u\colon u \ge u_0,\ \{u,\Phi_r(u)\} \in T(M)\} \;. \tag{5.3}$$

Each of the relationships $u_\ell(M) \le u_\ell(N)$ and $u_r(M) \le u_r(N)$ means that the curve $T(M)$ is located to the west of $T(N)$. Therefore, the curve $T(M)$ coincides with curves $T(M_1)$ and $T(M_2)$, where $M_1 = \{u_\ell(M), \Phi_\ell[u_\ell(M)]\}$ and $M_2 = \{u_r(M), \Phi_r[u_r(M)]\}$.

A set $F \subset \Omega(W)$ will be called (a,b)-*invariant*, if for every input $u(t)$ $(t \geq t_0)$ it follows from $a \leq u(t) \leq b$ and $\{u(t_0), x_0\} \in F$ that $\{u(t), W[t_0, x_0]u(t)\} \in F$ $(t \geq t_0)$.

Consider two fixed continuous inputs $u(t)$, $v(t)$ $(t \geq t_0)$ with values in $[a,b]$. Let the points $M_0 = \{u(t_0), x_0\}$ and $N_0 = \{v(t_0), y_0\}$ belong to an (a,b)-invariant set $F \subset \Omega(W)$, and moreover $M_0 \prec N_0$. Define

$$M(t) = \{u(t), W[t_0, x_0]u(t)\}, \quad N(t) = \{v(t), W[t_0, y_0]v(t)\} . \tag{5.4}$$

Lemma 5.1. Let $t_1 > t_0$. Then, alternatively, either

$$M_0 \prec M(t_1) \prec N_0, \; M_0 \prec N(t_1) \prec N_0 , \tag{5.5}$$

or $u_\ell[M(t_1)]$, $u_\ell[N(t_1)] \in [a,b]$ and

$$d_\ell[M(t_1), N(t_1)] \leq \|u(s) - v(s)\|_{t_0, t_1} , \tag{5.6}$$

or $u_r[M(t_1)]$, $u_r[N(t_1)] \in [a,b]$ and

$$d_r[M(t_1), N(t_1)] \leq \|u(s) - v(s)\|_{t_0, t_1} . \tag{5.7}$$

Proof. At the beginning it is convenient to pass to the canonical representation of the hysteron W; this permits to restrict the considerations to plays. The assertion of the lemma follows then quite simply in the case of monotone inputs (although the arguments are quite tedious - various combinations of increasing and decreasing inputs are to be analyzed). The transfer to piecewise monotone inputs is performed in a standard way. To complete the proof, one should pass to the limit; this is possible since the distances are upper semicontinuous. ∎

5.2. Estimates on differences of output signals

Consider an (a,b)-invariant set $F \subset \Omega(W)$. We shall characterize the proximity of the curves $T(M)$ and $T(N)$ that contain the points $M = \{u_M,x\}$ and $N = \{u_N,y\}$, by the values

$$\kappa(M,N;\delta) = \sup_{v_1,v_2 \in [a,b];\,|v_1-v_2|\le\delta} |T(v_1;M) - T(v_2;N)| \quad (\delta > 0)\,. \tag{5.8}$$

Define also

$$\kappa_\ell(\delta;F) = \sup_{M,N \in F\cap\Phi_\ell;\,|u_M-u_N|\le\delta} \kappa(M,N;\delta)\,, \tag{5.9}$$

$$\kappa_r(\delta;F) = \sup_{M,N \in F\cap\Phi_r;\,|u_M-u_N|\le\delta} \kappa(M,N;\delta)\,, \tag{5.10}$$

where the upper bounds over empty sets are assumed as equal to zero.

The following property is a consequence of Lemma 5.1.

Theorem 5.1. Suppose that the values of continuous inputs $u(t)$, $v(t)$ $(t \ge t_0)$ belong to the interval $[a,b]$, with the points $M_0 = \{u(t_0),x_0\}$ and $N_0 = \{v(t_0),y_0\}$ in an (a,b)-invariant set F. Then for any $t \ge t_0$ the estimate

$$|W[t_0,x_0]u(t) - W[t_0,y_0]v(t)| \le$$

$$\le \max\{\kappa_\ell[\delta(t);F],\kappa_r[\delta(t);F],\kappa[M_0,N_0;\delta(t)]\} \tag{5.11}$$

is satisfied with

$$\delta(t) = \|u(s) - v(s)\|_{t_0,t}\,. \tag{5.12}$$

∎

The following properties are conclusions from Theorem 5.1.

a. Suppose that the estimate

$$|T(u;\{u_0,x_0\}) - T(v;\{v_0,y_0\})| \le \lambda_1|u_0 - v_0| + \lambda_2|x_0 - y_0| +$$
$$+ \lambda_3|u - v| \quad , \tag{5.13}$$

holds for $u, v \in [a,b]$ and $\{u_0,x_0\}, \{v_0,y_0\} \in F$, and the intersections of the curves Φ_ℓ, Φ_r with F are graphs of some functions satisfying the Lipschitz condition with a constant λ_4 (clearly, $\lambda_4 \le \lambda_3$). Then, under the hypotheses of Theorem 5.1,

$$|W[t_0,x_0]u(t) - W[t_0,y_0]v(t)| \le$$
$$\le (\lambda_1 + \lambda_3)\|u(s) - v(s)\|_{t_0,t} + \lambda_2 \max\{|x_0 - y_0|, \lambda_4\|u(s) - v(s)\|_{t_0,t}\} . \tag{5.14}$$

In particular,

$$|W[t_0,x_0]u(t) - W[t_0,x_0]v(t)| \le (\lambda_1 + \lambda_3 + \lambda_2\lambda_4)\|u(s) - v(s)\|_{t_0,t} . \tag{5.15}$$

<u>b.</u> The constant $\lambda_1 + \lambda_3 + \lambda_2\lambda_4$ which enters Lipschitz condition (5.15), is not optimal. We shall give some more accurate estimates.

Denote by P the operator of orthogonal projection onto the abscissa axis in the plane $\{u,x\}$. Suppose that the set F does not contain any point of $\Phi_0 = \Phi_\ell \cap \Phi_r$. Assume that all the functions $\Pi(u;M)$ satisfy the Lipschitz condition on the set $P[F \cap \Pi(M)]$ with the same constant $\lambda_\Pi = \lambda_\Pi(F)$; moreover, for every $u_0 \in P[F \cap \Phi_\ell]$, $u_0 \ne b$,

$$\varlimsup_{u \to u_0 + 0} \left| \frac{\Phi_\ell(u) - \Pi(u\ ;\{u_0,\Phi_\ell(u_0)\})}{u - u_0} \right| \le \lambda_\Phi \quad , \tag{5.16}$$

and for every $u_0 \in P(F \cap \Phi_r)$, $u_0 \ne a$,

$$\varlimsup_{u \to u_0 - 0} \left| \frac{\Phi_r(u) - \Pi(u\ ;\{u_0,\Phi_r(u_0)\})}{u - u_0} \right| \le \lambda_\Phi \quad . \tag{5.17}$$

Furthermore, let for every $M_* = \{u_*,x_*\} \in F$ and $N_* = \{u_*,y_*\} \in F$

$$|\Pi(u;M_*) - \Pi(u;N_*)| \le \lambda_* |x_* - y_*| \quad (u \in P[F\cap\Pi(M_*)] \cap P[F \cap \Pi(N_*)]) \ . \tag{5.18}$$

Then for $\{u(t_0),x_0\}$, $\{v(t_0),y_0\} \in F$, $a \le u(t),v(t) \le b$ $(t \ge t_0)$, the estimate

$$|W[t_0,x_0]u(t) - W[t_0,y_0]v(t)| \le \lambda_\Pi \|u(s) - v(s)\|_{t_0,t} +$$
$$+ \lambda_* \max\{\lambda_\Phi \|u(s) - v(s)\|_{t_0,t}, \lambda_\Pi |u(t_0) - v(t_0)| +$$
$$+ |x_0 - y_0|\} \tag{5.19}$$

is true .

In particular, for every initial state $\{u_0,x_0\} \in F$, the operator $W[t_0,x_0]$ satisfies the Lipschitz condition

$$|W[t_0,x_0]u(t) - W[t_0,x_0]v(t)| \le (\lambda_\Pi + \lambda_*\lambda_\Phi) \|u(s) - v(s)\|_{t_0,t} \quad , \tag{5.20}$$

provided $u(t_0) = v(t_0) = u_0$, $a \le u(t),v(t) \le b$.

The estimate (5.19) is preserved if the set $F \cap \Phi_\ell \cap \Phi_r$ remains non-empty, but mes $P(F\cap\Phi_\ell\cap\Phi_r) = 0$. In the general case, the following estimate is true:

$$|W[t_0,x_0]u(t) - W[t_0,y_0]v(t)| \le \max\{Q,\lambda_0 \|u(s) - v(s)\|_{t_0,t}\} \ , \tag{5.21}$$

where Q is the right-hand side of inequality (5.19) and λ_0 is a real number such that

$$|\Phi_\ell(u) - \Phi_\ell(v)| \le \lambda_0 |u - v| \quad , \quad u, v \in P(F\cap\Phi_\ell\cap\Phi_r) \ .$$

As a consequence of (5.21), the estimate

$$|W[t_0,x_0]u(t) - W[t_0,x_0]v(t)| \le \max\{\lambda_0,\lambda_\Pi + \lambda_*\lambda_\Phi\} \|u(s) - v(s)\|_{t_0,t} \tag{5.22}$$

follows, in particular.

c. Let all functions $T(u;M)$ $(M \in F)$ have a joint continuity modulus $\mu_T(r)$ on the interval $[a,b]$. Assume that the following inequalities hold for all $v_1, v_2 \in [a,b]$:

$$|T(u;\{v_1,\Phi_\ell(v_1)\}) - T(u;\{v_2,\Phi_\ell(v_2)\})| \leq \mu(|v_1 - v_2|) \quad , \quad a \leq u \leq b \quad ;$$

$$|T(u;\{v_1,\Phi_r(v_1)\}) - T(u;\{v_2,\Phi_r(v_2)\})| \leq \mu(|v_1 - v_2|) \quad , \quad a \leq u \leq b \quad .$$

Then, under the hypotheses of Theorem 5.1,

$$|W[t_o,x_o]u(t) - W[t_o,y_o]v(t)| \leq \mu_T \; \|u(s) - v(s)\|_{t_o,t} +$$

$$+ \max\{\mu \; \|u(s) - v(s)\|_{t_o,t} \; , \; \|T(u;M_o) - T(u;N_o)\|_{a,b}\} \; . \tag{5.23}$$

The last inequality has a simpler form than the estimate (5.11), but it is less accurate, at the same time.

d. The estimate (5.23) admits a refinement. Suppose that for every $\{u_o,\Phi_\ell(u_o)\} = M_o \in F$, the estimate

$$|\Phi_\ell(u) - \Pi(u;M_o)| \leq \mu_\Phi(|u - u_o|) \quad , \quad u \in P[F \cap \Pi(M_o)]$$

takes place, whereas, for every $\{v_o,\Phi_r(v_o)\} = N_o \in F$, the estimate

$$|\Phi_r(u) - \Pi(u;N_o)| \leq \mu_\Phi(|u - v_o|) \quad , \quad u \in P[F \cap \Pi(N_o)]$$

holds. Assume that all functions $\Pi(u;M)$ $(M \in F)$ have the same continuity modulus $\mu_\Pi(r)$ over the relevant sets $P[F \cap \Pi(M)]$; suppose also that all functions $T(u;M)$ $(M \in F)$ have the same continuity modulus $\mu_T(r)$ over the interval $[a,b]$, and the estimate

$$|\Pi(u;M_*) - \Pi(u;N_*)| \leq \mu_*(|x_* - y_*|) \quad , \quad u \in P[F \cap \Pi(M_*)] \cap P[F \cap \Pi(N_*)]$$

holds for $M_* = \{u_*,x_*\} \in F$, $N_* = \{u_*,y_*\} \in F$.

Then, for the inputs $u(t)$, $v(t)$ such that $\{u(t_0),x_0\}$, $\{v(t_0),y_0\} \in F$ and $a \leq u(t), v(t) \leq b \quad (t \geq t_0)$, the estimate

$$|W[t_0,x_0]u(t) - W[t_0,y_0]v(t)| \leq$$

$$\leq \max\{Q,\mu_T(\|u(s) - v(s)\|_{t_0,t})\} \quad , \tag{5.24}$$

takes place with

$$Q = \mu_\Pi(\|u(s) - v(s)\|_{t_0,t}) +$$

$$+ \mu_*[\max\{\mu_\Phi(\|u(s) - v(s)\|_{t_0,t}), |x_0 - y_0| + \mu_\Pi(|u(t_0) - v(t_0)|)\}] \quad .$$

e. For several classes of hysterons (generalized plays, stops , etc.) the estimates (5.19), (5.20), (5.24) are optimal. For concrete hysterons, these estimates may assume quite simple form. For example, consider the hysteron depicted in Fig. 3.1. That hysteron describes a perfectly plastic Prandtl's fibre with elasticity modulus E . The estimate (5.19) assumes then the form

$$|W[t_0,x_0]u(t) - W[t_0,y_0]v(t)| \leq |x_0 + y_0| + 2E\|u(s) - v(s)\|_{t_0,t} \tag{5.25}$$

and the estimate (5.24) can be rewritten as

$$|W[t_0,x_0]u(t) - W[t_0,y_0]v(t)| \leq E\|u(s) - v(s)\|_{t_0,t} +$$

$$+ \max\{E\|u(s) - v(s)\|_{t_0,t}, |x_0 - y_0| + E|u(t_0) - v(t_0)|\} \quad . \tag{5.26}$$

6. Various input spaces

6.1. Statement of the problem

Until now, we have studied properties of the hysteron W with the operators

$$x(t) = W[t_0,x_0]u(t) \tag{6.1}$$

defined on the space C of continuous inputs. Moreover, those operators were treated as acting from the space C into the same space. In several situations it appears useful to consider (6.1) as an operator defined on some other space E of the inputs and acting either into the same space E or into some space E_1. If the space E comprises only continuous functions, we merely shall attempt to establish new properties of the operator (6.1). If, however, the space E contains also discontinuous functions, then the operator (6.1) must be reasonably extended onto the discontinuous inputs.

If (6.1) acts from E into E_1 ($E_1 \neq E$), then its properties (regularity, continuity, monotonicity, possibility of its approximation by various classes of operators, etc.) are hardly known. Nevertheless, some features have already become clear; we briefly discuss them in this chapter.

Recall that a space E is called *embedded* or *continuously embedded* in a space E_1, if any element of the space E simultaneously belongs to E_1, and moreover

$$\|x\|_{E_1} \leq \lambda \|x\|_E \quad (x \in E) \quad , \lambda\text{-finite constant.} \tag{6.2}$$

The operator I which assigns to any element $x \in E$ the same x viewed as an element of the space E_1, is called an *embedding operator*. If the embedding operator is compact then the space E is said to be *compactly embedded* in E_1.

Let us point out that the continuity of $W[t_0,x_0]$ treated as an operator acting from $C(t_0,t_1)$ into $C(t_0,t_1)$ (defined for all admissible inputs $u(t)$, i.e., inputs which satisfy the condition $\{u(t_0),x_0\} \in \Omega(W)$) does not yield the continuity of $W[t_0,x_0]$, treated as an operator in any space E embedded in $C(t_0,t_1)$, even in the case $W[t_0,x_0]\, E \subset E$.

6.2. Spaces of continuously differentiable functions

$C^1(t_0,t_1)$ is the Banach space of functions continuously differentiable on $[t_0,t_1]$, equipped with the norm

$$\|u(t)\|_{C^1(t_o,t_1)} = |u(t_o)| + \|u'(t)\|_{t_o,t_1}$$

where, as usual ,

$$\|v(t)\|_{t_o,t_1} = \|v(t)\|_{C(t_o,t_1)} \quad .$$

Operators (6.1) are defined in C^1 only in a special case, if the curves Φ_ℓ, Φ_r, $\Pi(M)$ are not only regular but also regularly connected (see Fig. 6.1).

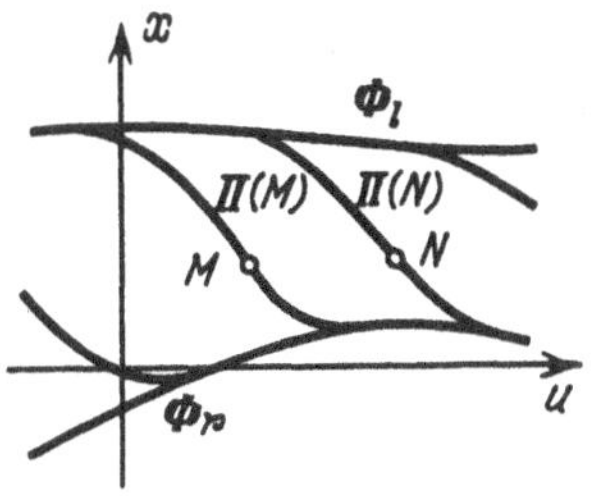

Fig. 6.1

For this reason, the spaces C^1 are not suitable for the study of hysterons such as a play, generalized play, stop, etc.

The same kind of problems occurs in the case of spaces of k-times continuously differentiable functions, with the norm

$$\|u(t)\|_{C^k(t_o,t_1)} = |u(t_o)| + \|u^{(k)}(t)\|_{t_o,t_1} \quad ,$$

as well as for spaces $S_p^k(t_o,t_1)$, $k > 1$, of functions having the finite norm

$$\|u(t)\|_{S_p^k(t_o,t_1)} = |u(t_o)| + \left\{ \int_{t_o}^{t_1} |u^{(k)}(t)|^p \, dt \right\}^{1/p} \quad ,$$

where $1 \leq p < \infty$.

The case of hysterons for which the curves $\Pi(M)$ are regularly connected with the curves Φ_ℓ, Φ_r has not been analyzed in a systematic way.

6.3. Play in the space S of absolutely continuous functions

In the space $S = S(t_0,t_1)$, one can introduce a norm according to the formula

$$\|u(t)\|_{S(t_0,t_1)} = |u(t_0)| + \int_{t_0}^{t_1} |u'(t)|\, dt \ .$$

Consider the generalized play $L(\Gamma_\ell,\Gamma_r)$. If the curves Γ_ℓ and Γ_r have no common points, then each operator $L[t_0,x;\Gamma_\ell,\Gamma_r]$ is defined in S (on a set of admissible inputs in S; this obvious restriction will not be mentioned in the further constructions).

The situation becomes more complicated if there are common points of the curves Γ_ℓ and Γ_r. A point $\{u_0,x_0\}$ of the intersection $\Gamma_\ell \cap \Gamma_r$ will be called L*-regular*, if there are $\delta > 0$ and $\lambda > 0$, such that

$$\Gamma_r(u_2) - \Gamma_r(u_1) \leq \lambda|u_2 - u_1^*| \quad (u_0 - \delta \leq u_1 \leq u_2 \leq u_0 + \delta) \ , \tag{6.3}$$

where

$$u_1^* = \min\{u_1, \max\{u\colon u_0 - \delta \leq u \leq u_0 + \delta, \ \Gamma_\ell(u) = \Gamma_r(u)\}\} \ . \tag{6.4}$$

The regularity property characterized by (6.3), (6.4) is not restrictive; it does obviously hold if at least one of the functions $\Gamma_\ell(u)$, $\Gamma_r(u)$ satisfies the Lipschitz condition in some neighbourhood of the point u_0.

Theorem 6.1. The operators $L[t_0,x_0;\Gamma_\ell,\Gamma_r]$ transform S into S if and only if all elements in the intersection of the curves Γ_ℓ and Γ_r are L-regular points. ∎

The next question concerns continuity of the operators $L[t_0,x_0:\Gamma_\ell,\Gamma_r]$.

Theorem 6.2. The operators $L[t_0,x_0;\Gamma_\ell,\Gamma_r]$ are continuous in S if and only if both functions $\Gamma_\ell(u)$ and $\Gamma_r(u)$ are locally absolutely continuous and all points in the intersection of their graphs are L-regular. ∎

If the curves Γ_ℓ and Γ_r have no common points and each of the functions

$$\frac{d}{du}\Gamma_\ell(u) \quad (-\infty < u < a_\ell), \quad \frac{d}{du}\Gamma_r(u) \quad (b_r < u < \infty) \tag{6.5}$$

satisfies the local Lipschitz condition, then also all operators $L[t_o,x_o;\Gamma_\ell,\Gamma_r]$ are locally Lipschitz continuous.

Note that the global Lipschitz continuity of the functions (6.5) does not imply the same property for the operators $L[t_o,x_o;\Gamma_\ell,\Gamma_r]$. Only if we have $\Gamma_\ell(u) = au+b_1$, $\Gamma_r(u) = au+b_2$, this implication is true, with the global Lipschitz constant $2a$.

If $\Gamma_\ell(u) \equiv \Gamma_r(u) \equiv \Gamma(u)$, then the generalized play degenerates to the superposition operator $Lu(u) = \Gamma[u(t)]$. In the degenerate case, the operator L is defined in the space S and is continuous there, if and only if the function $\Gamma(u)$ is locally Lipschitz continuous.

6.4. Hysteron in the space S

The canonical representation (4.8) of a hysteron W enables us to deduce its properties from the properties of the play $L(\Gamma_\ell,\Gamma_r)$ and the function $f(u,x)$. The question how to construct a representation of the hysteron which would comprehend the maximal available information on its properties, in spite of undoubtedly high interest remains almost completely open. Another important problem concerns methods of a direct determination of hysteron's properties from characteristics of the curves $\Pi(M)$, Φ_ℓ, Φ_r.

Suppose that the curves Φ_ℓ and Φ_r have empty intersection. Then the operators $W[t_o,x_o]$ are defined and locally bounded in S if and only if the functions $\Phi_\ell(u)$, $\Phi_r(u)$ are locally absolutely continuous and every sequence of the functions $\Pi(u;M_n)$, where $M_n \in \Omega(W)$ (see the axiom (a3)) and $M_n \to M \in \Omega(W)$, fulfils a uniform Lipschitz condition.

To any point $M \in \Omega_o(W)$ we shall assign a continuous function $\chi(u;M)$ (in addition to the functions $\Pi(u;M)$ and $T(u;M)$), defined on $-\infty < u < \infty$ (see Fig. 6.2):

$$\chi(u;M) = \begin{cases} \Phi_\ell[u_\ell(M)] & \text{at} \quad u \le u_\ell(M), \\ \Pi(u;M) & \text{at} \quad u_\ell(M) \le u \le u_r(M), \\ \Phi_r[u_r(M)] & \text{at} \quad u_r(M) \le u. \end{cases} \tag{6.6}$$

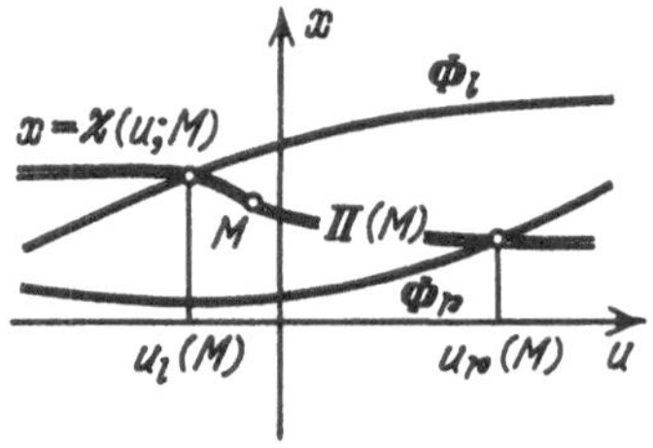

Fig. 6.2

<u>Theorem 6.3.</u> Assume that the curves Φ_ℓ and Φ_r have empty intersection. Let the operators $W[t_o,x_o]$ be locally bounded in S. Then the operators $W[t_o,x_o]$ are continuous in the space S if and only if the restriction of any sequence of the functions $\chi(u;M_n)$ to any interval $\{u_1,u_2\}$, where $M_n \in \Omega_o(W)$ and $M_n \to M \in \Omega(W)$, is a precompact set in $S(u_1,u_2)$. ∎

One can easily indicate conditions under which the operators $W[t_o,x_o]$ are Lipschitz continuous.

Consider, as an example, the plastic Prandtl's fibre (see Fig. 3.1). The corresponding operators $W[t_o,x_o]$ are defined in S and satisfy there the local Lipschitz condition with constant $2E$. As it has already been shown in Section 5.3, the operators $W[t_o,x_o]$, considered as acting in the space C, fulfil the Lipschitz condition with the same constant. Since S is embedded in C, $W[t_o,x_o]$ can also be regarded as operators from $S(t_o,t_1)$ into $C(t_o,t_1)$; remark that in the last case the Lipschitz condition holds with the constant E. Furthermore,

$$|W[t_o,x_o]u(t) - W[t_o,y_o]v(t)| \le |x_o - y_o| + E\int_{t_o}^{t} |u'(s) - v'(s)|\, ds \quad . \tag{6.7}$$

To prove (6.7), it is enough to use the relation

$$\frac{d}{dt}\,|W[t_o,x_o]u(t) - W[t_o,x_o]v(t)| \le E\,|u'(t) - v'(t)| \quad .$$

The general statements formulated in this section apply also to the case of curves Φ_ℓ and Φ_r which have common points, provided, in particular, the functions $\Phi_\ell(u)$ and $\Phi_r(u)$ fulfil the local Lipschitz condition in a certain neighbourhood of the set $P(\Phi_\ell \cap \Phi_r)$.

6.5. Hysterons in spaces H_α

The spaces $H_\alpha = H_\alpha(t_o,t_1)$ $(0 < \alpha \leq 1)$ are defined as the Banach spaces of continuous functions with the finite norm

$$\|u(t)\|_{H_\alpha} = |u(t_o)| + \sup_{t_o \leq \tau_1 < \tau_2 \leq t_1} \frac{|u(\tau_1) - u(\tau_2)|}{|\tau_1 - \tau_2|^\alpha} . \tag{6.8}$$

Let us confine ourselves to the following simple statement.

Theorem 6.4. Suppose that all the functions $\Pi(u;M)$ as well as the functions $\Phi_\ell(u)$, $\Phi_r(u)$ satisfy the same condition Lip γ:

$$|\Pi(u;M) - \Pi(v;M)|, \ |\Phi_\ell(u) - \Phi_r(u)|, \ |\Phi_r(u) - \Phi_r(v)| \leq$$
$$\leq \lambda(\rho)|u - v|^\gamma , \quad |u|, |v| \leq \rho ,$$

where $0 < \gamma \leq 1$, and assume

$$0 < \beta < \alpha\gamma , \quad \alpha \leq 1 . \tag{6.9}$$

Then all operators $W[t_o,x_o]$ map the space H_α into H_β and are continuous from H_α into H_β . ∎

Under the hypotheses of Theorem 6.4, $W[t_o,x_o]$ treated as operators from H_α into H_β satisfy the following Lipschitz condition Lip$(\gamma-\beta/\alpha)$:

$$\|W[t_o,x_o]u(t) - W[t_o,x_o]v(t)\|_{H_\beta} \leq \lambda(\rho)\{\|u(t) - v(t)\|_{H_\alpha}\}^{\gamma-\beta/\alpha} ,$$
$$(\|u(t)\|_{H_\alpha} , \|v(t)\|_{H_\alpha} \leq \rho) . \tag{6.10}$$

In particular, in the case of a stop the estimates (6.10) are true for $\gamma = 1$.

Necessary and sufficient conditions, ensuring that the operators $W[t_o,x_o]$ map H_α into H_β $(\beta < \alpha)$, are slightly more complicated, namely:

a. For every sequence of the points $M_n \in \Omega_o(W)$, such that $M_n \to M \in \Omega(W)$, the following equality holds:

$$\lim_{\delta\to 0} \sup_{n=1,2,\dots} \sup_{|u-v|\le\delta} \frac{|\Pi(u;M_n) - \Pi(v;M_n)|^\alpha}{|u-v|^\beta} = 0 .$$

b. The local continuity modulus $\mu(r)$ of the function $\Phi_\ell(u)$ (the same as for the function $\Phi_r(u)$) satisfies the condition

$$\lim_{r\to 0} \frac{[\mu(r)]^\alpha}{r^\beta} = 0$$

on every finite subinterval $[u_1,u_2]$ of its domain.

6.6. Discontinuous inputs

The formula (3.7) defines the output $x(t)$ of a hysteron W for any monotone (not necessarily continuous) input $u(t)$. The semigroup property enables us to determine the output (6.1) for any piecewise monotone (again possibly discontinuous) input. If the input $u(t)$ $(t_o \le t \le t_1)$ may be represented as the uniform limit of a sequence of piecewise monotone inputs $u_n(t)$ $(t_o \le t \le t_1)$, then the corresponding output (6.1) is the uniform limit of the sequence $x_n(t) = W[t_o,x_o]u_n(t)$ $(n = 1,2,\dots)$ (certainly, the conditions $u_n(t_o) = u(t_o)$ and $\{u(t_o),x_o\} \in \Omega(W)$ are to be fulfilled). Correctness of such a definition results from Theorem 6.5, formulated below.

The uniform limits $u(t)$ of sequences $u_n(t)$ (consisting of piecewise monotone functions) form a Banach space $\Xi = \Xi(t_o,t_1)$, with the norm defined by

$$\|u(t)\|_{\Xi(t_o,t)} = \sup_{t_o \le t \le t_1} |u(t)| . \tag{6.11}$$

Functions $u(t) \in \Xi$ may be discontinuous on at most countable sets;

moreover, for any discontinuity point τ of these functions the unilateral limits

$$u(\tau-0)=\lim_{\substack{t\to\tau\\ t<\tau}} u(t)\quad,\qquad u(\tau+0)=\lim_{\substack{t\to\tau\\ t>\tau}} u(t)$$

exist.

The space C is embedded in Ξ ; the space Ξ is not contained in L_∞ , since functions $u(t)$ and $v(t)$, which assume different values only in one point τ , do not coincide in the space Ξ . The space Ξ is non-separable.

Operators $W[t_0,x_0]$, defined in the space Ξ , have still the same properties as established in Chapters 3 and 5, where these operators have been treated as acting in the space C : the semigroup identity, the vibro-correctness property, etc., hold there. All these facts follow by Theorem 6.5.

Let Ξ_* be some separable subspace of the space $\Xi(t_0,t_1)$. Then the set T of all discontinuity points of the functions $u(t)\in\Xi_*$ different from t_0 and t_1 is countable; assume that $\tau_1, \tau_2, \ldots$ are elements of this set.

Denote by G the range of the strictly monotone function

$$s(t)=\begin{cases} t_0\ , & \text{if } t=t_0\ ,\\ 1+t+2\displaystyle\sum_{\{i:\tau_i<t\}} 2^{-i}\ , & \text{if } t\notin T\ ,\\ 1+t+2^{-i_0}+2\displaystyle\sum_{\{i:\tau_i<t\}} 2^{-i}\ , & \text{if } t=\tau_{i_0}\ ,\\ t_1+4\ , & \text{if } t=t_1 \end{cases} \tag{6.12}$$

defined on $[t_0,t_1]$. By definition, (6.12) has bilateral discontinuities in all points τ_i , and unilateral discontinuities at t_0 and t_1 . There exists an inverse $t=\sigma(s)$ of the function (6.12), defined on G .

For each function $u(t)\in\Xi_*$, let us define a continuous function $A\,u(s)$ at $t_0\le s\le t_1+4$. We construct it as equal $u[\sigma(s)]$ for $s\in G$ and then extend it by continuity onto the closure $\bar G$ of the set G . Finally we take $Au(s)$ linear on any interval in the complement to $\bar{G}$. The operator A , acting from $\Xi_*(t_0,t_1)$ into $C(t_0,t_1+4)$, is linear and isometric.

In turn, for a function $v(s)$ in the space $C(t_0,t_1+4)$ let us define the function $Bv(t) = v[s(t)]$; the linear operator B acts from $C(t_0,t_1+4)$ into Ξ , and $\|B\|_{C\to\Xi} = 1$.

By construction, the following relation is true for the pair of operators A and B .

<u>Theorem 6.5.</u> Let $u(t) \in \Xi_*$ and $\{u(t_0),x_0\} \in \Omega(W)$. Then

$$W[t_0,x_0]u(t) = B\,W[t_0,x_0]v(s) \quad , \tag{6.13}$$

where $v(s) = Au(s)$.

∎

To extend the operators onto any space of the Lebesgue type (or similar), we have to use another approach. Recall that an element of a Lebesgue space is not just a single function (pointwise defined) but rather represents the whole class of equivalent functions which coincide up to sets of measure zero.

Let us confine ourselves to the case of the monotone hysteron W (cf., Section 4.4), with the relevant functions $\Phi_\ell(u)$ and $\Phi_r(u)$ defined for all $u \in (-\infty,\infty)$. If there is a continuous input $u(t)$ in some class of measurable inputs different only on sets of measure zero, such that $\{u(t_0),x_0\} \in \Omega(W)$, then we assume the operator $W[t_0,x_0]$, defined on this class of inputs, has values $W[t_0,x_0]u(t)$. The operator $W[t_0,x_0]$, considered in any space L_p on such inputs , preserves the monotonicity property.

In many applications, the operators $W[t_0,x_0]$ should be extended onto the whole space L_p , with the resulting operators $W_L[t_0,x_0]$ still being monotone. Clearly, such extensions are not uniquely defined. Let us note that in the space L_∞ , a monotone extension of $W[t_0,x_0]$ can be constructed which itself is a continuous operator.

6.7. Hysteron in the space of functions with bounded variation

In the space $V = V(t_0,t_1)$ of functions with bounded variation, the norm is defined by the equality

$$\|u(t)\|_{V(t_0,t_1)} = |u(t_0)| + \mathrm{Var}_{t_0}^{t_1} u(t) \quad . \tag{6.14}$$

The space V is non-separable.

Certainly, the space V is embedded in Ξ ; thus the operators $W[t_0,x_0]$ are well-defined for inputs $u(t) \in V$ (provided $\{u(t_0),x_0\} \in \Omega(W)$).

The conditions which guarantee that operators $W[t_0,x_0]$ are defined and continuous in the space S , ensure the local Lipschitz continuity of these operators, etc., as given in Sections 6.3 and 6.4, coincide with the hypotheses providing analogous properties of $W[t_0,x_0]$ treated as operators from V into V . To prove it, let us first note that the continuity of the operators $W[t_0,x_0]$ in the space V implies their continuity also in the space S , and next make use of the following result, similar to Theorem 6.5.

Theorem 6.6. Let V_* be a separable subspace of the space V . Then for any input $u(t) \in V_*$, such that $\{u(t_0),x_0\} \in \Omega(W)$, the equality

$$W[t_0,x_0]u(t) = B_1 \, W[t_0,x_0]v(s) \tag{6.15}$$

holds with $v(s) = A_1 \, u(s)$, where A_1 is a linear, isometric operator from V_* into S ; B_1 is a linear non-expanding operator from S into V .

■

As one can easily see, hypotheses which ensure that the operators $W[t_0,x_0]$ act in the space V (without being continuous) are more general than the conditions providing that these operators act in S . In particular, if the curves Γ_ℓ and Γ_r have no common points then all the operators $L[t_0,x_0;\Gamma_\ell,\Gamma_r]$ (defining the corresponding generalized play) act not only from V into V , but also from Ξ into V .

6.8. Hysteron in Wiener spaces

Denote by $V_\beta = V\,(t_0,t_1)$ $(\beta > 1)$ the Banach space of functions $u(t)$ which have bounded variation $\mathrm{Var}_\beta \, u(t)$ on the interval $[t_0,t_1]$, where

$$\mathrm{Var}_\beta\, u(t) = \sup \sum_i |u(\tau_i) - u(\tau_{i-1})|^\beta \quad , \tag{6.16}$$

with the supremum taken over all divisions $t_0 = \tau_0 < \ldots < \tau_n = t_1$ of the interval $[t_0,t_1]$. The norm in Wiener space V_β is defined by the equality

$$\|u(t)\|_{V_\beta(t_0,t_1)} = |u(t_0)| + \{\mathrm{Var}_\beta\, u(t)\}^{1/\beta} \quad . \tag{6.17}$$

Clearly, $V_\beta \subset \Xi$; therefore the operators $W[t_0,x_0]$ are well-defined on the Wiener space V_β .

There are strong links between the Wiener and the Hölder spaces. For instance, if $u(t)$ is continuous and $u(t) \in V_\beta$, then

$$\|u(t)\|_{V_\beta} = \inf \|u[\sigma(t)]\|_{H_{1/\beta}} \quad ,$$

where the infimum is taken over all functions $\sigma(t)$ such that

$$\sigma(t_0) = t_0, \quad \sigma(t_1) = t_1, \quad \sigma'(t) > 0 \quad , \quad t_0 \leq t \leq t_1 \quad .$$

A comprehensive analysis of the operators $W[t_0,x_0]$ in the spaces V_β would certainly be of a mathematical interest.

At the very end, let us note that *for the plastic Prandtl's fibre the following estimate is true:*

$$\|W[t_0,x_0]u(t) - W[t_0,y_0]v(t)\|_{V_\beta(t_0,t_1)} \leq$$

$$\leq 2|x_0 - y_0| + 2^{1/\beta}\, E\{\mathrm{Var}_\beta[u(t) - v(t)]\}^{1/\beta} \quad . \tag{6.18}$$

Part 2. Identification theorem

A bride at a glance recognizes another bride. And they immediately find a common language, unknown to men and widows.

O'Henry

7. Identification problem

7.1. General identification problem

There are two substantially different approaches to the construction of mathematical relations that characterize the dynamic properties of a system W .

In the first approach, the system is considered as an assemblage of elementary components, with known mathematical characteristics and with specified rules that govern interactions between single elements. Such an approach is typical for physics, mechanics and chemistry to give a few examples. It is used at deriving various governing equations in mathematical physics, continuum mechanics, etc., where any system is treated as assembled of infinitesimal parts.

Alternatively, any system can be considered as a "black box". In such an approach, no information on the internal structure of the system is available. One only admits various classes of external actions as inputs for the "black box" and observes the produced effects (outputs). By an analysis of the relations between external actions and the corresponding reactions (i.e., an analysis of the experimental input-output relations), one constructs a mathematical model of the system. Such a treatment is specific for problems of automatic control, general system theory, etc. Further we shall use the name *mathematical identification method* for the "black box" approach.

This method can be reduced to solving the following three problems.

At first, it is necessary to develop a quantitative analysis of the experimental input-output relations and then specify properties of those relations that completely characterize a class of mathematical objects (equations, operators, etc.) which can be used for describing the "black box" under consideration.

There are no general recommendations concerning the choice of properties to be perceived. For this purpose, one can take various types of monotonicity, linearity or nonlinearity, character of the response to external actions periodic in time. In our further study, the "vibro--correctness" of the system we consider or, actually, lack of this property, i.e., the form of variations of the output observed in the case of small perturbations of the external action will play a special role.

Suppose that a class of mathematical objects with values of some parameters to be determined has been already chosen for the identification of a "black box".

The second identification problem consists in recovering the values of those parameters which correspond to a specific system under consideration. This problem turns out complicated already in the case where the number of parameters is finite and the parameters are scalars, because experimental data always contain errors of various nature. The case where either the number of parameters is infinite or some of the parameters are functions (i.e., elements of infinite dimensional spaces), is even more complicated. Thus, problems of the second group are usually solved in an approximate way, with an accuracy of the approximate solution determined by the proximity of the experimental input-output relations of the black box to the relevant correspondences in the constructed mathematical description (mathematical model) rather than by exactness of the recovery of the unknown parameters within the model.

The third problem consists in estimating an influence of the parameter perturbations within the mathematical model onto the input-output correspondences of the system.

If a black box represents merely an element of a certain complex system, then some additional hypotheses are often necessary for its mathematical description. Suppose, for instance, that reactions $u(t)$ of a black box W are external actions for another system W_1. Then the mathematical descriptions of the systems W and W_1 must be coordinated so that the reactions u(t) belong to the set of admissible inputs in the mathematical model of W_1 . Specially complicated is an analysis of feedback systems, even if they consist of parts (elements) of a simple nature. A formal combination of "trivial" mathematical descriptions of separate parts may lead to paradoxical conclusions for the whole system.

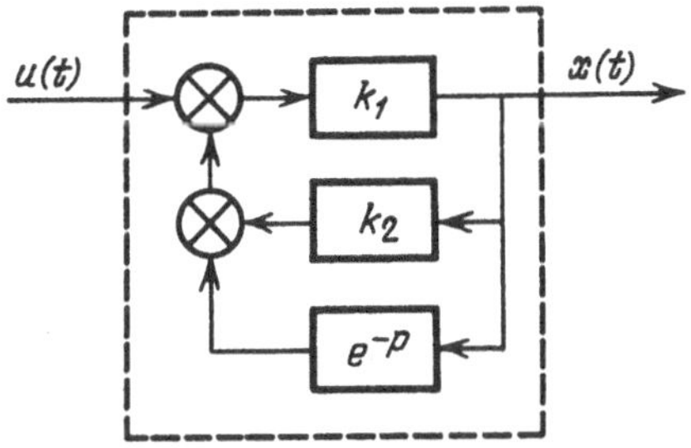

Fig. 7.1

As a classical example let us consider the "cybernetic" predictor shown in Fig. 7.1. In that system, k_1 and k_2 are some coefficients, symbol ⊗ denotes addition, e^{-p} - an element with pure delay (the output corresponding to an input z(t) is equal to z(t - 1)). If $k_1k_2 = 1$, then $x(t) = -u(t + 1)$ and this contradicts the principle of physical realizability!

7.2. Prehysteron

In this section we shall consider a deterministic, static transducer V with scalar input u and output x . For any feasible state $\{u_o,x_o\}$ of the transducer V at the time moment $t = t_o$ we shall assume that all piecewise monotone inputs u(t) $(t \geq t_o)$ are admissible if the initial condition $u(t_o) = u_o$ holds and, in addition, the corresponding outputs

$$x(t) = V[t_o,x_o]u(t) \qquad (t \geq t_o) \tag{7.1}$$

are continuous. By $\Omega(V)$ we shall denote the domain of feasible states of the transducer V .

Since V is deterministic, the input-output relation (7.1) will satisfy the semigroup identity

$$V[t_1,V[t_o,x_o]u(t_1)]u(t) = V[t_o,x_o]u(t) \quad (t_o \leq t_1 \leq t) \quad . \tag{7.2}$$

Therefore, any study of the operator (7.1) takes advantage of a possibly complete information on its behaviour over the class of all monotone inputs.

Let $u_n(t)$ $(t_o \leq t \leq t_1,\ n = 1, 2, \ldots)$ and $u_*(t)$ $(t_o \leq t \leq t_1)$ be continuous monotone inputs, satisfying the conditions

$$\{u_n(t_o),x_n\} \ , \ \{u_*(t_o),x_*\} \in \Omega(V) \quad , \tag{7.3}$$

where $x_n \to x_*$, and

$$\lim_{n\to\infty} \|u_n(s) - u_*(s)\|_{t_o,t_1} = 0 \quad . \tag{7.4}$$

If in these conditions

$$\lim_{n\to\infty} V[t_o,x_n]u_n(t_1) = V[t_o,x_*]u_*(t_1) \quad , \tag{7.5}$$

then the transducer V will be called *weakly correct*.

A deterministic, static and weakly correct transducer will be referred to as a *prehysteron*.

Obviously, any hysteron is also a prehysteron, the converse, however, is not true. Let, for example, two differential equations

$$\frac{dy}{du} = f(u,y), \quad \frac{dz}{du} = g(u,z) \tag{7.6}$$

with continuous right-hand sides be given ; let each initial condition $y(u_o) = y_o$ specify a unique solution $y = y(u;u_o,y_o)$ $(u_o \leq u < \infty)$ of the first equation in (7.6) and each initial condition $z(u_o) = z_o$ defines unique solution $z = z(u;u_o,z_o)$ $(-\infty < u \leq u_o)$ of the second equation in (7.6); then one may construct a prehysteron V with the whole plane as the domain of feasible states by defining the operators (7.1) on the monotone inputs so that

$$V[t_o,x_o]u(t) = \begin{cases} y[u(t);u(t_o),x_o] \quad , & \text{if } u(t) \text{ is increasing} \quad , \\ z[u(t);u(t_o),x_o] \quad , & \text{if } u(t) \text{ is decreasing} \quad . \end{cases}$$

Upon slight modification, the last example delivers a complete description of prehysteron. For this, the families of integrals defined for equations (7.6) are to be replaced by two families of the graphs of some continuous functions.

We shall denote by $K(u_0)$ intersection of the domain $\Omega(V)$ of all feasible states of the prehysteron V and the straight line $u = u_0$.

The set $K(u_0)$ may have a quite whimsical structure. If $\Omega(V)$ comprises two straight lines $x = x_1$ and $x = x_2$, then each $K(u_0)$ contains just two points.

<u>Lemma 7.1.</u> Let a prehysteron V be controllable. Then every set $K(u_0)$ is an interval.

<u>Proof.</u> Let $\{u_0, x_0\}$, $\{u_0, x_1\} \in K(u_0)$ and $u(t)$ $(t_0 \leq t \leq t_1)$ be a continuous, piecewise monotone input such that $u(t_0) = u(t_1) = u_0$, $V[t_0, x_0]u(t_1) = x_1$. By virtue of the weak correctness of prehysteron, the function

$$\varphi(\lambda) = V[t_0, x_0]u_\lambda(t_1) \qquad (0 \leq \lambda \leq 1) \quad ,$$

where

$$u_\lambda(t) = \lambda u(t) + (1 - \lambda)u_0 \qquad (t_0 \leq t \leq t_1; \quad 0 \leq \lambda \leq 1) \quad ,$$

is continuous, hence its values fill the interval $[\varphi(0), \varphi(1)]$, i.e., the interval $[x_0, x_1]$.

∎

7.3. <u>Basic identification theorem</u>

As it turns out, any prehysteron V is also a hysteron, if it exhibits a certain additional correctness property. This fact expresses the content of the basic theorem on identification.

If for every sequence of the states $\{a_n, x_n\} \in \Omega(V)$, convergent to some state $\{a_*, x_*\} \in \Omega(V)$, and for every sequence of real numbers b_n, convergent to some real b_*, the equality

$$\lim_{n \to \infty} \sup_{0 \leq t < \infty} |V[0, x_n](a_n + b_n \sin t) - V[0, x_*](a_* + b_* \sin t)| = 0 \qquad (7.7)$$

is valid, then the prehysteron V will be called *uniformly correct on harmonic inputs*.

The weak correctness of a prehysteron does not imply its uniform correctness on harmonic inputs; in turn, the uniform correctness on harmonic inputs does not guarantee the weak correctness.

Theorem 7.1. Let a controllable prehysteron V be uniformly correct on harmonic inputs. Then, in the case of piecewise monotone continuous inputs, the corresponding outputs (7.1) coincide with outputs of some hysteron.

By virtue of this theorem, any controllable prehysteron, uniformly correct on harmonic inputs, defined only on piecewise monotone inputs admits a canonical representation (cf., Chapter 4).

If for a controllable, deterministic and static transducer V all continuous inputs are admissible, and this transducer is vibro-correct, then due to Theorem 7.1 , V represents a hysteron.

A proof of Theorem 7.1 will be given in the next section.

7.4. Concluding remarks

a. The transducer considered in Theorem 7.1 was deterministic, static, controllable, weakly correct and uniformly correct on harmonic inputs. Neither of these properties follows from the others.

b. If a deterministic, autonomous transducer V is uniformly vibro--correct (in the natural sense) on the set of all piecewise monotone, continuous inputs, then in order to ensure that V is static it is sufficient that constant inputs produce constant outputs.

c. A deterministic transducer V is monocyclic if for any piecewise monotone input $u(t)$, continuous and periodic for $t \geq t_0$ $(u(t+T) = u(t)$ for $t \geq t_0)$, every output (7.1) is periodic with the same period for $t \geq t_0 + T$ (i.e., $x(t+T) = x(t)$ for $t \geq t_0 + T$). If a prehysteron is monocyclic, then it is uniformly correct on harmonic inputs. Therefore it follows from Theorem 7.1 that every monocyclic controllable prehysteron is a hysteron.

d. Suppose that for any initial state $\{u_0, x_0\} \in \Omega(V)$ and all piecewise monotone, continuous inputs $u_1(t)$, $u_2(t)$, $u_3(t)$ $(t_0 \leq t \leq t_1)$, satisfying the conditions

$$u_1(t_o) = u_2(t_o) = u_3(t_o) = u_o, \quad u_1(t_1) = u_2(t_1) = u_3(t_1)$$

and

$$u_1(t) \leq u_2(t) \leq u_3(t) \qquad (t_o \leq t \leq t_1) \quad ,$$

the inequality

$$\{V[t_o,x_o]u_2(t_1) - V[t_o,x_o]u_1(t)\}\{V[t_o,x_o]u_3(t_1) - V[t_o,x_o]u_2(t)\} \geq 0$$

holds. Then the prehysteron V is monocyclic.

8. Proof of Theorem 7.1

8.1. Singular points of the domain $\Omega(V)$

For any point $M_o = \{u_o,x_o\} \in \Omega(V)$ define the function

$$T(u;M_o) = V[t_o,x_o]w(t_1) \qquad (-\infty < u < \infty) \quad , \tag{8.1}$$

where $w(t)$ $(t_o \leq t \leq t_1)$ (at any fixed u) is an arbitrary monotone continuous input for which $w(t_o) = u_o$ and $w(t_1) = u$. The function (8.1) is well-defined because of the static character and weak correctness of the prehysteron V . The weak correctness of V implies also that every function (8.1) is jointly continuous with respect to both variables $u \in (-\infty,\infty)$, $M_o \in \Omega(V)$.

By $T(M_o)$ we shall denote the graph of function (8.1). That graph is then one of the defining curves (cf., Chapter 3) of the prehysteron V .

A point $M_* = \{u_*,x_*\} \in \Omega(V)$ will be called the *upper-singular point*, if there exists some $N = \{v,y\} \in \Omega(V)$ such that $M_* \in T(N)$ and

$$T(u;N) > T(u;M_*) \qquad (u \in (u_*,v)) \quad . \tag{8.2}$$

In this definition, the point N may be equally located to the left and to the right of M_* (see Figure 8.1) .

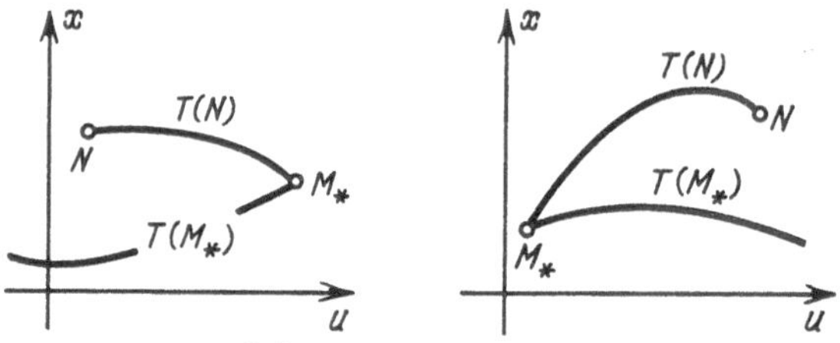

(a) (b)

Fig. 8.1

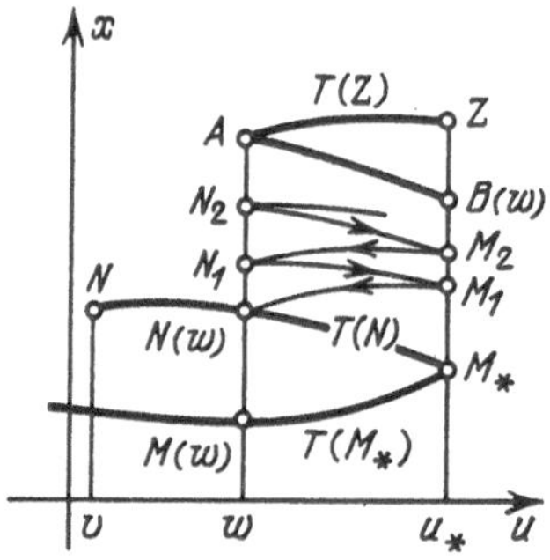

Fig. 8.2

In an analogous way one can define *lower singular points*.

By Lemma 7.1, for a prehysteron V that satisfies the hypotheses of Theorem 7.1, intersection $K(u_0)$ of the straight line $u = u_0$ with the domain $\Omega(V)$ of feasible states is an interval.

Lemma 8.1. Every upper (lower) singular point $M_* = \{u_*, x_*\} \in \Omega(V)$ is the upper (lower) end-point of the interval $K(u_*)$.

Proof. We shall confine ourselves to the situation depicted in Fig. 8.1,a. If the assertion of the lemma were false, there would exist a point $Z = \{u_*, z\}$ in $\Omega(V)$, with the ordinate $z > x_*$ (see Fig. 8.2).

Choose a point A with an abscissa $w \in [v, u_*]$ in the curve $T(Z)$, and denote by $B(w)$ the point with the abscissa u_* in the curve $T(A)$. Due to the continuity of the function (8.1), there exists a $w_0 \in [v, u_*]$

such that for $w_0 \leq w < u_*$ the parts of the curves $T(Z)$ and $T(A)$, contained between A and the points Z, $B(w)$, respectively, are located above the curve $T(N)$ (see Fig. 8.2).

Let us fix $w \in [w_0, u_*]$. Denote by P the mapping

$$P\{u_*,x\} = \{w,T(w;\{u_*,x\})\} \tag{8.3}$$

which transforms a segment M_*Z of the line $u = u_0$ into the line $u = w$; due to the semigroup property of the transducer V, it follows that $P(M_*Z)$ coincides with the segment $M(w)A$, where $M(w)$ denotes the point of $T(M_*)$ which has the abscissa w. Analogously, by Q we shall denote the mapping

$$Q\{w,x\} = \{u_*,T(u_*;\{w,x\})\} \tag{8.4}$$

which transforms $N(w)A$ of the straight line $u = w$ into the straight line $u = u_*$, where $N(w)$ is the point of $T(N)$ that in turn has the abscissa w. The mapping (8.4) transforms segment $N(w)A$ onto the segment $M_*B(w)$.

By M_1 we shall denote a point of the segment M_*Z which is transformed by the mapping (8.3) into the point $N(w)$; we shall assume that M_1 is located below the point $B(w)$. Then, in $N(w)A$ we determine a point N_1 which is transformed into M_1 by (8.4).

As the next step, in the segment M_1Z we select a point M_2 whose image by the mapping (8.3) is N_1; we shall again assume that M_2 is located below the point $B(w)$. Then there exists a point N_2 in N_1A whose image by the mapping (8.4) is M_2. If the described procedure can be infinitely continued, it will yield sequences of points N_k, M_k in the straight lines $u = w$ and $u = u_*$, satisfying the equalities $PM_k = N_{k-1}$ and $QN_k = M_k$. The sequences N_k and $M_k = \{u_*,x_k\}$ are convergent to some points $N^\square$ and $M^\square = \{u_*,x^\square\}$.

Consider the functions $V[0,x_k]u_w(t)$ and $V[0,x^\square]u_w(t)$, where

$$u_w(t) = \frac{1}{2}(u_* + w) + \frac{1}{2}(u_* - w)\cos t \quad (t \geq 0) . \tag{8.5}$$

By construction,

$$V[0,x_k]u_w(2k\pi) = x_* \quad (k = 1, 2, \ldots) . \tag{8.6}$$

Since $N^\square \in T(M^\square)$ and $M^\square \in T(N^\square)$ (by the continuity of the function (8.1)),

$$V[0,x^\square]u_w(2k\pi) = x^\square \neq x_* \quad (k = 1, 2, \ldots) \ . \tag{8.7}$$

The equalities (8.6) and (8.7) contradict the vibro-correctness of the prehysteron V on harmonic inputs.

Consequently, the procedure described in the previous section cannot be implemented at any value w. Therefore, for any $w \in [w_0,u_*)$, arbitrarily close to u_*, there exist finite sequences $N_1, N_2, \ldots, N_{n(w)}$ and $M_1, M_2, \ldots, M_{n(w)}$ of points in the straight lines $u = w$ and $u = u_*$, such that $PM_k = N_{k-1}$ and $QN_k = M_k$, with the point $M_{n(w)}$ belonging to the segment $ZB(w)$. Define

$$x(t;w) = V[0,x_{n(w)}]u_w(t) \quad (t \geq 0) \ ,$$

where $x_{n(w)}$ is the ordinate of $M_{n(w)}$ and $u_w(t)$ represents the input (8.5). By construction,

$$x[2n(w)\pi;w] \equiv x_* \quad (w_0 \leq w < u_*) \ .$$

On the other hand, $x_{n(w)} \to z$ at $w \to u_*$, and due to the uniform vibro-correctness on harmonic inputs, the equality

$$\lim_{w \to u_*} \sup_{t \geq 0} |x(t;w) - V[0,z]v_*(t)| = 0$$

is satisfied with $v_*(t) \equiv u_*$. Because $V[0,z]v_*(t) \equiv z \neq x_*$, we have again come to contradiction.

∎

Lemma 8.1 is the only point of the proof of Theorem 7.1 in which the uniform vibro-correctness of the prehysteron V on harmonic inputs is used.

<u>Lemma 8.2.</u> Let $M_0 = \{u_0,x_0\}$ be an interior point of the interval $K(u_0)$. Then in the curve $T(M_0)$ there exist singular points $M_1 = \{u_1,y\}$ and $M_2 = \{u_2,z\}$, such that $u_1 < u_0 < u_2$.

Proof. Choose a point $N_1 = \{u_o,x_1\}$ in the interval $K(u_o)$ so that $x_1 < x_o$. Because of the controllability of the prehysteron V, one can indicate such a piecewise monotone input $u_1(t)$ $(t_o \leq t \leq t_1)$ that $u_1(t_o) = u_1(t_1) = u_o$ and $V[t_o,x_o]u_1(t_1) = x_1$. Denote by τ the largest of the values $t \in [t_o,t_1]$ for which the point $\{u_1(t),V[t_o,x_o]u_1(t)\}$ belongs to the curve $T(M_o)$; since $x_1 < x_o$,

$$V[t_o,x_o]u_1(t) < T[u_1(t);M_o] \quad (\tau < t \leq t_1) . \tag{8.8}$$

Obviously, $u_1(\tau) \neq u_o$. Suppose, for instance, that the inequality $u_1(\tau) > u_o$ is satisfied. Then it follows from (8.8) that in a certain neighbourhood to the right of the point τ the input $u_1(\tau)$ is decreasing. Hence, $\{u_1(\tau),T[u_1(\tau);M_o]\}$ is the upper singular point.

Now, let us choose a point $N_2 = \{u_o,x_2\}$ of $K(u_o)$, such that $x_2 > x_o$, and construct a piecewise monotone, continuous input $u_2(t)$ $(t_o \leq t \leq t_2)$ for which $u_2(t_o) = u_2(t_2) = u_o$ and $V[t_o,x_o]u_2(t_2) = x_2$. By σ we shall denote the maximal $t \in [t_o,t_2]$, such that the point $\{u_2(t),V[t_o,x_o]u_2(t)\}$ belongs to $T(M_o)$. Then, $\{u_2(\sigma),T[u_2(\sigma);M_o]\}$ will be the lower singular point. It remains to prove that $u_2(\sigma) < u_o$.

Suppose the contrary. Obviously, $u_2(\sigma) \neq u_1(\tau)$. Let $u_o < u_1(\tau) < u_2(\sigma)$, for definiteness. Then the curve $u = u_2(t)$, $x = V[\sigma,T[u_2(\sigma);M_o]]u_2(t)$ $(\sigma < t \leq t_2)$ lies above the curve $T(M_o)$ and, in particular, its intersection with the straight line $u = u_1(\tau)$ is located above the upper singular point $\{u_1(\tau),T[u_1(\tau);M_o]\}$. This contradicts Lemma 8.1. ∎

8.2. Construction of curves $\Pi(M)$

Let us denote by Γ the set of those end-points of the intervals $K(u)$ $(-\infty < u < \infty)$ which belong to $\Omega(V)$. If $M \in \Omega(V) \setminus \Gamma$, then by $\Pi_o(M)$ we shall mean the maximal connected component of the curve $T(M)$ which includes M and is contained in $\Omega(V) \setminus \Gamma$. The curve $\Pi_o(M)$ represents the graph of some continuous function $\Pi_o(u;M)$, defined on a certain finite (due to Lemma 8.2) interval $(u_\ell(M), u_r(M))$.

Lemma 8.3. Suppose that the curves $\Pi_o(M_1)$ and $\Pi_o(M_2)$ do not coincide. Then these curves do not intersect.

Proof. Let $M_1 = \{u_1,x_1\}$, $M_2 = \{u_2,x_2\}$ and $u_1 \leq u_2$. At once, we can note that the functions $\Pi_o(u;M_1)$ and $\Pi_o(u;M_2)$ cannot assume the same values at all $u \in [u_1,u_2]$; otherwise (on account of the semigroup property) the curves $\Pi_o(M_1)$ and $\Pi_o(M_2)$ would completely coincide.

Suppose that the curves $\Pi_o(M_1)$ and $\Pi_o(M_2)$ have a common point $N = \{v,y\}$ where $v \in [u_1,u_2]$. Let, for definiteness, neither the interval $[v,u_2]$ degenerates to single point nor the functions $\Pi_o(u;M_1)$ and $\Pi_o(u;M_2)$ assume the same values in all points of the interval $[v,u_2]$. Then there exists such a common point $N_1 = \{v_1,y_1\}$ of the curves $\Pi_o(M_1)$, $\Pi_o(M_2)$ that $v_1 \in [v,u_2]$ and the functions $\Pi_o(u;M_1)$, $\Pi_o(u;M_2)$ assume different values on some interval (v_1,v_2) adjacent to v_1 from the right. N_1 will be either the upper or the lower singular point. This contradicts Lemma 8.2.

Let the functions $\Pi_o(u;M_1)$ and $\Pi_o(u;M_2)$ assume different values at every $u \in [u_1,u_2]$. Suppose that the curves $\Pi_o(M_1)$ and $\Pi_o(M_2)$ have a common point $N = \{v,y\}$, where $v > u_2$, for definiteness. Denote by σ_1 (respectively, by σ_2) the smallest $u \in [u_1,u_2]$ for which $T(u;N) = \Pi_o(u;M_1)$ (for which $T(u;N) = \Pi_o(u;M_2)$). One of the points

$$\{\sigma_1,T(\sigma_1;N)\} \ , \quad \{\sigma_2,T(\sigma_2;N)\} \ ,$$

does not belong to the line $u = u_2$. That point is singular and, by virtue of Lemma 8.1, cannot belong to the domain $\Omega(V) \setminus \Gamma$. We have again come to contradiction.

∎

Lemma 8.4. Let the input $u(t)$ $(t_* \leq t \leq t_1)$ satisfy the estimates

$$u_\ell(M) \leq u(t) \leq u_r(M) \qquad (t_* \leq t \leq t_1) \tag{8.9}$$

and

$$\{u(t_*),x_*\} \in \Pi(M) \ , \tag{8.10}$$

where $\Pi(M)$ is the curve obtained by addition of the points $\{u_\ell(M),T[u_\ell(M);M]\}$ and $\{u_r(M),T[u_r(M);M]\}$ to $\Pi_o(M)$. Then

$$V[t_*,x_*]u(t) = \Pi[u(t);M] \qquad (t_* \leq t \leq t_1) \ , \tag{8.11}$$

where $\Pi(u;M)$ denotes the restriction of function $T(u;M)$ to the interval $[u_\ell(M),u_r(M)]$.

Proof. It suffices to take advantage of Lemma 8.3 and the weak correctness of the prehysteron V .

∎

Lemma 8.5. Let $M = \{u_o,x_o\} \in \Omega(V)\setminus\Gamma$. Then each of the intervals $K[u_\ell(M)]$ and $K[u_r(M)]$ is non-degenerate (consists of more than one point); moreover, one of the points

$$N_\ell = \{u_\ell(M),\Pi[u_\ell(M);M]\}, \quad N_r = \{u_r(M),\Pi[u_r(M);M]\} \tag{8.12}$$

is the lower and the other is the upper end-point of the appropriate interval.

Proof. Suppose that the assertion of the lemma is false. For definiteness, let both points (8.12) be the upper end-points of the intervals $K[u_\ell(M)]$ and $K[u_r(M)]$ (see Fig. 8.3).

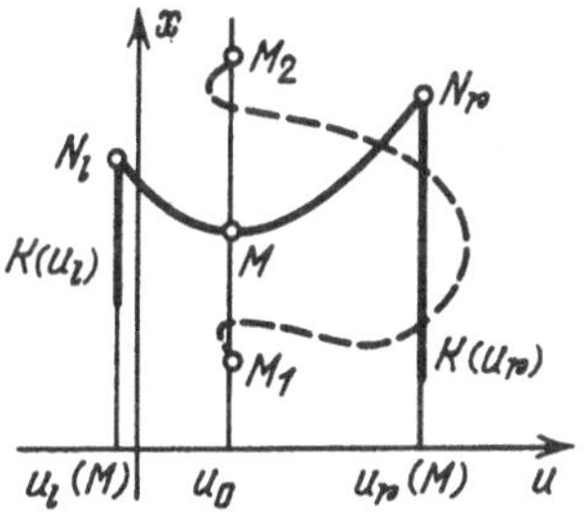

Fig. 8.3

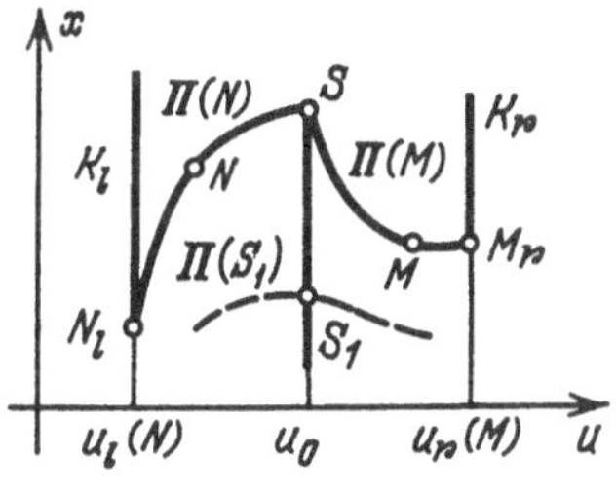

Fig. 8.4

Choose points $M_1 = \{u_o,x_1\}$ and $M_2 = \{u_o,x_2\}$ in the interval $K(u_o)$ as shown in Fig. 8.3. Since the prehysteron V is controllable, there exists a piecewise monotone, continuous input $u(t)$ $(t_o \le t \le t_1)$ for which $u(t_o) = u(t_1) = u_o$ and $V[t_o,x_1]u(t_1) = x_2$. For some values $t \in [t_o,t_1]$, the points of the curve

$$u = u(t), \quad x = V[t_o,x_1]u(t) \qquad (t_o \le t \le t_1)$$

(in Fig. 8.3 this curve is represented by the dashed line) belong to $\Pi(M)$; let t_* be the largest of those values. Then relations (8.9) and

(8.10) are fulfilled, and by (8.11) it follows that

$$V[t_0,x_1]u(t_1) = x_0 \quad .$$

We have come to contradiction. ∎

Now we pass to the main statement of this section.

Lemma 8.6. Let the curves $\Pi(M)$, $\Pi(N)$ $(M, N \in \Omega(V)\setminus\Gamma)$ be different. Then they have no common points and

$$[u_\ell(M) - u_\ell(N)][u_r(M) - u_r(N)] > 0 \quad . \tag{8.13}$$

Proof. For proving the first part of the assertion (on the empty intersection) it is enough (by Lemma 8.3) to show that none of the end-points of the curve $\Pi(M)$ is an end-point of the curve $\Pi(N)$. Due to Lemma 8.4, the left end-point of the curve $\Pi(M)$ cannot coincide with the left end-point of $\Pi(N)$, the same concerns the right end-points. Therefore, we only need to show that the right end-point of each of the curves under consideration does not coincide with the left end-point of the other curve.

Assume that the right end-point $S = \{u_0,x_0\}$ of the curve $\Pi(N)$ coincides with the left end-point of the curve $\Pi(M)$ (see Fig. 8.4). By Lemma 8.5, S is then also the lower or upper end-point of the non-degenerate interval $K(u_0)$; for definiteness assume that S is the upper end. By the same Lemma 8.5, the left end-point N_ℓ of the curve $\Pi(N)$ and the right end-point M_r of the curve $\Pi(M)$ are the lower ends of the intervals $K_\ell = K[u_\ell(N)]$ and $K_r = K[u_r(M)]$, respectively. Choose an interior point S_1 of the interval $K(u_0)$ and construct the curve $\Pi(S_1)$ (in Fig. 8.4, this curve is shown by the dashed line). It is geometrically clear that the abscissae of the end-points of the curve $\Pi(S_1)$ belong to the interval $[u_\ell(N),u_r(M)]$. Since the curve $\Pi(S_1)$ is situated below the union of the curves $\Pi(N)$ and $\Pi(M)$, both end-points of the curve $\Pi(S_1)$ are the lower ends of the appropriate intervals $K(u)$ (due to Lemma 8.4), and this contradicts Lemma 8.5. The first assertion of Lemma 8.6 has been proved.

Now suppose that

$$[u_\ell(M) - u_\ell(N)][u_r(M) - u_r(N)] = 0 \quad ,$$

and, for definiteness,

$$u_\ell(M) = u_\ell(N) = u_* \quad .$$

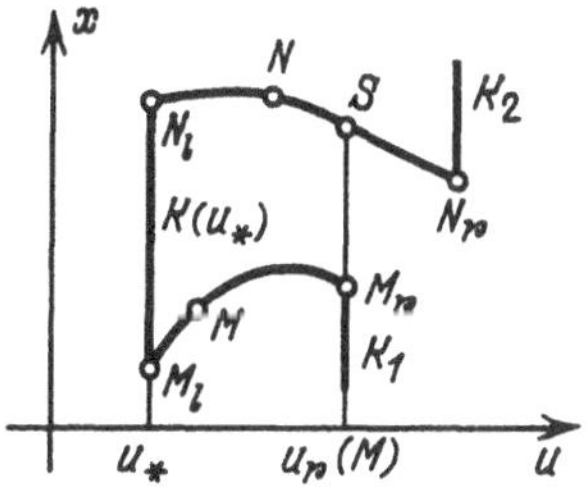

Fig. 8.5

As it has been already proved, the left end-points M_ℓ and N_ℓ of the curves $\Pi(M)$ and $\Pi(N)$ do not coincide. Therefore, M_ℓ and N_ℓ are the end-points of the interval $K(u_*)$; suppose that these points are located as shown in Figure 8.5. Then, by Lemma 8.5, it follows that the right end-point M_r of the curve $\Pi(M)$ is the upper end-point of the interval $K_1 = K[u_r(M)]$, while the right end-point N_r of the curve $\Pi(N)$ the lower end-point of the interval $K_2 = K[u_r(N)]$. Assume, for definiteness, that $u_r(M) \leq u_r(N)$; then the point S of the intersection of the curve $\Pi(N)$ with the straight line $u = u_r(M)$ does not belong to the interval K_1 . We have arrived at contradiction, therefore

$$[u_\ell(M) - u_\ell(N)][u_r(M) - u_r(N)] \neq 0 \quad . \tag{8.14}$$

It remains to prove that neither of intervals $[u_\ell(M),u_r(M)]$ and $[u_\ell(N),u_r(N)]$ is contained in the interior of the other. Suppose the converse; let

$$u_\ell(M) < u_\ell(N) < u_r(N) < u_r(M) \quad .$$

Then, by Lemma 8.5 ,

$$\{\Pi[u_\ell(N);M] - \Pi[u_\ell(N);N]\}\{\Pi[u_r(N);M] - \Pi[u_r(N);N]\} < 0 \quad .$$

In view of the last inequality, the curves $\Pi_0(M)$ and $\Pi_0(N)$ must have a common point, but this contradicts Lemma 8.3.

∎

8.3. Construction of curves Φ_ℓ, Φ_r

Let Ψ_ℓ and Ψ_r be respectively the sets of the left and right end--points of the curves $\Pi(M)$, where $M \in \Omega(V) \setminus \Gamma$. By Ψ_o we denote the set of the degenerate intervals $K(u)$ (singletons).

For any point $M_o = \{u_o, x_o\} \in \Omega(V)$, by $T_\ell(u;M_o)$ we denote the restriction of the function $T(u;M_o)$ to the interval $(-\infty, u_o]$; analogously we can define the function $T_r(u;M_o)$ $(u_o \leq u < \infty)$. The corresponding graphs will be denoted by $T_\ell(M_o)$ and $T_r(M_o)$, respectively.

Lemma 8.7. If the points $M = \{u_o, x_o\}$ and $N = \{v_o, y_o\}$ belong to the set $\Psi_\ell \cup \Psi_o$ and $v_o \leq u_o$, then the functions $T_\ell(u;M)$ and $T_\ell(u;N)$ assume the same values for $-\infty < u \leq v_o$.

Proof. By the semigroup property of the prehysteron V, we only have to show that

$$y_o = T(v_o;M) \quad . \tag{8.15}$$

First, let $M \in \Psi_o$. Suppose that the equality (8.15) is false. Then $N \neq M$ and hence $v_o < u_o$. The interval $K(v_o)$ contains not only the point N but also $M_1 = \{v_o, T(v_o;M)\}$ (see Fig. 8.6), therefore the point $N_\ell = N$ belongs to Ψ_ℓ and is the left end-point of a certain curve $\Pi(N_o)$. By Lemma 8.5, one of the end-points of the curve $\Pi(N_o)$ is the lower and the other - the upper end-point of the corresponding interval $K(u)$. Consequently, there are no points of the set Ψ_o with abscissae from the interval $[u_\ell(N_o), u_r(N_o)]$, i.e., the whole curve $\Pi(N_o)$ is located to the left of the point M. This means that the curve $x = T_\ell(u;M)$ has common points with the curve $\Pi(N_o)$. Therefore, Lemma 8.4 implies (8.15).

Now, let $M \in \Psi_\ell$. If $N \in \Psi_o$, then the equality (8.15) is obvious. We must only consider the case when also N belongs to Ψ_ℓ and $N \neq M$.

The points M and N are the left ends of some curves $\Pi(M_o)$ and $\Pi(N_o)$. In view of Lemma 8.6, the whole curve $\Pi(N_o)$ is located to the left of the right end-points M_r of the curve $\Pi(M_o)$. Therefore (see

Fig. 8.6), the curve $x = T_\ell(u;M_r)$ has common points with the curve $\Pi(N_o)$, where (by the same Lemma 8.6) these common points do not belong to $\Pi(M_o)$. This means, there exist common points of the curves $x = T_\ell(u;M_r)$ and $\Pi(N_o)$. Thus, by virtue of Lemma 8.4, (8.15) follows. ∎

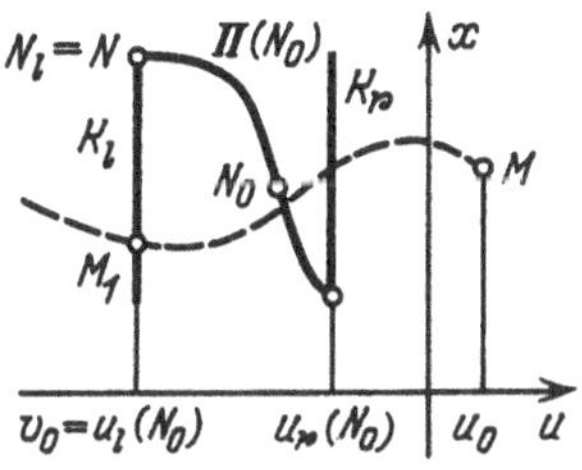

Fig. 8.6

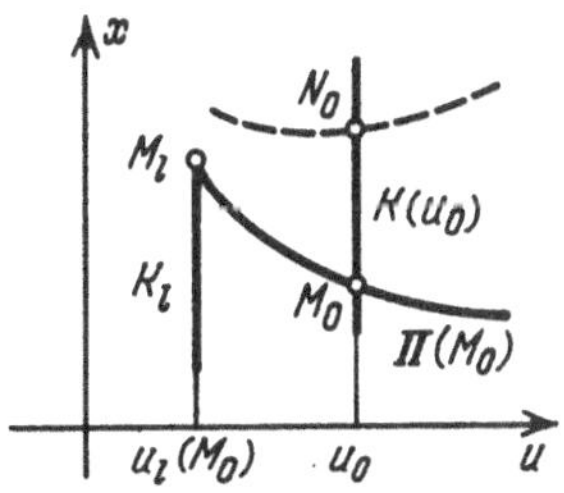

Fig. 8.7

Analogously, we can prove

Lemma 8.8. Let the points $M = \{u_o,x_o\}$ and $N = \{v_o,y_o\}$ belong to the set $\Psi_r \cup \Psi_o$ and $v_o \leq u_o$. Then the functions $T_r(u;M)$ and $T_r(u;N)$ assume the same values for $u_o \leq u < \infty$.

Due to Lemma 8.7, the union of the curves $T_\ell(M)$ $(M \in \Psi_\ell \cup \Psi_o)$ is the graph of a continuous function $\Phi_\ell(u)$ defined on the interval $-\infty < u < a_\ell$ (in particular, a_ℓ may be infinite); we shall denote the graph of the function $\Phi_\ell(u)$ by Φ_ℓ. Analogously, we can define the function $\Phi_r(u)$ $(b_r < u < \infty)$ and its graph Φ_r.

8.4. Completion of the proof of Theorem 7.1

Assume that the domain $\Omega(V)$ of the feasible states of a pre-hysteron V, considered in Theorem 7.1, as well as the curves $T(M)$, $\Pi(M)$, Φ_ℓ and Φ_r are the domain of the feasible states and the relevant curves of a certain hysteron W (cf., Chapter 3). We are going to show that the above domain and curves satisfy axioms (a1) ÷ (a5) of Section 3.2.

Axiom (a1) is inherent in Lemma 7.1.

To verify (a2), we shall prove that the curves Φ_ℓ and Φ_r consist of the end-points of the relevant intervals $K(u)$. Suppose the converse. For definiteness, assume there exists $M_o = \{u_o, \Phi_\ell(u_o)\}$ which is an interior point of the interval $K(u_o)$. By Lemma 8.4, the functions $\Phi_\ell(u)$ and $\Pi(u;M_o)$ assume equal values at

$$u_\ell(M_o) \leq u \leq \min\{u_r(M_o), a_\ell\} = c \quad .$$

Suppose (again, for definiteness) that the left end-point M_ℓ of the curve $\Pi(M_o)$ is the upper end (cf., Lemma 8.5) of the interval $K_\ell = K[u_\ell(M_o)]$. From this interval choose an interior point N_o located over the point M_o (see Fig. 8.7). By Lemma 8.6, the curves $\Pi(M_o)$ and $\Pi(N_o)$ have empty intersection (in Fig. 8.7, the curve $\Pi(N_o)$ is represented by the dashed line); therefore the left end-point N_ℓ of the curve $\Pi(N_o)$ cannot belong to Φ_ℓ what contradicts definition of the curve Φ_ℓ .

Let $u_o \in (b_r, a_\ell)$ and the interval $K(u_o)$ be non-degenerate. To complete the verification of axiom (a2), it remains to show that the points $M_o = \{u_o, \Phi_\ell(u_o)\}$ and $N_o = \{u_o, \Phi_r(u_o)\}$ are different. Suppose that these points coincide with the upper end (for concreteness) of the interval $K(u_o)$. Choose then an interior point S_o of $K(u_o)$ and construct the curve $\Pi(S_o)$. All the intervals $K(u)$ at $u_\ell(S_o) \leq u \leq u_r(S_o)$ are non-degenerate; their upper end-points belong simultaneously to the curves Φ_ℓ and Φ_r (this follows by the continuity of the functions $\Phi_\ell(u)$ and $\Phi_r(u)$, and in view of the already proved part of axiom (a2)); in particular, points $\{u_\ell(S_o), \Pi[u_\ell(S_o);S_o]\}$ and $\{u_r(S_o), \Pi[u_r(S_o); S_o]\}$ represent the upper ends of the intervals $K[u_\ell(S_o)]$ and $K[u_r(S_o)]$, respectively. This contradicts Lemma 8.5 .

Denote by $\Omega_o(W)$ union of all the curves $\Pi(M)$ where $M \in \Omega(V) \setminus \Gamma$, and define $\Omega_1(W) = \Omega_o(W) \cup \Phi_\ell \cup \Phi_r$. To verify axiom (a3), we only must prove that $\Omega(V)$ coincides with $\Omega_1(V)$. To this purpose, because of the controllability of the prehysteron V it is enough to show that each curve

$$u = u(t), \quad x = V[t_o, x_o]u(t) \qquad (t_o \leq t \leq t_1) \quad ,$$

where $\{u(t_o), x_o\} \in \Omega_1(W)$, which corresponds to a piecewise monotone continuous input $u(t)$ $(t_o \leq t \leq t_1)$, is entirely contained in $\Omega_1(W)$. For this, in turn, it suffices to prove (because the operators $V[t_o, x_o]$, defining prehysteron V , satisfy the semigroup identity) that $T(M_0) \subset \Omega_1(W)$ at every $M_0 \in \Omega_1(W)$. For $M_0 \in \Omega_0(W)$, the inclusion $T(M_0) \subset \Omega_1(W)$ is a consequence of Lemma 8.6; this inclusion is obvious if $M_0 \in \Phi_\ell \cap \Phi_r$.

Now suppose that $M_0 \in \Omega_1(W)$, but $M_0 \notin \Omega_0(W)$ and $M_0 \notin \Phi_\ell \cap \Phi_r$. For definiteness we shall assume that $M_0 = \{u_0,x_0\} \in \Phi_r$ and $M_0 \notin \Phi_\ell$. Define

$$u_* = \sup\{u: u < u_0,\ T(u,M_0) \notin \Phi_r\} \ .$$

If $u_* = -\infty$, then $T(u;M_0) = \Phi_r(u)$ at all values $u \in (-\infty,\infty)$; hence we only have to consider the case of finite u_* (u_* may be equal to u_0). The inclusion $T(M_0) \subset \Omega_1(W)$ is evident if there exists $u_1 \in [u_*,u_0]$ for which the point $\{u_1,T(u_1;M_0)\}$ either is the right end of some curve $\Phi(M)$ or belongs to Φ_ℓ . Otherwise, the point $M_* = \{u_*,T(u_*;M_0)\}$ belongs to Φ_r and, as it has been established at the verification of axiom (a2), M_* is one of the end-points of the non-degenerate interval $K(u_*)$; for concreteness, let it be the upper end (see Fig. 8.8). Because $b_r < u_*$ and $\Phi_r(u)$ is continuous, we can indicate an interval $(u_1,u_*]$, adherent to u_* from the left, on which the function $\Phi_r(u)$ is defined, with every point $\{u,\Phi_r(u)\}$ at $u \in (u_1,u_*]$ being the upper end of the appropriate interval $K(u)$. In Figure 8.8 , union of the above-mentioned intervals is indicated by hatching. This set is contained in $\Omega_1(W)$. The part of the curve $T(M_0)$ which is contained in the hatched domain in Figure 8.8, lies in $\Omega_1(W)$, as well. In this part of the curve (in Fig. 8.8 it is shown by the double line)there are points of $\Omega_0(W)$ and, moreover, points of $\Omega(V)\setminus\Gamma$. Therefore the whole curve $T(M)$ is contained in $\Omega_1(W)$. Axiom (a3) has been proved.

Axiom (a4) is satisfied due to Lemma 8.6 .

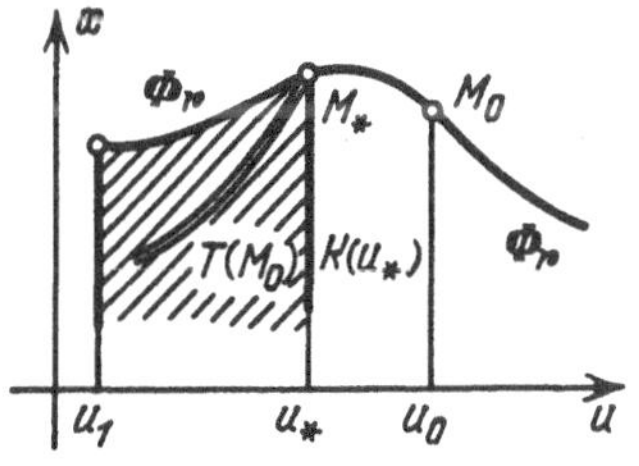

Fig. 8.8

For verifying axiom (a5), we shall restrict ourselves to establishing existence of the minimum (3.5). Let $u < a_\ell$ and $u \notin A_\ell \cup A_0$. If at least one interior point $N = \{u_1, T(u_1;M)\}$ of some interval $K(u_1)$, where $u_1 > u$, belongs to the curve $T(M)$, then the value (3.5) is given by equality $\gamma_r(u) = u_\ell(N)$. If, however, all the points $\{v, T(v;M)\}$ at $v > u$ are the ends of the relevant intervals $K(v)$, then (by definition of the number a_ℓ) among them there must be points of the set $\Phi_\ell \cap \Phi_r$. Hence (3.5) exists and equals $\min\{v: v > u, v \in A_0\}$.

Since the domain $\Omega(V)$ of the feasible states, the set $\Omega_0(W)$, the curves Φ_ℓ, Φ_r and the family of curves $\Pi(M)$ satisfy axioms (a1) - (a5), they also define (cf., Chapter 3) a certain hysteron W. By the same construction, for each piecewise monotone, continuous input $u(t)$, the output of hysteron W coincides with the output of the prehysteron V (provided that the initial states of hysteron W and prehysteron V are equal).

The proof of Theorem 7.1 has been completed.

∎

9. α-identifiability

9.1. Statement of the problem

Let V_1 and V_2 be two prehysterons. Assume that for any point $\{u_0, x_1\} \in \Omega(V_1)$ there exists a point $\{u_0, x_2\} \in \Omega(V_2)$ such that at any piecewise monotone, continuous input $u(t)$ $(t \geq t_0)$, satisfying the condition $u(t_0) = u_0$, the inequality

$$|V_1[t_0, x_1]u(t) - V_2[t_0, x_2]u(t)| \leq \alpha \quad (t \geq t_0) \tag{9.1}$$

holds, where α is some fixed positive number.

Analogously, assume that for each point $\{u_0, x_2\} \in \Omega(V_2)$ there exists such a point $\{u_0, x_1\} \in \Omega(V_1)$ that at any piecewise monotone, continuous input $u(t)$ which satisfies the condition $u(t_0) = u_0$, the same inequality (9.1) takes place. Then the prehysterons V_1 and V_2 will be called *α - close*.

If the operators $V_1[t_0, x_1]$ and $V_2[t_0, x_2]$ can be extended by continuity onto arbitrary continuous inputs which satisfy condition $u(t_0) = u_0$ (for example, if V_1 and V_2 are hysterons), then the α - closeness of the prehysterons V_1 and V_2 implies that (9.1) is true for all appropriate continuous inputs $u(t)$.

Now, we introduce a special class of hysterons, with simpler structure, further referred to as *normal hysterons*. We are going to establish a characterization for the class of prehysterons that are α-close (for any $\alpha > 0$) to some normal hysteron.

9.2. Normal hysteron

Let Ω be a closed domain, bounded by the graphs Φ_ℓ and Φ_r of two continuous functions $\Phi_\ell(u)$ and $\Phi_r(u)$, defined on the whole axis $-\infty < u < \infty$. Assume that for each point $M \in \Omega$ there is a unique curve $\Pi(M)$ which contains this point and is the graph of a continuous function $\Pi(u;M)$ defined on the finite interval $[u_\ell(M), u_r(M)]$. Let the end-points

$$M_\ell = \{u_\ell(M), \Pi[u_\ell(M);M]\} ,$$
$$M_r = \{u_r(M), \Pi[u_r(M);M]\} \tag{9.2}$$

of the curve $\Pi(M)$ lie on Φ_ℓ and Φ_r , respectively, whereas all the remaining points of $\Pi(M)$ belong neither to Φ_ℓ nor to Φ_r . The curve $\Pi(M)$ degenerates to a point, if $M \in \Phi_\ell \cap \Phi_r$.

The curves Φ_ℓ, Φ_r and the system of curves $\Pi(M)$ define a certain hysteron W ; that hysteron will be referred to as *normal* (cf., Section 4.5). In the case of a normal hysteron there is no necessity to provide the general, cumbersome characterization of the operators $W[t_0,x_0]$ as exposed in Chapter 3. In this case, for a monotone input $u(t)$ the corresponding output is

$$W[t_0,x_0]u(t) = \begin{cases} \Phi_\ell[u(t)] , & \text{if } u(t) \le u_\ell(M_0) , \\ \Pi[u(t);M_0] , & \text{if } u_\ell(M_0) \le u(t) \le u_r(M_0) , \\ \Phi_r[u(t)] , & \text{if } u_r(M_0) \le u(t) , \end{cases} \tag{9.3}$$

where $M_0 = \{u(t_0), x_0\}$. After this, as in Chapter 3, for passing to piecewise monotone, continuous inputs the semigroup identity must be used, while a limit construction is necessary for passing to arbitrary inputs.

The ordinary play and stop represent examples of normal hysterons. If W is an arbitrary hysteron and the functions $\Phi_\ell(u)$, $\Phi_r(u)$ are defined for all $u \in (-\infty,\infty)$, then W is not necessarily normal, since the curves Φ_ℓ and Φ_r may contain parts which do not consist of the end-

-points of the curves $\Pi(M)$.

A prehysteron will be called *locally controllable* if there exists a non-decreasing, positive function $\eta(u)$ $(u \geq 0)$ such that for any two initial states $\{u_0,x_0\}$, $\{u_1,x_1\} \in \Omega(W)$ one can choose a piecewise monotone, continuous input $u(t)$ for which

$$u(t_0) = u_0, \quad u(t_1) = u_1, \quad |u(t)| \leq \eta(|u_0| + |u_1|) \quad (t_0 \leq t \leq t_1)$$

and

$$W[t_0,x_0]u(t_1) = x_1 \quad .$$

The local controllability of a prehysteron is equivalent to the fact that the functions $\Phi_\ell(u)$ and $\Phi_r(u)$ in its characterization are defined for all $u \in (-\infty,\infty)$. In particular, any *normal hysteron is locally controllable*.

9.3. Theorem on α-identification

Let us consider a locally controllable prehysteron V_1 which is uniformly correct on harmonic inputs. Due to Theorem 4.1, that prehysteron admits the canonical representation

$$V_1 = f[u,L(\Gamma_\ell^1,\Gamma_r^1)] \quad . \tag{9.4}$$

We shall consider the function $f(u,z)$, jointly continuous and strongly monotone in z for every fixed u , as defined on the whole plane $-\infty < u,\ z < \infty$. Since the prehysteron V_1 is locally controllable, the functions $\Phi_\ell(u)$ and $\Phi_r(u)$ in its definition are defined for all $u \in (-\infty,\infty)$; therefore, the functions $\Gamma_\ell^1(u)$ and $\Gamma_r^1(u)$ are defined for all u , as well.

Assign a number $\alpha > 0$ and denote by $\delta(u)$ $(-\infty < u < \infty)$ a continuous, positive function such that by virtue of the relations

$$\Gamma_r^1(u) \leq x, \ x_0 \leq \Gamma_\ell^1(u), \quad |x - x_0| \leq \delta(u)$$

it follows that $|f(u,x) - f(u,x_0)| < \alpha$. Further, let us construct strictly increasing functions $\Gamma_\ell^2(u)$, $\Gamma_r^2(u)$ $(-\infty < u < \infty)$ for which

$$\Gamma_r^2(u) \le \Gamma_r^1(u) \le \Gamma_\ell^1(u) \le \Gamma_\ell^2(u) \qquad (-\infty < u < \infty)$$

and

$$|\Gamma_\ell^1(v) - \Gamma_\ell^2(v)| < \delta(u) \qquad (\Gamma_r^2(u) \le \Gamma_r^2(v) \le \Gamma_\ell^2(u)) \quad ,$$

$$|\Gamma_r^1(v) - \Gamma_r^2(v)| < \delta(u) \qquad (\Gamma_r^2(u) \le \Gamma_\ell^2(v) \le \Gamma_\ell^2(u)) \quad .$$

Having given the function $f(u,z)$ and the play $L(\Gamma_\ell^2,\Gamma_r^2)$ we shall construct the hysteron

$$V_2 = f[u,L(\Gamma_\ell^2,\Gamma_r^2)] \quad ; \tag{9.5}$$

(9.5) is then a normal hysteron.

Theorem 9.1. The normal hysteron (9.5) is α-close to the prehysteron (9.4).

Proof. To each point $\{u_o,x_1\} \in \Omega(V_1)$ let us associate the same point, treated as a point $\{u_o,x_2\} \in \Omega(V_2)$; then the inequalities (9.1) are fulfilled. To each point $\{u_o,x_2\} \in \Omega(V_2)$ let us associate the closest point $\{u_o,x_1\}$ in the intersection of the domain $\Omega(V_1)$ of the feasible states of the prehysteron V_1 with the straight line $u = u_o$; then the inequalities (9.1) are satisfied, too.

∎

9.4. A remark

The construction of the hysteron (9.5), given in Section 9.3, has a general character. By employing that construction, we may construct (via an appropriate approximation of one function $f(u,z)$ of two variables and two functions $\Gamma_\ell^1(u)$, $\Gamma_r^1(u)$ of one variable) hysterons with various additional properties, α-close to the prehysteron V .

10. Approximate construction of hysteron

10.1. Distance between hysterons

Consider two hysterons (prehysterons) W_1 and W_2 with different, in general, domains $\Omega(W_1)$ and $\Omega(W_2)$ of feasible states. The value

$$\theta(W_1,W_2) = \sup_{u(t)} \sup_{x_o} \inf_{y_o} \|W_1[t_o,x_o]u(t) - W_2[t_o,y_o]u(t)\|_{t_o,t_1} \quad (10.1)$$

will be called a *deviation of the hysteron* W_1 *from hysteron* W_2. The numbers x_o and y_o in the right-hand side of the formula (10.1) are to be chosen so that, at least for $u = u_o$, the points $\{u_o,x_o\}$ and $\{u_o,y_o\}$ belong to the domains $\Omega(W_1)$ and $\Omega(W_2)$, respectively. The functions $u(t)$ $(t_o \le t \le t_1)$, for which the upper bound is taken in (10.1), have to satisfy the condition $u(t_o) = u_o$.

A *distance between the hysterons* W_1 *and* W_2 will be defined by the formula

$$\rho(W_1,W_2) = \max\{\theta(W_1,W_2),\ \theta(W_2,W_1)\}. \quad (10.2)$$

The distance (10.2) may be infinite.

Deviation (10.1) and distance (10.2) are reminiscent of Hausdorff's deviation of one set to another and of Hausdorff's distance between two sets. The distance defined by formula (10.2) satisfies all axioms of metric.

If $\rho(W_1,W_2) \le \alpha$, then the hysterons (prehysterons) W_1 and W_2 are, certainly, α-close.

By $T_1(u;M)$, $\Phi_\ell^1(u)$, $\Phi_r^1(u)$ and $T_2(u;M)$, $\Phi_\ell^2(u)$, $\Phi_r^2(u)$ we shall denote the functions that enter definitions of the hysterons (prehysterons) W_1 and W_2.

Theorem 10.1. Let functions $\Phi_\ell^1(u)$, $\Phi_r^1(u)$ and $\Phi_\ell^2(u)$, $\Phi_r^2(u)$ be defined for all $u \in (-\infty,\infty)$. Then the equality

$$\rho(W_1,W_2) = \max\{\chi_\ell(W_1,W_2),\ \chi_r(W_1,W_2)\} \quad (10.3)$$

holds, where

$$\chi_\ell(W_1,W_2) = \sup_{-\infty < u,v < \infty} |T_1[u;\{v,\Phi_\ell^1(v)\}] - T_2[u;\{v,\Phi_\ell^2(v)\}]| \quad (10.4)$$

and

$$\chi_r(W_1,W_2) = \sup_{-\infty<u,v<\infty} |T_1[u;\{v,\Phi_r^1(v)\}] - T_2[u;\{v,\Phi_r^2(v)\}]| \quad . \tag{10.5}$$

Proof. The inequality

$$\rho(W_1,W_2) \geq \max\{\chi_\ell(W_1,W_2),\ \chi_r(W_1,W_2)\}$$

follows immediately from definitions (10.4) and (10.5). Therefore, we only have to prove the inequality

$$\rho(W_1,W_2) \leq \max\{\chi_\ell(W_1,W_2),\ \chi_r(W_1,W_2)\} \quad , \tag{10.6}$$

or, equivalently,

$$\theta(W_1,W_2) \leq \max\{\chi_\ell(W_1,W_2),\ \chi_r(W_1,W_2)\} \tag{10.7}$$

and

$$\theta(W_2,W_1) \leq \max\{\chi_\ell(W_1,W_2),\ \chi_r(W_1,W_2)\} \quad . \tag{10.8}$$

We shall carry out the proof of the inequality (10.7).

It will be shown if to each point $M_o = \{u_o,x_o\} \in \Omega(W_1)$ we assign a certain point $N_o = \{u_o,y_o\} \in \Omega(W_2)$ such that at any continuous input $u(t)$ $(t_o \leq t \leq t_1)$, satisfying the condition $u(t_o) = u_o$, the inequality

$$|W_1[t_o,x_o]u(\tau) - W_2[t_o,y_o]u(\tau)| \leq \max\{\chi_\ell(W_1,W_2),\ \chi_r(W_1,W_2)\} \tag{10.9}$$

is satisfied for all $\tau \in [t_o,t_1]$. Clearly, it is enough to prove that (10.9) is satisfied in the case of piecewise monotone, continuous inputs $u(t)$.

The ordinate of the point N_o will be defined by the equality (see Fig. 10.1)

$$y_o = T_2[u_o;\{u_\ell^1(M_o),\Phi_\ell^2[u_\ell^1(M_o)]\}] \quad . \tag{10.10}$$

Then, by (10.4), the inequality

$$|T_1(u;M_o) - T_2(u;N_o)| \leq \chi_\ell(W_1,W_2) \quad (-\infty < u < \infty) \tag{10.11}$$

will be satisfied.

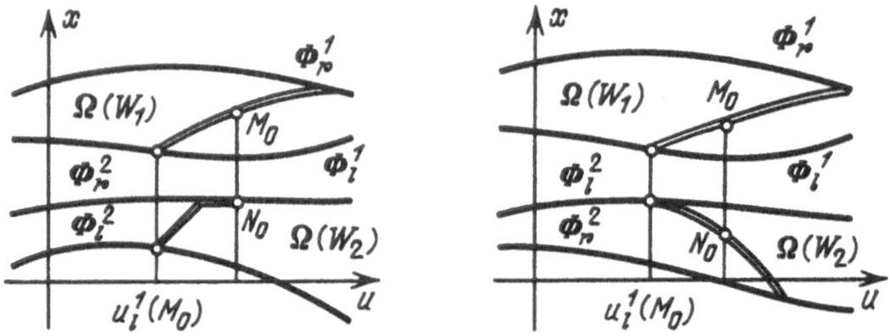

Fig. 10.1

Assume that the points $t_o = s_o < s_1 < \dots < s_n = \tau$ divide $[t_o,\tau]$ into the monotonicity intervals of function $u(t)$. We are going to show that at each fixed i , the inequality

$$|T_1(u;M_{i-1}) - T_2(u;N_{i-1})| \leq$$

$$\leq \max\{\chi_\ell(W_1,W_2),\ \chi_r(W_1,W_2)\} \quad (-\infty < u < \infty) \quad , \tag{10.12}$$

where

$$M_i = \{u(s_i), W_1[t_o,x_o]u(s_i)\}, \quad N_i = \{u(s_i), W_2[t_o,y_o]u(s_i)\} \quad , \tag{10.13}$$

implies the estimate

$$|T_1(u;M_i) - T_2(u;N_i)| \leq$$

$$\leq \max\{\chi_\ell(W_1,W_2),\ \chi_r(W_1,W_2)\} \quad (-\infty < u < \infty) \quad . \tag{10.14}$$

Then, by virtue of (10.11), the estimate

$$|T_1[u;\{u(\tau),W_1[t_o,x_o]u(\tau)\}] - T_2[u;\{u(\tau),W_2[t_o,y_o]u(\tau)\}]| \leq$$

$$\leq \max\{\chi_\ell(W_1,W_2),\ \chi_r(W_1,W_2)\} \quad (-\infty < u < \infty)$$

holds, and it yields (10.9) if we take $u = u(\tau)$.

Now, let the inequality (10.12) be satisfied. For concreteness, let us assume that $u(t)$ is a non-decreasing function on the interval $[s_{i-1},s_i]$. Then the number $u(s_i)$ fulfils one of the following four conditions:

$$u(s_i) < \min\{u_r^1(M_{i-1}),\ u_r^2(N_{i-1})\} \quad , \tag{10.15}$$

$$u_r^1(M_{i-1}) \leq u(s_i) < u_r^2(N_{i-1}) \quad , \tag{10.16}$$

$$u_r^2(N_{i-1}) \leq u(s_i) < u_r^1(M_{i-1}) \quad , \tag{10.17}$$

$$\max\{u_r^1(M_{i-1}),\ u_r^2(N_{i-1})\} \leq u(s_i) \quad . \tag{10.18}$$

If (10.15) is true, inequality (10.14) coincides with (10.12). If (10.16) holds, we define the auxiliary points

$$M_* = \{u_r^2(N_i),\ \Phi_r^1[u_r^2(N_i)]\} \quad ,$$

$$N_* = \{u_r^2(N_i),\ \Phi_r^2[u_r^2(N_i)]\} = \{u_r^2(N_i),\ \Phi_r^2[u_r^2(N_{i-1})]\}$$

(cf., Fig. 10.2). Due to (10.5), the difference of functions $T_1(u;M_*)$ and $T_2(u;N_*)$ can be estimated from above by $\chi_r(W_1,W_2)$, and hence for proving (10.14) it is sufficient to note that $T_2(u;N_{i-1}) = T_2(u;N_i)$, and the curve $T_1(M_i)$ is located between curves $T_1(M_{i-1})$ and $T_1(M_*)$. The case (10.17) can be considered in an analogous way. At last, if (10.18) holds, inequality (10.14) follows directly from the definition of $\chi_r(W_1,W_2)$.

∎

Theorem 10.1 admits a natural extension onto the cases where the functions $\Phi_\ell^1(u)$, $\Phi_r^1(u)$, $\Phi_\ell^2(u)$, $\Phi_r^2(u)$ (or some of them) are not defined for all u.

The form of (10.2) becomes remarkably simple for the plays $L_1 = L(\Gamma_\ell^1,\Gamma_r^1)$, $L_2 = L(\Gamma_\ell^2,\Gamma_r^2)$. Then,

$$\rho(L_1,L_2) = \max\{\sup_u |\Gamma_\ell^1(u) - \Gamma_\ell^2(u)| \ ,\ \sup_u |\Gamma_r^1(u) - \Gamma_r^2(u)|\} \quad . \tag{10.19}$$

Let E_1, E_2 and $2h_1$, $2h_2$ be the parameters of stops W_1 and W_2 (cf., Fig. 3.1), where

$$h_1 \leq h_2 \quad . \tag{10.20}$$

Then

$$\rho(W_1,W_2) = \begin{cases} h_2 - h_1, & \text{if } E_1 \le E_2 \text{ and } h_1E_2 \le h_2E_1, \\ h_1 + h_2 - 2h_2 E_1/E_2, & \text{if } E_1 \le E_2 \text{ and } h_1E_2 \ge h_2E_1, \\ h_2 + h_1 - 2h_1 E_2/E_1, & \text{if } E_1 \ge E_2 \ . \end{cases} \tag{10.21}$$

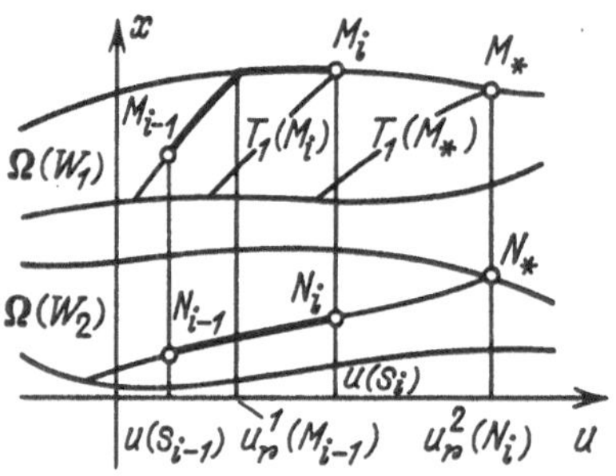

Fig. 10.2

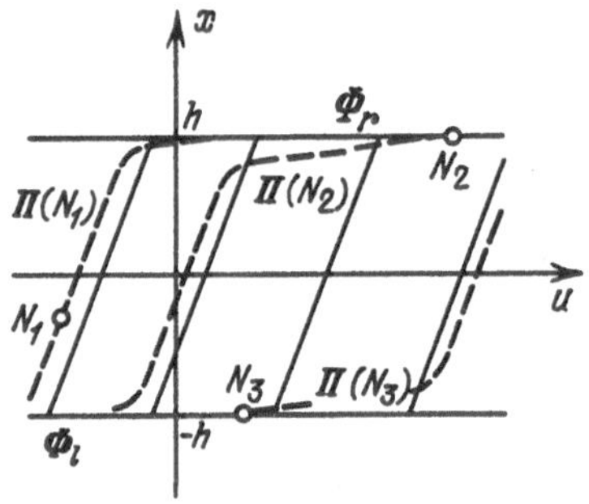

Fig. 10.3

10.2. Bounded inputs

Choose some positive number R and define, analogously to (10.1), the value

$$\theta(W_1,W_2;R) = \sup_{u(t)} \sup_{x_0} \inf_{y_0} \|W_1[t_0,x_0]u(t) - W_2[t_0,y_0]u(t_0)\|_{t_0,t_1} \ , \tag{10.22}$$

where the outer upper bound is to be taken over all inputs $u(t)$ subject to the constraint

$$|u(t)| \le R \quad (t_0 \le t \le t_1) \ . \tag{10.23}$$

We also define the distance

$$\rho(W_1,W_2;R) = \max\{\theta(W_1,W_2;R),\ \theta(W_2,W_1;R)\} \ , \tag{10.24}$$

by proceeding analogously to (10.2).

Similarly to Theorem 10.1, we can prove

Theorem 10.2. Let an interval $[-R,R]$ be contained in the domain of all the functions $\Phi_\ell^1(u)$, $\Phi_r^1(u)$, $\Phi_\ell^2(u)$, $\Phi_r^2(u)$. Suppose that for each point $M_o = \{u_o,x_o\} \in \Omega(W_1)$, where $|u_o| \leq R$, at least one of the numbers $u_\ell^1(M_o)$, $u_r^1(M_o)$ belongs to $[-R,R]$ and, analogously, for each point $N_o = \{u_o,y_o\} \in \Omega(W_2)$, where $|u_o| \leq R$, at least one of the numbers $u_\ell^2(N_o)$, $u_r^2(N_o)$ belongs to $[-R,R]$. Then the equality

$$\rho(W_1,W_2;R) = \max\{\chi_\ell(W_1,W_2;R),\ \chi_r(W_1,W_2;R)\} \tag{10.25}$$

is satisfied with

$$\chi_\ell(W_1,W_2;R) = \max_{|u|,|v| \leq R} |T_1[u;\{v,\ \Phi_\ell^1(v)\}] - T_2[u;\{v,\ \Phi_\ell^2(v)\}]| \tag{10.26}$$

and

$$\chi_r(W_1,W_2;R) = \max_{|u|,|v| \leq R} |T_1[u;\{v,\ \Phi_r^1(v)\}] - T_2[u;\{v,\ \Phi_r^2(v)\}]| \ . \tag{10.27}$$

Let us note that $R_1 < R_2$ implies the inequalities

$$\rho(W_1,W_2;R_1) \leq \rho(W_1,W_2;R_2) \leq \rho(W_1,W_2) \ .$$

For plays, the distance $\rho(L_1,L_2;R)$ is defined by the right--hand side of (10.19), with the inner upper bounds to be taken over the whole interval $[-R,R]$. For stops, at sufficiently large R the distance $\rho(W_1,W_2;R)$ coincides with $\rho(W_1,W_2)$ (see also the formula (10.21)).

A natural question arises whether it is possible to ensure the closeness of hysterons W_1 and W_2 in either of the metrics (10.2) and (10.24), only by assuming that for each function $T_1(u;M_1)$ which defines the hysteron W_1 there is some close to it (in the uniform norm) function $T_2(u;M_2)$, defining for the hysteron W_2, and conversely, for each function $T_2(u;N_2)$, defining for W_2, there is some close to it function $T_1(u;N_1)$, defining for W_1. Unfortunately, the answer is no . A construction of the relevant example follows immediately from Figure 10.3. In that figure, W_1 is an ordinary play and W_2 is a hysteron with the same set of feasible states (for W_2, the curves

Φ_ℓ^2 and Φ_r^2 are represented by the straight lines $x = -h$ and $x = h$) with the curves $\Pi_2(N)$ indicated by dashed lines in Figure 10.3 .

10.3. Frames of hysterons

Let W be a certain hysteron generated by the curves Φ_ℓ, Φ_r and the family of curves $\Pi(M)$. As usual, we shall denote by $T(M)$ the defining curves of W .

Let us choose positive numbers R and δ_o . We shall consider two following finite sets of functions

$$\theta_\ell^1(u),\ \theta_\ell^2(u),\ \dots,\ \theta_\ell^n(u)\ ,\quad (-R - 2\delta_o \leq u \leq R + 2\delta_o) \tag{10.28}$$

$$\theta_r^1(u),\ \theta_r^2(u),\ \dots,\ \theta_r^m(u)\ ,\quad (-R - 2\delta_o \leq u \leq R + 2\delta_o)\,. \tag{10.29}$$

Their graphs will be denoted by

$$\theta_\ell^1,\ \theta_\ell^2,\ \dots,\ \theta_\ell^n;\ \ \theta_r^1,\ \theta_r^2,\ \dots,\ \theta_r^m\ . \tag{10.30}$$

In each graph θ_ℓ^i $(i = 1, 2, \dots, n)$, a certain point $M_i = \{u_i, \theta_\ell^i(u_i)\}$ is chosen and, similarly, in each graph θ_r^j $(j = 1, 2, \dots, r)$, a certain point $N_j = \{v_j, \theta_r^j(v_j)\}$ (in Fig. 10.4, the continuous lines represent graphs of the first two of functions (10.28), with points M_1 and M_2 marked, while the dashed lines refer to the corresponding functions (10.29), with points N_1 and N_2 marked). We shall assume that the inequalities

$$-R - 2\delta_o < u_1 < -R - \delta_o < u_2 < u_3 < \dots < R + \delta_o < u_n < R + 2\delta_o \tag{10.31}$$

and

$$-R - 2\delta_o < v_1 < -R - \delta_o < v_2 < v_3 < \dots < R + \delta_o < v_m < R + 2\delta_o \tag{10.32}$$

are satisfied.

The system Ξ of curves (10.30) with selected points M_i and N_j will be referred to as a δ*-frame* $(0 < \delta \leq \delta_o)$ *of the hysteron* W

on the interval $[-R,R]$, if for each curve θ_ℓ^i with the marked point M_i there is a point $M_i^* = \{u_i^*, \Phi_\ell(u_i^*)\}$ such that

$$|u_i^* - u_i| < \delta \ , \quad \|\theta_\ell^i(u) - T(u;M_i^*)\|_{-R,R} < \delta \ ,$$

and, analogously, for each curve θ_r^j with the marked point N_j there is a point $N_j^* = \{v_j^*, \Phi_r(v_j^*)\}$ such that

$$|v_j^* - v_j| < \delta \ , \quad \|\theta_r^j(u) - R(u;N_j^*)\|_{-R,R} < \delta \ .$$

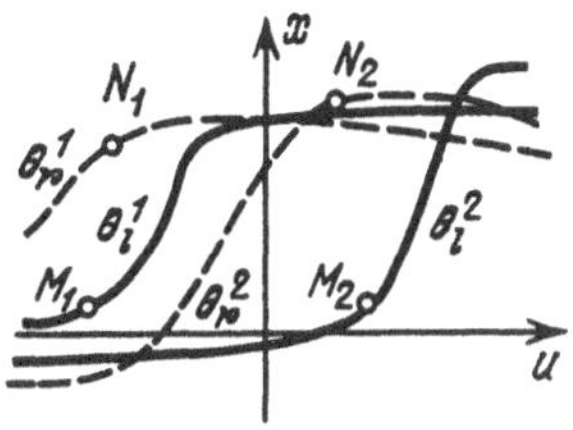

Fig. 10.4

Let w be any fixed number in the interval $[-R,R]$. Among the abscissae of the points M_i choose the maximal value u_{i_1} and the minimal value u_{i_2}, such that

$$u_{i_1} + \delta \leq w \leq u_{i_2} - \delta$$

(this can be done in view of (10.31)), and put

$$\gamma_\ell(w) = \|\theta_\ell^{i_1}(u) - \theta_\ell^{i_2}(u)\|_{-R-2\delta_0, R+2\delta_0} \ .$$

Analogously, let us choose the maximal and minimal abscissae v_{j_1} and v_{j_2} of the points N_j, such that

$$v_{j_1} + \delta \leq w \leq v_{j_2} - \delta \ ,$$

and put

$$\gamma_r(w) = \|\theta_r^{j_1}(u) - \theta_r^{j_2}(u)\|_{-R-2\delta_0,R+2\delta_0} .$$

The number

$$\gamma(\Xi) = \sup_{w\in[-R,R]} \max\{\gamma_\ell(w), \gamma_r(w)\}$$

will be called a *density of the frame* Ξ .

Consider now two hysterons W_1 and W_2 which fulfil the hypotheses of Theorem 10.2. By that theorem, we can conclude the following

Theorem 10.3. Suppose that the hysterons W_1 and W_2 have a common δ-frame with the density $\gamma(\Xi)$. Then the estimate

$$\rho(W_1,W_2;R) \leq 2\delta + \gamma(\Xi)$$

holds for the distance (10.24) between the hysterons W_1 and W_2 . ∎

It is clear that Theorem 10.3 plays a principal role at an approximate construction of the hysteron W . By virtue of this theorem, that problem can be reduced at first to a construction, based only on experimental data, of a finite number of curves forming the δ-frame Ξ , and then to a construction (using mathematical methods) of a certain hysteron W_1 which also has the δ-frame Ξ . For some special problems, the above scheme has been implemented; its detailed development (organization of a special experiment, processing of experimental results, optimal synthesis of a hysteron from its δ-frame , etc.) is to be accomplished in future.

10.4. Approximation by operators different from hysterons

Until now, in this chapter we have discussed problems concerning the relations of proximity for pairs of hysterons. A natural question arises about a possibility of approximating hysterons by operators of some different structure. This problem has been given only a fragmentary treatment. Below we formulate a few simple results on the possibility of constructing approximations to ordinary plays and stops in the form of solutions to some suitable differential equations.

Let $L = L(\Gamma_\ell, \Gamma_r)$ be a play with the generating lines

$$\Gamma_\ell(u) = Eu + Eh, \quad \Gamma_r(u) = Eu - Eh \quad (-\infty < u < \infty) \quad ,$$

where E and h are some positive numbers. For L, the domain of its feasible states is characterized by the inequality

$$|Eu - x| \leq Eh \quad .$$

Consider the differential equation

$$\frac{dx}{dt} = kg[Eu(t) - x] \quad , \tag{10.33}$$

where

$$g(z) = \begin{cases} z + h \; , & \text{for } z \leq -h \; , \\ 0 \; , & \text{for } |z| \leq h \; , \\ z - h \; , & \text{for } z \geq h \end{cases} \tag{10.34}$$

and k is any positive number. The function (10.34) is Lipschitz continuous, with Lipschitz constant 1 . Thus, for any continuous function $u(t)$ $(t_0 \leq t \leq t_1)$ and each initial state

$$x(t_0) = x_0 \quad , \tag{10.35}$$

the equation (10.33) admits a unique solution

$$x(t) = P_k[t_0, x_0]u(t) \quad (t_0 \leq t \leq t_1) \quad . \tag{10.36}$$

Operators (10.36) have good properties: they are defined on $C(t_0,t_1)$, continuous in that space, as well as monotone with respect to the input $u(t)$, monotone with respect to the initial value x_0 , etc. The operators (10.36) may be used as approximations to plays, because of

<u>Theorem 10.4.</u> Let $\{u_0, x_0\} \in \Omega(L)$. Suppose that a continuous input $u(t)$ $(t_0 \leq t \leq t_1)$ satisfies the condition $u(t_0) = u_0$. Then

$$\lim_{k \to \infty} \|P_k[t_0, x_0]u(t) - L[t_0, x_0]u(t)\|_{t_0, t_1} = 0 \quad . \tag{10.37}$$

Although a geometrical sense of (10.37) is quite simple, its detailed proof turns out rather awkward.

∎

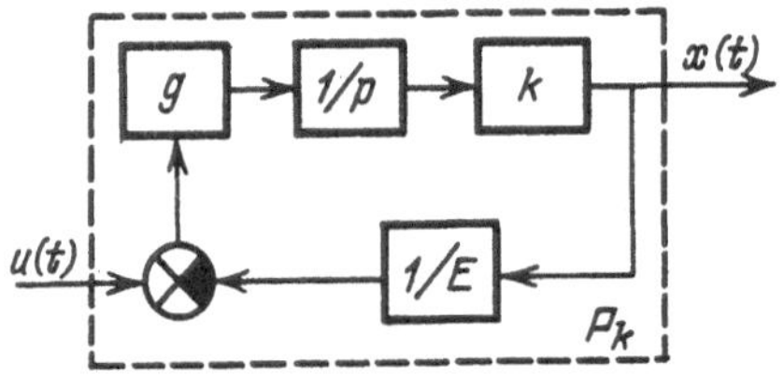

Fig. 10.5

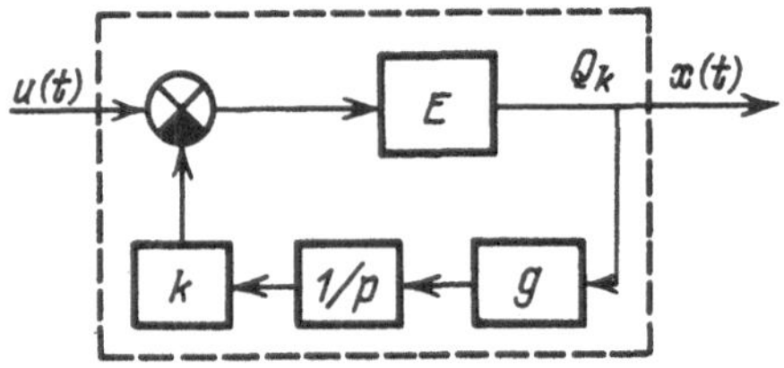

Fig. 10.6

The operators (10.36) describe input-output relations of the transducer P_k whose block-diagram is shown in Fig. 10.5. In that diagram, g denotes a static element whose output w(t) is dependent on the input z(t) according to the equality

$$w(t) = g[z(t)] \ .$$

1/E and k are coefficients, 1/p is the integrator whose output w(t), corresponding to the input z(t), is given by the equality

$$w(t) = x_o + \int_{t_o}^{t} z(s)\, ds \ ,$$

with x_o representing the initial state.

Now let us turn to the study of the stop (see Fig. 3.1) with parameters E and 2h . To approximate it, consider the operators

$$x(t) = Q_k[t_o, x_o]u(t) \ , \tag{10.38}$$

whose values are characterized as solutions of the differential equations

$$\frac{d}{dt}[x - u(t)] = -kg(x) \quad ,$$

subject to the initial condition (10.35).

Because of the identity

$$Q_k[t_o,x_o + u(t_o)]u(t) + P_k[t_o,x_o]u(t) \equiv u(t) \quad ,$$

Theorem 10.4 implies the following

Theorem 10.5. Let $\{u_o,x_o\} \in \Omega(W)$. Suppose that a continuous input $u(t)$ $(t_o \leq t \leq t_1)$ satisfies the condition $u(t_o) = u_o$. Then

$$\lim_{k\to\infty} \|Q_k[t_o,x_o]u(t) - W[t_o,x_o]u(t)\|_{t_o,t_1} = 0 \quad . \tag{10.39}$$

■

The operators (10.38) describe input-output relations of the transducer Q_k whose block-diagram is depicted in Figure 10.6.

At the end, let us note that the limits (10.37) and (10.39) are uniform with respect to equi-continuous classes of admissible input signals.

Part 3. Vibro-correct differential equations and variable hysterons

... have a look, Balloon,
that's our future ...
J. Hašek

11. Necessary condition of vibro-correctness

11.1. Integrator

In the first four chapters of this part, we are concerned with transducers governed by some special differential equations. At the end, by using the established properties of those transducers, a study of hysterons with time-dependent characteristics will be developed.

For such hysterons, input-output relations are constructed in two steps: first, the output corresponding to a piecewise monotone continuous input is to be determined; then, a possibility of performing the limit passage to arbitrary continuous inputs is to be shown.

In this part, we shall consider transducers with outputs defined at first for smooth inputs (for instance, continuously differentiable), and afterwards, by taking a limit extended to arbitrary continuous inputs.

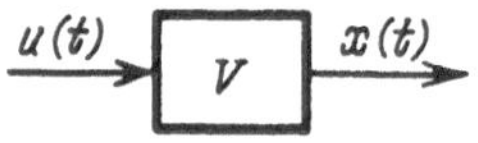

Fig. 11.1

In all the constructions, we shall consider transducers V with scalar inputs $u(t)$ $(t \geq t_0)$ (see Figure 11.1); the output $x(t)$ $(t \geq t_0)$ can be vector-valued, with values in R^N. The state of the transducer V at time instant $t = t_0$ will be represented by a point $\{u_0, x_0\} \in \in R^{N+1}$, where $u_0 = u(t_0)$ and

$$x_0 = x(t_0) \quad . \tag{11.1}$$

For the time-being, we shall not specify any class of the admissible inputs.

We shall assume that at an initial state $\{u_0, x_0\}$ and for a continuously differentiable input $u(t)$ $(t \geq t_0)$ which satisfies the condition $u(t_0) = u_0$, the output $x(t) = x_0$ is defined as a solution of the differential equation

$$\frac{dx}{dt} = f[t, x, u(t), u'(t)] \quad , \tag{11.2}$$

subjected to the initial condition (11.1). If the function $f(t,x,u,v)$ is continuous, then the solution

$$x(t) = V[t_0, x_0]u(t) \tag{11.3}$$

does exist. If $f(t,x,u,v)$ is locally Lipschitz continuous with respect to x (alternatively, it satisfies the Osgood condition or any analogous hypothesis), then the solution (11.3) is unique. If $f(t,x,u,v)$ satisfies natural growth conditions in x, then the solution (11.3) is defined at all $t \geq t_0$, etc.

We shall say the equation (11.2) is *vibro-correct*, if for each $x_* \in R^N$ and any continuous input $u_*(t)$ $(t \geq t_0)$ there exists an interval $[t_0, t_1]$ such that for each continuously differentiable (belonging to the space $C^1 = C^1(t_0, \infty)$) input $u(t)$ which fulfils the condition

$$\|u(t) - u_*(t)\|_{t_0, t_1} = \max_{t_0 \leq t \leq t_1} |u(t) - u_*(t)| \leq \delta_0 \quad , \tag{11.4}$$

where δ_0 is a fixed positive number, the functions (11.3) are defined for all $t \in [t_0, t_1]$ and the equality

$$\lim_{\delta \to 0} \sup_{u(t),v(t) \in D(\delta)} \| V[t_0,x_0]u(t) - V[t_0,x_0]v(t) \|_{t_0,t_1} = 0$$

holds with

$$D(\delta) = \{u(t): u(t) \in C^1, \ \|u(t) - u_*(t)\|_{t_0,t_1} \le \delta\} .$$

In other words, the equation (11.2) is vibro-correct if for each $x_0 \in R^N$ and any continuous input $u_*(t)$ $(t \ge t_0)$ there is a continuous function $x_*(t)$ defined on a certain interval $[t_0,t_1]$, such that the equality

$$\lim_{n\to\infty} \|u_n(t) - u_*(t)\|_{t_0,t_1} = 0 ,$$

where $u_n(t) \in C^1(t_0,t_1)$, implies that

$$\lim_{n\to\infty} \|V[t_0,x_0]u_n(t) - x_*(t)\|_{t_0,t_1} = 0 .$$

If only continuously differentiable inputs are taken into consideration, then the vibro-correctness of equation (11.2) means, in particular, that the transducer V is stable with respect to noises of small amplitude (rather than small in the norm of C^1) at the input. Vibro-correctness of the equation (11.2) is equivalent to a possibility of extending the operators $V[t_0,x_0]$ by continuity from the set of continuously differentiable inputs onto the set of all continuous inputs. This extension preserves (or determines) the input-output relations of transducer V, if all continuous inputs are admissible and if the transducer is stable with respect to noises of small amplitude. Even if the transducer V is physically stable with respect to such noises but the input-output relations on the smooth inputs are described by equation (11.2) which is not vibro-correct, then such a description is obviously incorrect!

We are interested in specifying conditions which guarantee the vibro-correctness. In Chapters 11 - 13, the function $f(t,x,u,v)$ is assumed to be jointly continuous with respect to $t, u, v, \in R^1$, $x \in R^N$.

11.2. Simple examples

The equation (11.2) is evidently vibro-correct, if its right-hand side does not depend explicitly upon $u'(t)$, or, equivalently, if it has the form

$$\frac{dx}{dt} = f[t,x,u(t)] \quad , \tag{11.5}$$

where $f(t,x,u)$ is locally Lipschitz continuous with respect to x.

As a second example let us consider the scalar differential equation

$$\frac{dx}{dt} = a[u(t)]u'(t) + b[u(t)] \quad . \tag{11.6}$$

For smooth $u(t)$, the solution (11.3), satisfying initial condition (11.1), admits the representation

$$V[t_o,x_o]u(t) = x_o \exp \int_{u(t_o)}^{u(t)} a(\xi)d\xi + \int_{t_o}^{t} b[u(\tau)] \exp \int_{u(\tau)}^{u(t)} a(\xi)d\xi d\tau \quad . \tag{11.7}$$

Thus, the equation (11.6) is vibro-correct.

As a last example let us consider the following scalar equation

$$\frac{dx}{dt} = [u'(t)]^2 \quad . \tag{11.8}$$

This equation is not vibro-correct, since, for example, in the case of the inputs $u_n(t) = \frac{1}{n} \sin nt$, the corresponding solutions

$$x_n(t) = V[0,0]u_n(t) = \frac{1}{2} t + \frac{1}{4n} \sin 2nt$$

do not converge uniformly to the solution $x_*(t) = V[0,0]u_*(t) \equiv 0$, where $u_*(t) = 0$ on any of the intervals $[0,t_1]$.

11.3. Necessary condition of vibro-correctness

The vibro-correctness of equation (11.2) cannot be guaranteed by any smoothness of the function $f(t,x,u,v)$. In particular, this follows from the example (11.8). It turns out that *linearity of the right-hand side of the equation with respect to* $u'(t)$, i.e., the special form of f,

$$f(t,x,u,v) \equiv \varphi(t,x,u)v + \psi(t,x,u) \quad , \tag{11.9}$$

is the necessary condition of vibro-correctness. As a matter of fact, condition (11.9) follows from some properties of equations (11.2), much weaker than vibro-correctness.

If only constant functions $u_*(t)$ are admissible in the definition of vibro-correctness (cf., Section 11.1), then the equation (11.2) will be called *vibro-correct on constant inputs*.

Theorem 11.1. If the equation (11.2) is vibro-correct on constant inputs, then $f(t,x,u,v)$ assumes the form (11.9).

Proof. Let us fix numbers t_o, $u_o \in R^1$ and a point $x_o \in R^N$. Then, at $v \neq 0$, we can define the function

$$a(v) = \frac{1}{v}\,[f(t_o,x_o,u_o,v) - f(t_o,x_o,u_o,0)]$$

with values in R^N. For proving the theorem, it is enough to show that $a(v_1) = a(v_2)$, if v_1, $v_2 \neq 0$ and have different sign.

Suppose the converse; then there exist such $v_1 < 0$ and $v_2 > 0$ that

$$a(v_1) \neq a(v_2) \quad . \tag{11.10}$$

In view of (11.10), one can find a linear functional ℓ defined on R^N, such that

$$\begin{aligned} &v_2\ell[f(t_o,x_o,u_o,v_1) - f(t_o,x_o,u_o,0)] - \\ &- v_1\ell[f(t_o,x_o,u_o,v_2) - f(t_o,x_o,u_o,0)] = 2c > 0 \quad . \end{aligned} \tag{11.11}$$

Below we shall consider smooth and continuous, piecewise smooth inputs $u(t)$ $(t \geq t_o)$ for which

$$|u(t) - u_o| \leq 1, \quad |u'(t)| \leq |v_1| + |v_2| \quad . \tag{11.12}$$

By Schauder's theorem, solutions of the differential equation (11.2) with the initial condition (11.1) that correspond to the above inputs, are obviously defined on the interval $[t_o,t_o+h]$, where $h = \{1,M^{-1}\}$ and

$$M = \max\{|f(t,x,u,v)|: |t-t_o|,\ |x-x_o|,\ |u-u_o| \leq 1;\ |v| \leq |v_1| + |v_2|\}.$$

Let us choose an arbitrary positive integer n. We decompose the interval $[t_o, t_o+h]$ into 2^n different parts by the points $\sigma_o = t_o, \sigma_1, \sigma_2, \ldots, \sigma_{2^n} = t_o + h$; further, we divide each interval $[\sigma_{i-1}, \sigma_i]$ by a point τ_i in the ratio $v_2 : |v_1|$. Therefore,

$$t_o = \sigma_o < \tau_1 < \sigma_1 < \ldots < \sigma_{i-1} < \tau_i < \sigma_i < \ldots < \sigma_{2^n} = t_o + h \quad (11.13)$$

and

$$\sigma_i - \sigma_{i-1} = 2^{-n} h, \quad v_1(\tau_i - \sigma_i) = v_2(\tau_i - \sigma_{i-1}) \quad . \quad (11.14)$$

Denote by $u_n(t)$ the continuous input which assumes the same value u_o in all the points σ_i, in all the points τ_i has the value

$$u_o + 2^{-n} \frac{|v_1 v_2| h}{|v_1| + v_2} ,$$

and on each of the intervals $[\sigma_{i-1}, \tau_i]$ and $[\tau_i, \sigma_i]$ is linear (see Fig. 11.2). On each interval $[\tau_i, \sigma_i]$ the derivative of the function $u_n(t)$ is equal to v_1, while on each interval $[\sigma_{i-1}, \tau_i]$ this derivative is equal to v_2. For sufficiently large values n, functions $u_n(t)$ satisfy conditions (11.12). Hence, the corresponding solutions of the equation (11.2), subjected to the initial condition (11.1), are defined on the interval $[t_o, t_o + h]$.

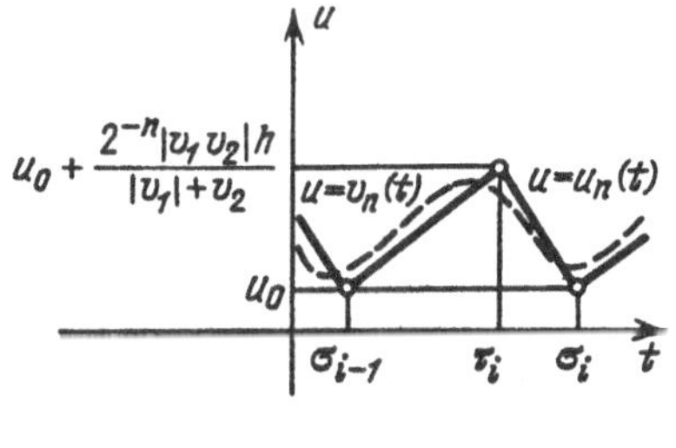

Fig. 11.2

Let us fix n and construct a sequence of inputs $w_k(t)$, smooth on $[t_o, t_o + h]$ (in Fig. 11.2 represented by the dashed line), which fulfil the condition (11.12):

$$|w_k(t) - u_0| \leq 1 \quad , \quad |w_k'(t)| \leq |v_1| + v_2 \quad ,$$

are uniformly convergent to $u_n(t)$ and their derivatives converge to $u_n'(t)$ in the norm of the space L_1 :

$$\lim_{k\to\infty} \int_{t_0}^{t_0+h} |w_k'(t) - u_n'(t)|dt = 0 \quad .$$

Let $V[t_0,x_0]w_k(t)$ be one of the solutions of the equation (11.2) at $u(t) = w_k(t)$; since the sequence $V[t_0,x_0]w_k(t)$ is compact in $C = C(t_0,t_0+h)$, it can be considered as uniformly convergent. The limit function is thus a solution of equation (11.2) at $u(t) = u_n(t)$, and it satisfies condition (11.1). This solution will be denoted by

$$x_n(t) = V[t_0,x_0]u_n(t) \quad . \tag{11.15}$$

By $v_n(t)$ we shall denote an element of the sequence $w_k(t)$, such that

$$\|v_n(t) - u_n(t)\|_{t_0,t_0+h} < \frac{1}{n} \tag{11.16}$$

and

$$\|V[t_0,x_0]v_n(t) - V[t_0,x_0]u_n(t)\|_{t_0,t_0+h} < \frac{1}{n} \quad . \tag{11.17}$$

By construction, the sequence $u_n(t)$ is uniformly convergent to the constant input $u_*(t) \equiv u_0$, hence (11.16) implies also the uniform convergence of the sequence of continuously differentiable functions $v_n(t)$ to $u_*(t)$. Therefore, due to the vibro-correctness of equation (11.2) on constant inputs, the equality

$$\lim_{n\to\infty} \|V[t_0,x_0]v_n(t) - V[t_0,x_0]u_*(t)\|_{t_0,t_0+h_1} = 0$$

holds at some $h_1 \in (0,h]$; hence, by (11.17),

$$\lim_{n\to\infty} \|V[t_0,x_0]u_n(t) - V[t_0,x_0]u_*(t)\|_{t_0,t_0+h_1} = 0 \quad . \tag{11.18}$$

Let us introduce the notations

$$\alpha_n(t) = f[t,x_n(t),u_n(t),v_1] - f\{t,V[t_o,x_o]u_*(t),u_o,0\} ,$$

$$\beta_n(t) = f[t,x_n(t),u_n(t),v_2] - f\{t,V[t_o,x_o]u_*(t),u_o,0\}$$

and choose $h_2 = 2^{-r}h \in (0,h_1)$ (this can be done on account of (11.11)) so that

$$v_2\ell[\alpha_n(\eta)] - v_1\ell[\beta_n(\xi)] \geq c > 0 \quad (t_o \leq \xi,\eta \leq t_o + h_2) , \tag{11.19}$$

where ℓ is a certain linear functional defined on R^N .

With this construction we have completed the preparatory work; now we pass to the proof of Theorem 11.1.

We shall use the decomposition (11.13). For any sufficiently large n , the interval $[t_o,t_o+h_2]$ comprises the first 2^{n-r} intervals $[\sigma_{i-1},\sigma_i]$. By the obvious equality

$$x_n(t) - V[t_o,x_o]u_*(t) =$$

$$= \int_{t_o}^{t} \{f[s,x_n(s),u_n(s),u_n'(s)] - f[s,V[t_o,x_o]u_*(s),u_o,0]\}ds ,$$

it follows that

$$x_n(t_o+h_2) - V[t_o,x_o]u_*(t_o+h_2) = \sum_{i=0}^{2^{n-r}} \left[\int_{\sigma_{i-1}}^{\tau_i} \beta_n(s)ds + \int_{\tau_i}^{\sigma_i} \alpha_n(s)ds\right]$$

and, further,

$$\ell\{x_n(t_o+h_2) - V[t_o,x_o]u_*(t_o+h_2)\} =$$

$$= \sum_{i=1}^{2^{n-r}} \left\{ \int_{\sigma_{i-1}}^{\tau_i} \ell[\beta_n(s)]ds + \int_{\tau_i}^{\sigma_i} \ell[\alpha_n(s)]ds \right\} . \tag{11.20}$$

On the other hand, by the mean-value theorem applied to the above integrals for $i = 1, 2, \ldots, 2^{n-r}$, there are ξ_i, $\eta_i \in [\sigma_{i-1},\sigma_i]$ such that

$$\int_{\sigma_{i-1}}^{\tau_i} \ell[\beta_n(s)]ds + \int_{\tau_i}^{\sigma_i} \ell[\alpha_n(s)]ds = \ell[\beta_n(\xi_i)]\ (\tau_i - \sigma_{i-1}) +$$

$$+ \ell[\alpha_n(\eta_i)]\ (\sigma_i - \tau_i) = \frac{2^{-n}h}{|v_1| + v_2} \{-v_1\ell[\beta_n(\xi_i)] + v_2\ell[\alpha_n(\eta_i)]\} .$$

Therefore, (11.19) implies the estimate

$$\int_{\sigma_{i-1}}^{\tau_i} \ell[\beta_n(s)]ds + \int_{\tau_i}^{\sigma_i} \ell[\alpha_n(s)]ds \geq \frac{2^{-n}hc}{|v_1| + v_2} \quad (i = 1, \ldots, 2^{n-r}) . \tag{11.21}$$

(11.20) and (11.21) yield the inequality

$$\ell\{x_n(t_o + h_2) - V[t_o,x_o]u_*(t_o + h_2)\} \geq \frac{2^{-r}hc}{|v_1| + v_2}$$

which contradicts (11.18). ∎

11.4. Vibro-correctness in a point

Theorem 11.1 admits various generalizations and reinforcements. The modifications of its hypotheses in the case of a function $f(t,x,u,v)$ defined only on some bounded domain are clear. The relevant modifications in the case of operators $V[t_o,x_o]$ continuous in some point $u_*(t)$ $(t \geq t_o)$ of the space C are more complicated.

We shall say that equation (11.12) is *vibro-correct in the point* $\{t_o,x_o,u_*(t)\}$ if there exists an interval $[t_o,t_o+h]$ such that for any sequence $u_n(t)$ of continuously differentiable inputs, uniformly convergent to $u_*(t)$ on $[t_o,t_o+h]$, the corresponding sequence $V[t_o,x_o]u_n(t)$ is uniformly convergent on $[t_o,t_o+h]$, too.

Suppose that the equation (11.2) is vibro-correct in the point $\{t_o,x_o,u_*(t)\}$ and in every right neighbourhood of the point t_o there exists an interval in which the input $u_*(t)$ is Lipschitz continuous. Already in such a case, the function $f[t_o,x_o,u_*(t_o),v]$ is linear with respect to v .

The latter assertion admits the following converse. If $u_*(t)$ does not satisfy the Lipschitz condition on any interval contained in a certain right neighbourhood of the point t_0, then for each x_0 one can construct an equation (11.2), vibro-correct in the point $\{t_0,x_0,u_*(t)\}$, with the right-hand side $f[t_0,x_0,u_*(t_0),v]$ nonlinear with respect to the last variable.

It is interesting to note that there exist equations (11.2) with the right-hand sides highly nonlinear with respect to v, which are vibro-correct on large classes of inputs. For example, the scalar equation

$$\frac{dx}{dt} = g(x)|u'(t)| + h[t,x,u(t),u'(t)] \quad ,$$

with a strictly decreasing function $g(x)$ such that $g(0) = 0$,

$$\int_{-\infty}^{\infty} \frac{dx}{g(x)} < \infty \quad ,$$

and with any continuous, uniformly bounded function $h(t,x,u,v)$, is vibro-correct in all points $\{t_0,x_0,u_*(t)\}$ where $u_*(t)$ is a continuous input with infinite variation on every interval.

12. Sufficient condition of vibro-correctness

12.1. Main result

Due to Theorem 11.1, only equations of the form

$$\frac{dx}{dt} = \varphi[t,x,u(t)]u'(t) + \psi[t,x,u(t)] \tag{12.1}$$

can be vibro-correct. As in Chapter 11, the solutions $x(t)$ of equation (12.1), subject to the initial condition

$$x(t_0) = x_0 \quad , \tag{12.2}$$

will be denoted by

$$x(t) = V[t_0,x_0]u(t) \quad , \tag{12.3}$$

with the operators $V[t_0,x_0]$ defined on all smooth inputs $u(t)$ $(t \geq t_0)$.

Directly by the definition of vibro-correctness it follows that if (12.1) is a vibro-correct equation, then for each fixed smooth input it admits a unique solution which satisfies initial condition (12.2). This implies that the continuity of $\varphi(t,x,u)$ and $\psi(t,x,u)$ does not suffice for ensuring vibro-correctness of the relevant equation (12.1).

To ensure the vibro-correctness, some additional hypotheses concerning a regularity of the functions $\varphi(t,x,u)$ and $\psi(t,x,u)$ must be introduced.

Theorem 12.1. Let a function $\varphi(t,x,u)$ be differentiable with respect to t and x and functions

$$\varphi(t,x,u), \quad \varphi'_t(t,x,u), \quad \varphi'_x(t,x,u), \quad \psi(t,x,u) \tag{12.4}$$

be continuous with respect to all their arguments and locally Lipschitz continuous in x. Then the equation (12.1) is vibro-correct.

The above theorem will be proved in Section 12.4.

If $x = \{\xi_1,\ldots,\xi_N\}$ and $\varphi(t,x,u)$ is a vector-function with components $\varphi_i(t,\xi_1,\ldots,\xi_N,u)$, then $\varphi'_t(t,x,u)$ is a vector-function with components $\frac{\partial}{\partial t}\varphi_i(t,\xi_1,\ldots,\xi_N,u)$ and $\varphi'_x(t,x,u)$ is a matrix with elements $\frac{\partial}{\partial \xi_j}\varphi(t,\xi_1,\ldots,\xi_N,u)$.

The authors do not know any intuitive necessary and sufficient vibro-correctness condition for the equation (12.1).

It would be of interest to find conditions which are not covered by Theorem 12.1 but suffice for the vibro-correctness of the scalar equation

$$\frac{dx}{dt} = \varphi[t,u(t)]u'(t) + \psi[t,x,u(t)] \tag{12.5}$$

(in (12.5), in contrast to (12.1), the coefficient at $u'(t)$ is independent of x).

For the vibro-correctness of equation

$$\frac{dx}{dt} = \varphi[x,u(t)]u'(t) \tag{12.6}$$

it is sufficient that the function $\varphi(x,u)$, jointly continuous with respect to all variables, is locally Lipschitz continuous in x .

12.2. An auxiliary equation

Assume that the function $\varphi(t,x,u)$ fulfils the hypotheses of Theorem 12.1. Let us consider the differential equation

$$\frac{dx}{du} = \varphi(t,x,u) \quad , \tag{12.7}$$

where t plays the role of a parameter. For any fixed t and each initial condition

$$x(u_o) = x_o \quad , \tag{12.8}$$

equation (12.7) admits a unique solution

$$x(u) = Q(u,u_o,x_o,t) \quad . \tag{12.9}$$

We shall assume that it is impossible to extend the solution (12.9) out of a finite or infinite open interval $(a(t,u_o,x_o),\ b(t,u_o,x_o))$ including u_o . Let the equality

$$\lim_{u\to a(t,u_o,x_o)} [\,|u| + |x(u)|\,] = \lim_{u\to b(t,u_o,x_o)} [\,|u| + |x(u)|\,] = \infty \tag{12.10}$$

hold . The right-hand side of the equality (12.9) defines a *shift operator* (in time, from u_o to u) along trajectories of the equation (12.7).

By the definition of shift operator, the identities

$$Q(u_o,u_o,x_o,t) \equiv x_o \tag{12.11}$$

and

$$Q[u_2,u_1,Q(u_1,u_o,x_o,t),t] \equiv Q(u_2,u_o,x_o,t) \tag{12.12}$$

are implied (the second of these identities makes sense provided that the shifts $Q(u_1,u_o,x_o,t)$ and $Q(u_2,u_o,x_o,t)$ are defined).

The derivatives of function $Q(u,u_0,x_0,t)$ with respect to u, u_0, x_0 and t will be denoted by

$$Q_1(u,u_0,x_0,t),\ Q_2(u,u_0,x_0,t),\ Q_3(u,u_0,x_0,t),\ Q_4(u,u_0,x_0,t), \quad (12.13)$$

respectively. The derivatives (12.13) (by virtue of general theorems on the differentiability of the solutions of differential equations with respect to initial data and parameters) exist and are continuous (jointly in all arguments). $Q_3(u,u_0,x_0,t)$ has the form of a matrix, whereas Q_1, Q_2 and Q_4 are vectors. Below we establish several identities for the derivatives (12.13).

The identity

$$Q[v,u,Q(u,v,z,t),t] \equiv z \quad (12.14)$$

follows by virtue of (12.11) and (12.12). By differentiating (12.14) with respect to u, we get

$$Q_2[v,u,Q(u,v,z,t),t] + Q_3[v,u,Q(u,v,z,t),t]Q_1(u,v,z,t) \equiv 0 . \quad (12.15)$$

At the same time, in view of the definition of the operator (12.9),

$$Q_1(u,v,z,t) \equiv \varphi[t,Q(u,v,z,t),u] , \quad (12.16)$$

and, due to (12.15),

$$Q_2[v,u,Q(u,v,z,t),t] + Q_3[v,u,Q(u,v,z,t),t]\varphi[t,Q(u,v,z,t),u] \equiv 0 . \quad (12.17)$$

As a consequence of (12.16) and (12.14), it follows that

$$Q_1[u,v,Q(v,u,x,t),t] \equiv \varphi(t,x,u) . \quad (12.18)$$

By differentiating (12.14) with respect to t, we arrive at the identity

$$Q_4[u,v,Q(v,u,x,t),t] + Q_3[u,v,Q(v,u,x,t),t]Q_4(v,u,x,t) \equiv 0 . \quad (12.19)$$

In turn, the differentiation of (12.14) with respect to z gives us

$$Q_3[u,v,Q(v,u,x,t),t]Q_3(v,u,x,t) \equiv I \quad . \tag{12.20}$$

12.3. A substitution

Now we return to the study of the equation (12.1). We shall look for the solutions $x(t)$ of that equation which have the form

$$x(t) = Q[u(t),v,z(t),t] \quad , \tag{12.21}$$

where $Q(u,v,z,t)$ is defined by (12.9) and v denotes a fixed real number. The representation (12.21) remains always true. To prove this, it is enough to set (on account of (12.12))

$$z(t) = Q[v,u(t),x(t),t] \quad . \tag{12.22}$$

In this representation, functions $x(t)$ and $u(t)$ are defined on the same non-empty intervals, whenever $v = u(t_o)$.

If $x(t)$ is a solution of the equation (12.1), then by differentiating (12.22), we get the relation

$$\frac{dz}{dt} = Q_2[v,u(t),x(t),t]u'(t) + Q_3[v,u(t),x(t),t]\,\frac{dx}{dt} + Q_4[v,u(t),x(t),t],$$

which can be treated as a differential equation determining $z(t)$, provided that, instead of $x(t)$, the representation (12.21) has been substituted into the right-hand side. After this substitution, the equation assumes the form

$$\frac{dz}{dt} = Q_2\{v,u(t),Q[u(t),v,z(t),t],t\}u'(t) +$$

$$+ Q_3\{v,u(t),Q[u(t),v,z(t),t],t\}\varphi\{t,Q[u(t),v,z(t),t],u(t)\}u'(t) +$$

$$+ Q_3\{v,u(t),Q[u(t),v,z(t),t],t\}\psi\{t,Q[u(t),v,z(t),t],u(t)\} +$$

$$+ Q_4\{v,u(t),Q[u(t),v,z(t),t],t\} \quad ,$$

and, due to (12.17),

$$\frac{dz}{dt} = F[t,z,u(t);v] \quad , \tag{12.23}$$

where

$$F(t,z,u;v) = Q_4[v,u,Q(u,v,z,t),t] + \\ + Q_3[v,u,Q(u,v,z,t),t] \quad \psi[t,Q(u,v,z,t),u] \ . \tag{12.24}$$

On account of (12.19), the function (12.24) admits the following equivalent representation

$$F(t,z,u;v) = Q_3[v,u,Q(u,v,z,t),t]\{\psi[t,Q(u,v,z,t),u] - Q_4(u,v,z,t)\} \ . \tag{12.25}$$

A function (12.21) satisfies the initial condition (12.2) if and only if

$$z(t_o) = Q[v,u(t_o),x_o,t_o] \ . \tag{12.26}$$

By (12.11), it follows then that for

$$v = u(t_o) \tag{12.27}$$

we should seek that solution

$$z(t) = W[t_o,x_o]u(t) \tag{12.28}$$

of the equation (12.23), which fulfils the initial condition

$$z(t_o) = x_o \ . \tag{12.29}$$

Now assume that the function $u(t)$ is continuously differentiable and $z(t)$ represents the solution of equation (12.23) subject to the initial condition (12.26). The function (12.21) will then be continuously differentiable in an appropriate neighbourhood of the point $t = t_o$.

Upon differentiating equality (12.21) with respect to t, then inserting the right-hand side of equation (12.23) instead of $\frac{dz}{dt}$ into the resulting equality and eventually substituting expression (12.22) in place of z, by virtue of identities (12.14)-(12.19) it will follow that $x(t)$ is the searched solution of equation (12.1).

Hence, upon substituting relation (12.21), the integration of equation

(12.1) with the right-hand side including $u'(t)$ has been reduced to integrating equation (12.23) whose right-hand side does not depend explicitly on $u'(t)$.

12.4. Proof of Theorem 12.1

In view of the hypotheses of Theorem 12.1, the function (12.24) satisfies a local Lipschitz condition with respect to z. Thus, at $v = u(t_o)$ and any continuous function $u(t)$, the equation (12.23) admits a unique solution satisfying the initial condition (12.29); we shall preserve the notation (12.28) for that solution.

Consider a fixed triple $\{t_o, x_o, u_*(t)\}$, where $u_*(t)$ is a continuous function. Let $x_n \to x_o$ and $\|u_n(t) - u_*(t)\|_{t_o-1, t_o+1} \to 0$. Then, on a certain interval $|t - t_o| \le h$, the solutions

$$z_n(t) = W[t_o, x_n]u_n(t)$$

of the equation (12.23) are defined, with

$$\lim_{n\to\infty} \|W[t_o,x_n]u_n(t) - W[t_o,x_o]u_*(t)\|_{t_o-h, t_o+h} = 0 . \tag{12.30}$$

If h is sufficiently small, then (12.30) implies the equality

$$\lim_{n\to\infty} \|Q\{u_n(t), u_n(t_o), W[t_o,x_n]u_n(t), t\} -$$
$$- Q\{u_*(t), u_*(t_o), W[t_o,x_o]u_*(t), t\}\|_{t_o-h, t_o+h} = 0 ,$$

which in turn yields the assertion of Theorem 2.1.

∎

The hypotheses of Theorem 12.1 on the existence and properties of functions (12.4) were necessary for the existence of unique (at $t \ge t_o$) solutions to equations (12.23). For some particular classes of the equations (12.1), the substitution (12.21) can be realized under more general hypotheses without loss of the uniqueness of solutions of the corresponding equations (12.23). A detailed analysis of those more general situations was not performed. In the case of equation (12.6), the corresponding equation (12.23) assumes the simplest possible form

$$\frac{dz}{dt} = 0 \quad .$$

12.5. Lemma on differential inequalities

In the sequel we shall make use of the following property.

Let $y_{min}(t)$, $y_{max}(t)$ $(0 \leq t \leq \delta)$ be respectively the minimal and maximal solutions of the initial-value problem

$$\frac{dy}{dt} = L(t,y), \quad y(0) = 0 \quad ,$$

where the scalar function $L(t,y)$ is continuous.

Lemma 12.1. Assume that a function $x(t)$ $(0 \leq t \leq \delta)$ satisfies the conditions

$$\frac{dx}{dt} \geq L[t,x(t)] + \alpha(t), \quad x(0) = 0 \quad , \tag{12.32}$$

where $\alpha(t)$ is positive at $0 < t \leq \delta$ and

$$\int_0^t \alpha(s)ds > y_{max}(t) - y_{min}(t) - \int_0^t \beta(s)ds \qquad (0 < t \leq \varepsilon) \quad , \tag{12.33}$$

with a certain positive number ε and

$$\beta(t) = \min_{y_{min}(t) \leq x \leq y_{max}(t)} \{L(t,x) - L[t,y_{min}(t)]\} \quad .$$

Then $x(t) > y_{max}(t)$ $(0 < t \leq \delta)$.

Proof. By (12.32) and positiveness of the function $\alpha(t)$ it follows (in view of standard theorems on differential inequalities) that

$$x(t) > y_{min}(t) \qquad (0 < t \leq \delta) \quad .$$

Suppose that the assertion of the lemma is false. Then on a certain interval $(0,\delta_1] \subset (0,\delta]$ the estimates $y_{min}(t) < x(t) < y_{max}(t)$ would hold, and therefore

$$L[t,x(t)] - L[t,y_{min}(t)] \geq \beta(t) \quad (0 \leq t \leq \delta_1) .$$

Consequently,

$$\frac{d}{dt}[x(t) - y_{min}(t)] \geq L[t,x(t)] + \alpha(t) - L[t,y_{min}(t)] \geq \alpha(t) + \beta(t)$$

and further

$$x(t) - y_{min}(t) \geq \int_0^t \alpha(s)ds + \int_0^t \beta(s)ds \quad (0 \leq t \leq \delta_1) .$$

The last estimate contradicts the inequality (12.33).

∎

12.6. Vibro-correctness on smooth inputs

We now return to the analysis of the equation (12.1). In a natural way, we can introduce the notion of vibro-correctness for the case of a fixed input $u_*(t)$ and for some special classes of inputs. In this section, we shall formulate conditions which guarantee the vibro-correctness on all smooth (in particular, continuously differentiable) inputs. Clearly, such conditions will be less restrictive than the hypotheses of Theorem 12.1. It is sufficient to ensure the applicability of substitution (12.21) and uniqueness of solutions to equations (12.23) in the case of smooth inputs $u(t)$.

Theorem 12.2. Let functions (12.4) be jointly continuous with respect to all arguments and the function $\psi(t,x,u)$ satisfy the unilateral Lipschitz condition

$$(\psi(t,x,u) - \psi(t,y,u), x - y) \leq L(\rho)|x - y|^2 \quad (|t|, |x|, |y|, |u| \leq \rho) \tag{12.34}$$

with respect to x. Then the equation (12.1) is vibro-correct on the smooth inputs.

∎

Under the hypotheses of Theorem 12.1, as it is clear by its proof, both the vibro-correctness forwards in time t and vibro-correctness backwards in time take place. Theorem 12.1 is solely concerned with the forwards vibro-correctness.

In connection with Theorem 12.2 a natural question arises: does the uniqueness of solutions to the equations (12.23) with continuous $u(t)$ follow by the uniqueness of the appropriate solutions for all smooth $u(t)$? The answer is no .

As an example consider the scalar equation

$$\frac{dx}{dt} = h(t,x) + |u(t) - \sqrt{|t|}\,| , \tag{12.35}$$

where

$$h(t,x) = \begin{cases} 0 , & \text{if } t \leq 0 , \\ 2t , & \text{if } 0 \leq t , \ t^2 \leq x , \\ 2x/t , & \text{if } 0 \leq t , \ -t^2 \leq x \leq t^2 , \\ -2t , & \text{if } 0 \leq t , \ x \leq -t^2 . \end{cases}$$

At the continuous input $u_*(t) = \sqrt{|t|}$ $(t \geq 0)$, equation (12.35) admits continuum of solutions $x = kt^2$ $(k \in [-1,1])$ which satisfy the initial condition $x(0) = 0$. At the same time, for a smooth input $u(t)$ each initial condition determines a unique solution. In the case of a non-zero initial value, the uniqueness is evident; if however $x(0) = 0$, then one has to use Lemma 12.1. By virtue of that lemma, every solution satisfies the estimate $x(t) \geq t^2$, hence the derivative of the difference of two solutions is equal to zero and coincidence of the initial values implies coincidence of the solutions.

A second natural question arises: does the non-uniqueness of solutions to the equation (12.23) imply the lack of vibro-correctness of the equation (12.1) ? Again, the answer is no .

As an example one can again take the equation (12.35), since Lemma 12.1 implies that for every smooth input $u(t)$ any solution $x(t)$ corresponding to the homogeneous initial condition satisfies the estimate $x(t) > t^2$ at $t > 0$.

Results analogous to Theorems 12.1 and 12.2 can be established for scalar equations (12.1), with functions $\psi(t,x,u)$ discontinuous in x , if their solutions are understood, for example, in Filippov's sense.

13. Vibro-solutions

13.1. Definition

We continue the study of the equation (12.1),

$$\frac{dx}{dt} = \varphi[t,x,u(t)]u'(t) + \psi[t,x,u(t)] \quad . \tag{13.1}$$

We shall assume that the equation (13.1) is vibro-correct.

Let $u_*(t)$ $(t \geq t_o)$ be a fixed continuous input. By definition, there are maximal τ (possibly infinite) and function $x_*(t)$ defined on $[t_o,\tau]$, such that for every $t_1 \in (t_o,\tau)$, since

$$\lim_{n\to\infty} \|u_n(t) - u_*(t)\|_{t_o,t_1} = 0 \tag{13.2}$$

where $u_n(t) \in C^1(t_o,t_1)$, it follows that

$$\lim_{n\to\infty} \|V[t_o,x_o]u_n(t) - x_*(t)\|_{t_o,t_1} = 0 \quad . \tag{13.3}$$

We shall call the function $x_*(t)$ a *vibro-solution of the equation* (13.1), subject to the initial condition

$$x_*(t_o) = x_o \tag{13.4}$$

and *corresponding to the input* $u_*(t)$. The name "vibro-solution" will be preserved also for every restriction of the function $x_*(t)$ to any interval of the form $[t_o,\tau_1) \subset [t_o,\tau)$ or $[t_o,\tau_1] \subset [t_o,\tau)$.

We shall assume that the hypotheses of Theorem 12.1 are fulfilled, i.e., the functions

$$\varphi(t,x,u), \quad \varphi'_t(t,x,u), \quad \varphi'_x(t,x,u), \quad \psi(t,x,u) \tag{13.5}$$

are continuous and satisfy the local Lipschitz condition with respect to x. In this case, for each vibro-solution $x_*(t)$ $(t_o \leq t < \tau)$ either $\tau = \infty$ or $|x_*(t)| \to \infty$ at $t \to \tau$.

The vibro-solution $x_*(t)$ is characterized by

$$x_*(t) = V[t_o,x_o]u_*(t) \quad (t_o \leq t < \tau) \quad . \tag{13.6}$$

Such a notation is justified since in the case of the smooth inputs the vibro-solutions and usual solutions coincide.

At the study of vibro-solutions one should remember that the right--hand side of the equation (13.1) corresponding to $x = x_*(t)$ and $u = u_*(t)$ may make no sense for any value t. However, if the input $u_*(t)$ is differentiable at $t = t_*$, then also the vibro-solution $x_*(t)$ is differentiable at $t = t_*$ and satisfies equation (13.1) in this point in the usual sense. If the input $u_*(t)$ is absolutely continuous, then also the function (13.6) is absolutely continuous and satisfies the integral equation

$$x(t) = x_0 + \int_{t_0}^{t} \varphi[s,x(s),u_*(s)]u_*'(s)ds + \int_{t_0}^{t} \psi[s,x(s),u_*(s)]ds \quad . \quad (13.7)$$

It is important to notice that for the operators (13.6) the semigroup identity

$$V[t_0,x_0]u_*(t) = V[t_1,V[t_0,x_0]u_*(t_1)]u_*(t) \qquad (13.8)$$

holds for all t such that either its right- or left-hand side is defined. The operators (13.6) are in the natural sense continuous with respect to small perturbations of the initial values x_0, inputs $u_*(t)$ (small in the uniform metric) and the right-hand sides of the equation (this holds provided that the variations of all functions (13.5) are small).

All the properties of vibro-solutions we have just mentioned can be concluded from the following by-product of the proof of Theorem 12.1.

Theorem 13.1. Assume that the hypotheses of Theorem 12.1 are satisfied, the function $u_*(t)$ $(t \geq t_0)$ is continuous and $x_*(t)$ is of the form (13.6). Let $z_*(t)$ $(t_0 \leq \tau_0 \leq t \leq \tau_1)$ be the solution of equation (12.23), with the initial condition

$$z(\tau_0) = Q[v,u_*(\tau_0),x_*(\tau_0),\tau_0] \quad . \qquad (13.9)$$

Then $\tau_1 < \tau$ and

$$V[t_0,x_0]u_*(t) = Q[u_*(t),v,z_*(t),t] \quad (\tau_0 \leq t \leq \tau_1) \quad . \qquad (13.10)$$ ∎

Without hypotheses of Theorem 12.1 fulfilled, it is even not clear if the characterization (13.6) makes sense (for some smooth inputs $u(t)$, the corresponding vibro-solutions may then be defined on a smaller interval than it was for the standard solutions). In general, vibro-correctness of equation (13.1) does not guarantee any continuous dependence of the vibro-solutions on initial values, there are no analogues of the semigroup identity, etc. To construct suitable examples, one may use equation (12.35), perhaps slightly modified.

13.2. Global vibro-correctness

If every vibro-solution (at an arbitrary continuous input) is defined for all $t \geq t_0$, then we shall say that the equation (13.1) is *globally vibro-correct*.

If the hypotheses of Theorem 12.1 hold, then the vibro-solution can be constructed according to equality (13.10). By Theorem 13.1, the equation (13.1) is globally vibro-correct if, first, all solutions of equation (12.7) are defined at $u \in (-\infty,\infty)$ and, secondly, any solution of equation (12.23) is defined at all $t \geq t_0$. By employing such a scheme, one may easily derive various sufficient conditions of the global vibro-correctness.

For example, consider the equation

$$\frac{dx}{dt} = \varphi[x,u(t)]u'(t) \quad . \tag{13.11}$$

For its global vibro-correctness it is sufficient that, beside the hypotheses of Theorem 12.1, the estimate

$$|\varphi(x,u)| \leq c(u)(1 + |x|)$$

holds with a continuous function $c(u)$.

Theorem 13.2. Let the hypotheses of Theorem 12.1 be satisfied. Assume that the estimates

$$|\varphi(t,x,u)| \leq c(t,u)(1 + |x|) \quad , \tag{13.12}$$

$$|\varphi_t'(t,x,u)|, \ |\psi(t,x,u)| \leq c(t,u)\ p(|x|) \quad , \tag{13.13}$$

$$|\varphi_x'(t,x,u)| \leq c(t,u)r(|x|) \tag{13.14}$$

hold with a continuous function $c(t,u)$ and functions $p(w)$, $r(w)$ which satisfy the condition *)

$$\int^{\infty} e^{-r(w)} \frac{dx}{p(w)} = \infty \quad . \tag{13.15}$$

Then the equation (13.1) is globally vibro-correct.

Proof. Condition (13.12) guarantees that solutions of equation (12.7) admit global extensions. In turn, by virtue of conditions (13.12)-(13.15), it is then possible to construct global extensions to the solutions of equations (12.23). ∎

If the function $p(w)$ is chosen in the form $p(w) = 1 + w$, then according to (13.15) the asymptotic growth of $r(w)$ must be of lower order than for $\ln(1 + w)$ (for example, the function $r(w)$ can be taken in the form $\ln \ln(e + w)$). In the case of $p(w) = 1 + w^{\alpha}$ with $\alpha \in [0,1)$, equality (13.15) will be satisfied if, for example, the estimate (13.14) has the form

$$|\varphi_x'(t,x,u)| \leq c_1(t,u) \ln(e + |x|)$$

(for proving it, one should take $r(w) = \ln(e + w^{1-\alpha})$).

As well-known, hypotheses that ensure the existence of global extensions to solutions of ordinary differential equations almost always can be formulated as unilateral conditions for the right-hand sides rather than as estimates on asymptotic growth. It would also be useful to derive hypotheses which ensure the global vibro-correctness of equation (13.1). Such hypotheses may be found in the form of unilateral bounds for the vector-functions $\varphi(t,x,u)$, $\psi(t,x,u)$ and their derivatives, as well as in the form of combinations of some norm estimates and such bounds.

We now give a criterion for global vibro-correctness restricted to the

*) The symbol $\int^{\infty} g(w)dw = \infty$ means that, for any a, the integral $\int_a^{\infty} g(w)dw$ is infinite.

case of scalar equations.

Theorem 13.3. Let the hypotheses of Theorem 12.1 be satisfied and the estimate (13.12) hold. Assume that the function $\varphi_t'(t,x,u)$ does not change its sign in each of the domains $x \leqq -\xi(t,u)$ and $x \geqq \eta(t,u)$, where $\xi(t,u)$, $\eta(t,u)$ are continuous and positive. Assume that $\psi(t,x,u) \geqq 0$ at $x \leqq -\xi(t,u)$ and $\psi(t,x,u) \leqq 0$ at $x \geqq \eta(t,u)$. Then the scalar equation (13.1) is globally vibro-correct.

Proof. Due to Theorem 13.1, it will suffice to construct for any continuous input $u(t)$ $(t \geqq t_0)$ and each $t_1 > t_0$ some numbers R_-, R_+, v_-, v_+ such that

$$F[t,z,u(t);v_-] \geqq 0 \qquad (t_0 \leqq t \leqq t_1,\ z \leqq R_-) \ ,$$

$$F[t,z,u(t);v_+] \leqq 0 \qquad (t_0 \leqq t \leqq t_1,\ z \geqq R_+) \ ,$$

where the function $F(t,z,u;v)$ is defined by (12.24). The numbers R_-, R_+, v_-, v_+ can be defined by

$$R_- = \inf\{Q[u,v,-\xi(t,u),t]:\ t_0 \leqq t \leqq t_1,\ m \leqq u,\ v \leqq M\} \ ,$$

$$R_+ = \sup\{Q[u,v,\eta(t,u),t]:\ t_0 \leqq t \leqq t_1,\ m \leqq u,\ v \leqq M\} \ ,$$

$$v_- = \begin{cases} m, & \text{if } \varphi_t'(t,x,u) \leqq 0 \text{ for } x \leqq -\xi(t,u) \ , \\ M, & \text{if } \varphi_t'(t,x,u) \geqq 0 \text{ for } x \leqq -\xi(t,u) \ , \end{cases}$$

$$v_+ = \begin{cases} m, & \text{if } \varphi_t'(t,x,u) \geqq 0 \text{ for } x \geqq \eta(t,u) \ , \\ M, & \text{if } \varphi_t'(t,x,u) \leqq 0 \text{ for } x \geqq \eta(t,u) \ , \end{cases}$$

where

$$m = \min\{u(t):\ t_0 \leqq t \leqq t_1\} \ , \quad M = \max\{u(t):\ t_0 \leqq t \leqq t_1\} \ .$$

∎

13.3. Inputs on finite time interval

In this section, we confine ourselves to the case of globally vibro-correct equation (13.1). We assume the hypotheses of Theorem 12.1 are fulfilled. The operators (13.6), as already mentioned, are defined over any $C(t_0,t_1)$ and are continuous there. Those operators are stable with respect to small perturbations of the initial value x_0 and small (in the metric of C^1) perturbations of the functions $\varphi(t,x,u)$ and $\psi(t,x,u)$. If the function $\psi(t,x,u)$ is locally Lipschitz continuous with respect to u, then $V[t_0,x_0]$ treated as an operator in $C(t_0,t_1)$ is locally Lipschitz continuous, too.

If the operators (13.6) are monotone (or have some property reminiscent of monotonicity), their study becomes much simpler. We shall say that *equation* (13.1) *is monotone*, if for any x_0 from

$$u(t_0) = v(t_0), \quad u(t) \leq v(t) \quad (t \geq t_0) \tag{13.16}$$

it follows that

$$V[t_0,x_0]u(t) \leq V[t_0,x_0]v(t) \quad (t \geq t_0) \quad . \tag{13.17}$$

Here and in the sequel, inequalities between vectors are to be understood component-wise. To the knowledge of the authors, there are no conditions which would directly guarantee the monotonicity of equations (13.1) having a general form. However, one may easily establish such conditions in various special cases.

Consider the equation

$$\frac{dx}{dt} = \varphi[t,u(t)]u'(t) + \psi[t,x,u(t)] \quad . \tag{13.18}$$

It is sufficient for ensuring its monotonicity that the vector-functions $\varphi(t,u)$ and $-\varphi_t'(t,u)$ have non-negative components, the vector-function $\psi(t,x,u)$ has all components non-decreasing with respect to u and the matrix $\psi_x'(t,x,u)$ is non-negative out of the diagonal *).

If (13.1) is a scalar equation, the following more general statement is true: for monotonicity of (13.1) it suffices that functions $\varphi(t,x,u)$ and $-\varphi_t'(t,x,u)$ are non-negative and that $\psi(t,x,u)$ is non-decreasing in u.

*) A matrix $M = [m_{ij}]$ is called non-negative (positive) out of the diagonal, if $m_{ij} \geq 0$ $(m_{ij} > 0)$ at $i \neq j$.

13.4. Inputs on infinite time interval

We are going to continue the study of the globally vibro-correct equation (13.1). The following questions will be of interest for us: what conditions guarantee that for the inputs $u(t)$ which are bounded on $[t_0,\infty)$, also the corresponding vibro-solutions are bounded; what properties have $V[t_0,x_0]$, viewed as operators in the space $C(t_0,\infty)$ of bounded continuous vector-functions; what additional properties are characteristic for the outputs that correspond to periodic inputs, etc. ?

Consider equation (13.11). If

$$|\varphi(x,u)| \leq c(u)\ (1 + |x|) \quad ,$$

then every operator $V[t_0,x_0]$ acts in $C(t_0,\infty)$ and is locally Lipschitz continuous in this space. Vector-functions $V[t_0,x_0]u(t)$ are periodic in the case of periodic inputs $u(t)$, they are almost periodic for almost periodic $u(t)$, etc.

Analogous properties are exhibited by solutions of the equation

$$\frac{dx}{dt} = \varphi[t,x,u(t)]u'(t) \tag{13.19}$$

whose right-hand side is dependent on t, if for instance,

$$|\varphi_t'(t,x,u)| \leq a(t,r) \quad (|u| \leq r) \tag{13.20}$$

and for every $r > 0$

$$\int_{t_0}^{\infty} a(t,r)dt < \infty \quad . \tag{13.21}$$

The operators $V[t_0,x_0]$ generated by equation (13.18), act in the space $C(t_0,\infty)$, if for example

$$|\varphi(t,u)| \leq \alpha_1(r) \quad (t \geq t_0,\ |u| \leq r) \tag{13.22}$$

and

$$(\psi(t,x,u),x) \leq \alpha_2(r) - \alpha_3(r)(x,x) \quad (t \geq t_0,\ |u| \leq r) \tag{13.23}$$

(here and in the sequel, all the functions $\alpha_i(r)$ are continuous and positive). Under these hypotheses, the estimate

$$\overline{\lim_{t\to\infty}} \ |V[t_0,x_0]u(t)| \leq \rho(r) < \infty \qquad (13.24)$$

is true for any x_0 and $|u(t)| \leq r$. Operators $V[t_0,x_0]$ are locally Lipschitz continuous in the space $C(t_0,\infty)$ if, in addition to (13.22) and (13.23), the condition

$$(\psi_x'(t,x,u)y,y) \leq$$

$$\leq -\alpha_4(r)[|\varphi_t'(t,u)| + |\psi_u'(t,x,u)|](y,y) \quad (t \geq t_*;\ |u|,|x| \leq r) \qquad (13.25)$$

is satisfied for some fixed t_*. Now suppose that, instead of (13.25), the following more restrictive condition

$$(\psi_x'(t,x,u)y,y) \leq$$

$$\leq -\alpha_4(r)[1 + |\varphi_t'(t,u)| + |\psi_u'(t,x,u)|](y,y), \quad (t \geq t_*; |u|,|x| \leq r) \qquad (13.26)$$

is satisfied. Then $V[t_0,x_0]$ is "stable" in the following sense: for each positive r and ε there exists a number $\delta > 0$ such that

$$\|u(t)\|_{t_0,\infty},\ \|v(t)\|_{t_0,\infty} \leq r, \quad \overline{\lim_{t\to\infty}} \ |u(t) - v(t)| \leq \delta \qquad (13.27)$$

imply the relation

$$\overline{\lim_{t\to\infty}} \ |V[t_0,x_0]u(t) - V[t_0,y_0]v(t)| < \varepsilon \quad (x_0,\ y_0 \in R^N) \ . \qquad (13.28)$$

Assume now that the functions $\varphi(t,u)$ and $\psi(t,x,u)$ in the right-hand side of equation (13.18) satisfy the relations

$$\lim_{t\to\infty} \max_{|u|\leq h} |\varphi(t,u) - \bar{\varphi}(t,u)| = \lim_{t\to\infty} \max_{|u|\leq r} |\varphi_t'(t,u) - \bar{\varphi}_t'(t,u)| = 0 \ ,$$

$$\lim_{t\to\infty} \max_{|u|,|x|\leq r} |\psi(t,x,u) - \bar{\psi}(t,x,u)| = 0$$

at every $r > 0$, where $\bar{\varphi}(t,u)$ and $\bar{\psi}(t,x,u)$ are T-periodic with respect to t. If, in addition, the conditions (13.22), (13.23) and (13.25) are fulfilled, then for each asymptotically T-periodic input $u(t)$ the corresponding output $V[t_0,x_0]u(t)$ is also asymptotically T-periodic.

Let us give a simple condition which will guarantee that solutions of (13.19) are asymptotically periodic. Assume that for $t\to\infty$, the function $\varphi(t,x,u)$ is uniformly convergent on every set $|x|,|u| \le r$ to a function $\bar{\varphi}(x,u)$, locally Lipschitz continuous with respect to x. Suppose that every matrix $\varphi'_x(t,x,u)$ is definite out of diagonal and, besides, the vector-function $\varphi'_t(t,x,u)$ has a fixed sign at $t \ge t_*$. Then for any asymptotically T-periodic (asymptotically almost periodic) input $u(t)$, the corresponding output $V[t_0,x_0]u(t)$ is either unbounded or asymptotically T-periodic (asymptotically almost periodic). This statement follows, for example, if the function (12.24) has constant sign at an appropriate choice of v.

Operators $V[t_0,x_0]$, generated by the scalar equation (13.1), act in the space $C(t_0,t_1)$, provided that hypotheses of Theorem 13.3 are satisfied and, in addition, the estimates

$$|\varphi(t,x,u)| \le \alpha_5(r)(1+|x|) \qquad (t \ge t_*,\ \ |u| \le r)\ , \tag{13.29}$$

$$\xi(t,u),\ \eta(t,u) \le \alpha_6(r) \qquad (t \ge t_*,\ \ |u| \le r) \tag{13.30}$$

hold. If, moreover, the function $\psi(t,x,u)$ is strictly positive at $t \ge t_*$ and $x \le -\xi(t,u)$, as well as strictly negative at $t \ge t_*$ and $x \ge \eta(t,u)$, then for $|u(t)| \le r$ and any x_0 the estimate (13.24) holds.

In conclusion, let us consider inputs which are close to constants $u_*(t) \equiv u_*$. By Theorem 13.1, (at $v = u_*$) properties of the corresponding outputs are reminiscent of those for solutions of the differential equation

$$\frac{dx}{dt} = \psi(t,x,u_*) \ .$$

For example, if $\psi(t,x,u) = \psi(x,u)$ and x_* is an asymptotically stable equilibrium point for the equation

$$\frac{dx}{dt} = \varphi(x,u_*) \ , \tag{13.31}$$

then for each $\varepsilon > 0$ there is $\delta > 0$ such that

$$\|u(t) - u_*(t)\|_{t_0,\infty} < \delta \quad \text{and} \quad |x_0 - x_*| < \delta$$

imply the estimate

$$\overline{\lim_{t\to\infty}} |V[t_0,x_0]u(t) - x_*| < \varepsilon \quad .$$

If, moreover, $|u(t) - u_*| \to 0$ as $t \to \infty$, then

$$\lim_{t\to\infty} |V[t_0,x_0]u(t) - x_*| = 0 \quad .$$

Analogous assertions remain true if the equation (13.31) is globally stable, admits an asymptotically stable limit cycle, etc. Analogously, one can consider the outputs corresponding to inputs which are close to a fixed smooth periodic function.

14. Equations with constraints

14.1. Equations with discontinuous right-hand sides

Below we shall consider scalar differential equations

$$\frac{dx}{dt} = f(t,x) \tag{14.1}$$

with the right-hand side discontinuous as a function of x . The function $x(t)$ will be called a *solution* (*classical solution*) of equation (14.1) if it is absolutely continuous and fulfils the equation almost everywhere.

Let the right-hand side of the equation (14.1) have the form

$$f(t,x) = \begin{cases} \gamma_-'(t) \ , & \text{if} \quad x < \gamma_-(t) \\ \max\{\gamma_-'(t),\ \varphi(t,x)\} \ , & \text{if} \quad x = \gamma_-(t) \ , \\ \varphi(t,x) \ , & \text{if} \quad \gamma_-(t) < x < \gamma_+(t) \\ \min\{\gamma_+'(t),\ \varphi(t,x)\} \ , & \text{if} \quad x = \gamma_+(t) \ , \\ \gamma_+'(t) \ , & \text{if} \quad x > \gamma_+(t) \ , \end{cases} \tag{14.2}$$

where $\varphi(t,x)$ $(-\infty < t,x < \infty)$ is continuous as a function of two variables, it satisfies the local Lipschitz condition with respect to x, and the functions $\gamma_-(t)$, $\gamma_+(t)$ $(-\infty < t < \infty)$ are absolutely continuous and fulfil the inequality

$$\gamma_-(t) < \gamma_+(t) \quad (-\infty < t < \infty) \ . \tag{14.3}$$

The solutions of equation (14.1) with the right-hand side (14.2) will be called *solutions of the equation*

$$\frac{dx}{dt} = \varphi(t,x) \tag{14.4}$$

subject to the constraints $\gamma_-(t)$ *and* $\gamma_+(t)$. In Figure 14.1, some of these solutions are shown by the thickened lines; the fine lines contained in the domain

$$\Omega = \{\{t,x\}: \gamma_-(t) \leq x \leq \gamma_+(t)\} \ , \tag{14.5}$$

represent the integral curves of equation (14.4). The integral curves which lie outside the domain (14.5) have the form of graphs of either of functions $x = \gamma_-(t) - c$ or $x = \gamma_+(t) + c$, where $c > 0$.

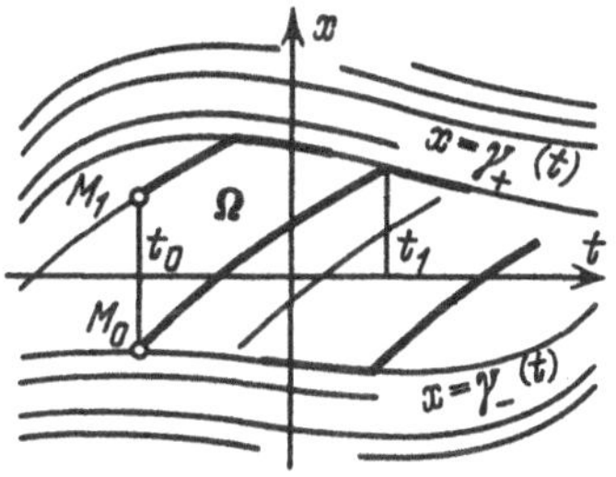

Fig. 14.1

In order to determine some special solutions of equation (14.4) with constraints (similarly as for the general equation (14.1)), we impose the initial condition

$$x(t_o) = x_o \ . \tag{14.6}$$

In other words, the special solutions that satisfy the condition (14.6), are solutions of the integral equation

$$x(t) = x(t_o) + \int_{t_o}^{t} f[s,x(s)]ds \quad . \tag{14.7}$$

The equation (14.1) in a general form, with the right-hand side discontinuous with respect to x, may have no solutions. However, because the function $\varphi(t,x)$ is locally Lipschitz continuous with respect to x, *for any initial condition* (14.6) *there exists a unique solution*

$$x(t) = q(t;t_o,x_o) \quad (t \geq t_o) \tag{14.8}$$

of the equation (14.4) *with constraints* $\gamma_-(t)$ *and* $\gamma_+(t)$.

Proof. To prove the above assertion, it is convenient to pass from the equation (14.1) with the right-hand side (14.2) to the differential inclusion (equation in contingences)

$$\frac{dx}{dt} \in F(t,x) \quad . \tag{14.9}$$

In (14.9), $F(t,x)$ is a single-valued function only if $x \neq \gamma_-(t)$ and $x \neq \gamma_+(t)$; it is then equal to $f(t,x)$. Otherwise, $F(t,x)$ is multi-valued and the segment $[\gamma_-'(t),\varphi[t,\gamma_-(t)]]$ represents its value at $x = \gamma_-(t)$, whereas $[\gamma_+'(t),\varphi[t,\gamma_+(t)]]$ contains its values at $x = \gamma_+(t)$. By virtue of the classical Filippov's theorems, equation (14.9) admits a unique solution at any initial condition. It remains to show that this solution satisfies also (14.1). Details are left to the reader. ∎

The solutions (14.8) are defined only "forwards in time t ". Directly from their definition it follows that

$$q(t_o;t_o,x_o) \equiv x_o \tag{14.10}$$

and the semigroup identity

$$q(t;t_o,x_o) \equiv q[t;t_1,q(t_1;t_o,x_o)] \quad (t_o \geq t_1 \geq t) \tag{14.11}$$

holds. The last identity gives rise to the definition

$$U(t_o,t)x = q(t;t_o,x) \tag{14.12}$$

of the *shift operator from* t_o *to* t *along the trajectories of equation* (14.3) *with constraints* $\gamma_-(t)$ *and* $\gamma_+(t)$. Any operator (14.12) is single-valued but it is non-invertible. It may assume the same values for different x (for example, if the points $M_o = \{t_o,x_o\}$ and $M_1 = \{t_o,x_1\}$ are located as in Fig. 14.1, then $q(t_1;t_o,x_o) = q(t_1;t_o,x_1)$. In terms of the shift operator, the equality (14.11) assumes the form

$$U(t_1,t)\,U(t_o,t_1) = U(t_o,t) \quad (t_o \leq t_1 \leq t) \ . \tag{14.13}$$

14.2. Arbitrary continuous constraints

Beside the equation (14.4) with constraints $\gamma_-(t)$ and $\gamma_+(t)$, let us consider the equation

$$\frac{dx}{dt} = \varphi^1(t,x) \tag{14.14}$$

with some constraints $\gamma_-^1(t)$ and $\gamma_+^1(t)$. At the same time, beside (14.5), we shall consider the domain

$$\Omega^1 = \{\{t,x\}:\ \gamma_-^1(t) \leq x \leq \gamma_+^1(t)\} \ . \tag{14.15}$$

By $q^1(t;t_o,x_o)$ we shall denote the solution of the problem (14.14), (14.6) subject to the constraints $\gamma_-^1(t)$ and $\gamma_+^1(t)$. We shall assume the equations (14.4) and (14.14), as well as the domains (14.5) and (14.15) are close to each other:

$$|\varphi(t,x) - \varphi^1(t,x)| \leq \varepsilon_1 \quad (t_o \leq t \leq t_1,\ \ -\infty < x < \infty) \ , \tag{14.16}$$

$$|\gamma_-(t) - \gamma_-^1(t)|,\ |\gamma_+(t) - \gamma_+^1(t)| \leq \varepsilon_2 \quad (t_o \leq t \leq t_1) \ . \tag{14.17}$$

Lemma 14.1. Assume that the continuous functions $\varphi(t,x)$ and $\varphi^1(t,x)$ satisfy the condition (14.16), both functions are locally Lipschitz continuous with respect to x and one of them satisfies the global*) Lipschitz condition in x , with a constant ℓ . Suppose that absolutely continuous functions $\gamma_-(t)$, $\gamma_+(t)$ and $\gamma_-^1(t)$, $\gamma_+^1(t)$ satisfy the condition

*) Consider, for instance, the function $\varphi(t,x)$. This function satisfies the global Lipschitz condition if $|\varphi(t,x) - \varphi(t,y)| \leq \ell|x-y|$ for $t_o \leq t \leq t_1;\ \gamma_-(t) - \varepsilon_2 \leq x,y \leq \gamma_+(t) + \varepsilon_2$.

(14.17). Then for $\{t_o,x_o\} \in \Omega$, $\{t_o,y_o\} \in \Omega^1$, the estimates

$$|q(t;t_o,x_o) - q^1(t;t_o,y_o)| \leq$$

$$\leq \frac{e^{\ell(t-t_o)} - 1}{\ell}\varepsilon_1 + e^{\ell(t-t_o)}\max\{|x_o - y_o|, \varepsilon_2\} \quad (t_o \leq t \leq t_1) \tag{14.18}$$

hold.

Proof. It is enough to note that the graphs of the functions

$$x(t) = q(t;t_o,x_o), \quad y(t) = q^1(t;t_o,y_o) \quad (t_o \leq t \leq t_1) \tag{14.19}$$

are contained in the domains (14.5) and (14.15), respectively, and therefore

$$\frac{d}{dt}|x(t) - y(t)| \leq \varepsilon_1 + \ell|x(t) - y(t)|$$

at all those values t for which $|x(t) - y(t)| \geq \varepsilon_2$.

∎

In the next step, we are going to construct solutions to the equation (14.4) with continuous constraints $\gamma_-(t)$ and $\gamma_+(t)$ which satisfy the inequality

$$\gamma_-(t) \leq \gamma_+(t) \quad (-\infty < t < \infty) \tag{14.20}$$

but are not necessarily absolutely continuous. In this case, it may be impossible to pass to equation (14.1) with the right-hand side (14.2).

Consider sequences $\gamma_-^n(t)$, $\gamma_+^n(t)$ $(n = 1, 2, \ldots)$ of absolutely continuous functions which satisfy the condition

$$\gamma_-^n(t) < \gamma_-(t) \leq \gamma_+(t) < \gamma_+^n(t) \quad (-\infty < t < \infty;\ n = 1, 2, \ldots)$$

and are convergent to $\gamma_-(t)$ and $\gamma_+(t)$, uniformly on all finite intervals $[t_o,t_1]$:

$$\lim_{n\to\infty} \|\gamma_-^n(t) - \gamma_-(t)\|_{t_o,t_1} = \lim_{n\to\infty} \|\gamma_+^n(t) - \gamma_+(t)\|_{t_o,t_1} . \tag{14.21}$$

By

$$x_n(t) = q_n(t;t_o,x_o) \quad (t \geq t_o) \tag{14.22}$$

we shall denote the solutions of equation (14.4) with absolutely continuous constraints $\gamma_-^n(t)$, $\gamma_+^n(t)$. If $\{t_o,x_o\} \in \Omega$, then due to Lemma 14.1, the sequence (14.22) converges to some continuous function uniformly over finite intervals $[t_o,t_1]$. Thus we can define the *solutions of equation* (14.4) *with constraints* $\gamma_-(t)$ *and* $\gamma_+(t)$ by the equality

$$x(t) = q(t;t_o,x_o) = \begin{cases} x_o + \gamma_-(t) - \gamma_-(t_o), & \text{if} \quad x_o < \gamma_-(t), \\ \lim\limits_{n\to\infty} q_n(t;t_o,x_o), & \text{if} \quad \gamma_-(t) \leq x_o \leq \gamma_+(t), \\ x_o + \gamma_+(t) - \gamma_+(t_o), & \text{if} \quad x_o > \gamma_+(t) . \end{cases} \tag{14.23}$$

The new definition is correct because in the case of absolutely continuous constraints it gives rise to the solution described in Section 14.1. By this definition, the semigroup identity (14.11) holds for the solution (14.23). Lemma 14.1 implies a continuous dependence of the solutions (14.23) on the initial values $\{t_o,x_o\} \in \Omega$. Given the solutions (14.23), one can define the shift operators (14.12), etc.

Sometimes it is useful to give a direct characterization (without any limit passage employed) of the solutions $q(t;t_o,x_o)$ to equation (14.4) with non-smooth constraints $\gamma_-(t)$, $\gamma_+(t)$.

Introduce the notation

$$x(t) = p(t;t_o,x_o) \tag{14.24}$$

for the standard solution of equation (14.4) (without constraints), satisfying the initial condition (14.6). A continuous function $z(t)$ $(t \geq t_o)$ will be called (*)-*proper*, if $\gamma_-(t) \leq z(t) \leq \gamma_+(t)$ and the relation

$$\gamma_-(t) \leq p[t;\tau_1,z(\tau_1)] \leq \gamma_+(t) \quad (\tau_1 \leq t \leq \tau_2)$$

implies that $z(t) = p[t;\tau_1,z(\tau_1)]$ on the whole interval $[\tau_1,\tau_2]$. As it can be easily seen, for initial data $\{t_o,x_o\} \in \Omega$, (14.23) *is the* *only (*)-proper function with the graph in* Ω at all $t \geq t_0$. Therefore, the solution $x(t)$ $(t \geq t_0)$ of equation (14.4) with constraints $\gamma_-(t)$ and $\gamma_+(t)$, satisfying the initial condition (14.6) with $\{t_0,x_0\} \in \Omega$, can be defined as the unique (*)-proper function which satisfies the condition (14.6). Obviously, one must prove the existence of functions which are proper in the above sense.

A notion of the solution to equation (14.4) with constraints can also be introduced in the case of discontinuous (but Borel) functions $\varphi(t,x)$ which are locally bounded and satisfy locally a unilateral Lipschitz condition. It is convenient to use Filippov's definition for the solutions of the corresponding equations.

In all the constructions within Sections 14.1 and 14.2, the solutions $x(t)$ of equation (14.4) with the constraints $\gamma_-(t)$ and $\gamma_+(t)$ were defined only at $t \geq t_0$. Therefore all those constructions remain valid if the right-hand side of the equation (14.4) and both constraints are defined only for $t \geq t_0$.

The estimate (14.18) does not depend on any parameter characterizing the absolute continuity of constraints. Due to Lemma 14.1, upon passing to the limit we can thus conclude the following stronger result.

<u>Lemma 14.2.</u> Let the continuous functions $\varphi(t,x)$ and $\varphi^1(t,x)$ satisfy the hypotheses of Lemma 14.1. Assume that the continuous functions $\gamma_-(t)$, $\gamma_+(t)$ and $\gamma_-^1(t)$, $\gamma_+^1(t)$ satisfy the condition (14.17). Then the estimates (14.18) hold at $\{t_0,x_0\} \in \Omega$, $\{t_0,y_0\} \in \Omega^1$.

∎

14.3. Vibro-correct equations with constraints

Let us consider an equation

$$\frac{dx}{dt} = f[t,x,u(t),u'(t)] \tag{14.25}$$

with the right-hand side dependent on the input $u(t)$. We shall assume that the function $f(t,x,u,v)$ $(-\infty < t,x,u,v < \infty)$ is jointly continuous with respect to all variables and satisfies the local Lipschitz condition with respect to x. Let, in addition to the equation (14.25), two families of continuous functions $\gamma_-(t,u)$ and $\gamma_+(t,u)$, such that

$$\gamma_-(t,u) \leq \gamma_+(t,u) \quad (-\infty < t,\ u < \infty) \quad , \tag{14.26}$$

be defined. Then, for any continuously differentiable function $u(t)$ $(t \geq t_0)$ and each initial condition (14.6) there exists a solution of the equation (14.25) subject to the constraints $\gamma_-[t,u(t)]$ and $\gamma_+[t,u(t)]$; we shall denote this solution by

$$x(t) = W[t_0,x_0]u(t) \quad (t \geq t_0) \quad . \tag{14.27}$$

We shall consider the functions $u(t)$ $(t \geq t_0)$ as inputs of a certain transducer W and the operators (14.27) as the corresponding input-output operators. The pairs $\{u,x\}$ will represent the states of that transducer. It will be useful to assume that the set $\Omega(W,t)$ of feasible states is indeed t-dependent and comprises the pairs for which

$$\gamma_-(t,u) \leq x \leq \gamma_+(t,u) \quad .$$

Operators (14.27) have been defined on continuously differentiable inputs $u(t)$. A natural question that arises concerns the possibility of extending (14.27) by continuity onto all continuous inputs, to operators which are continuous in the whole space $C(t_0,t_1)$. If such an extension exists, then the equation (14.25) considered together with constraints $\gamma_-(t,u)$ and $\gamma_+(t,u)$, or equivalently the transducer W, is referred to as *vibro-correct*. Due to Theorem 11.1, for the vibro-correctness it is necessary that the equation (14.25) has the form

$$\frac{dx}{dt} = \varphi[t,x,u(t)]u'(t) + \psi[t,x,u(t)] \quad . \tag{14.28}$$

For denoting the transducer W and the operators $W[t_0,x_0]$, we shall sometimes use more detailed symbols $W(\varphi,\psi;\gamma_-,\gamma_+)$ and $W[t_0,x_0;\varphi,\psi,\gamma_-,\gamma_+]$, respectively.

Theorem 14.1. Assume that the functions

$$\varphi(t,x,u),\ \ \varphi'_t(t,x,u),\ \ \varphi'_x(t,x,u),\ \ \psi(t,x,u) \tag{14.29}$$

are continuous and satisfy the local Lipschitz condition with respect to x . Then the equation (14.28) is vibro-correct for all constraints which satisfy condition (14.26).

Proof. Let a continuous input $u_*(t)$ $(t \geq t_0)$ and an initial state $\{u_*(t_0), x_0\} \in \Omega(W, t_0)$ be given. Furthermore, suppose that a sequence of continuously differentiable inputs $u_n(t)$ $(t \geq t_0)$ has been defined so that

$$\{u_n(t_0), x_0\} \in \Omega(W, t_0) \quad (n = 1, 2, \ldots) \quad ,$$

and, uniformly on every interval $[t_0, t_1]$,

$$\lim_{n\to\infty} \|u_n(t) - u_*(t)\|_{t_0, t_1} = 0 \quad . \tag{14.30}$$

To prove the theorem, we must show that the corresponding sequence of the outputs $W[t_0, x_0]u_n(t)$ converges to a certain function $x_*(t)$ uniformly on every interval $[t_0, t_1]$:

$$\lim_{n\to\infty} \|W[t_0, x_0]u_n(t) - x_*(t)\|_{t_0, t_1} = 0 \quad . \tag{14.31}$$

In all further constructions, we shall make use only of those values of the functions (14.29) which correspond to such values of the arguments that

$$\gamma_-(t,u) \leq x \leq \gamma_+(t,u) \quad \text{and} \quad t_0 \leq t \leq t_1 \quad .$$

Therefore, the functions (14.29) can be considered as globally Lipschitz continuous with respect to x . In particular, we can assume that for every t the solution (cf., Section 12.2)

$$x(u) = Q(u, u_0, x_0, t)$$

of the equation

$$\frac{dx}{du} = \varphi(t, x, u) \quad ,$$

satisfying the initial condition $x(t_0) = x_0$, is defined for all $u \in (-\infty, \infty)$. Consider the equation (cf., (12.23))

$$\frac{dz}{dt} = F[t,z,u(t);v] \tag{14.32}$$

with the right-hand side (12.24). By construction, each function

$$z_n(t) = Q\{v,u_n(t),W[t_o,x_o]u_n(t),t\} \tag{14.33}$$

is then a (*)-proper solution of equation (14.32) subject to the constraints

$$Q[v,u,\gamma_-(t,u)], \quad Q[v,u,\gamma_+(t,u),t] \quad . \tag{14.34}$$

Consequently, each function (14.33) is a solution of equation (14.32) subject to constraints (14.34). The right-hand side of (14.32) is now no longer dependent on any derivative of the input $u(t)$. Hence, due to Lemma 14.1 and according to (14.30), the sequence (14.33) is convergent in $C(t_o,t_1)$ to some function $z_*(t)$:

$$\lim_{n\to\infty} \|z_n(t) - z_*(t)\|_{t_o,t_1} = 0 \quad . \tag{14.35}$$

But, on account of (14.33) (cf., Chapter 12),

$$W[t_o,x_o]u_n(t) = Q[u_n(t),v,z_n(t),t] \quad ,$$

Thus, due to (14.35) it follows that the equality (14.31) is satisfied with

$$x_*(t) = Q[u_*(t),v,z_*(t),t] \quad .$$

∎

Due to Theorem 14.1, operators (14.27) can be considered as defined on all continuous inputs. Functions $W[t_o,x_o]u(t)$ $(t \geq t_o)$ will be referred to as *vibro-solutions of the equation* (14.28) *subject to the constraints* $\gamma_-(t,u)$ *and* $\gamma_+(t,u)$.

We shall denote those vibro-solutions by $V[t_o,x_o]u(t)$. Then, for any continuous input $u(t)$, the function $W[t_o,x_o]u(t)$ can be defined (see the definition of a (*)-proper function in Section 14.2) as the unique continuous function $x(t)$ such that

$$\gamma_-[t,u(t)] \leq x(t) \leq \gamma_+[t,u(t)]$$

and the relation

$$\gamma_-[t,u(t)] \leq V[\tau_0,x(\tau_0)]u(t) \leq \gamma_+[t,u(t)] \qquad (t_0 \leq \tau_0 \leq t \leq \tau_1)$$

implies the equality

$$x(t) = V[\tau_0,x(\tau_0)]u(t) \qquad (\tau_0 \leq t \leq \tau_1) \quad .$$

An analogous approach may also be applied in other constructions where a certain deterministic transducer is converted into another one just by imposing constraints.

14.4. Properties of vibro-solutions to equations with constraints

As it follows from the construction, operators $W[t_0,x_0]$ act in the space $C(t_0,t_1)$ and are continuous there, if functions (14.29) are continuous and satisfy local Lipschitz condition with respect to x. If, in addition, the local Lipschitz condition with respect to u is fulfilled by the function $\psi(t,x,u)$, as well as by the constraints $\gamma_-(t,u)$ and $\gamma_+(t,u)$, then (see Lemma 14.2) the operators $W[t_0,x_0]$ are locally Lipschitz continuous in $C(t_0,t_1)$.

Under hypotheses of Theorem 14.1 (without additional assumptions), the operators $W[t_0,x_0]$ satisfy the Lipschitz condition with respect to the initial value x_0.

In a number of applications, monotonicity of the operators $W[t_0,x_0]$ plays an important role. From $x_0 \leq y_0$ it follows then that

$$W[t_0,x_0]u(t) \leq W[t_0,y_0]u(t) \qquad (t \geq t_0) \quad .$$

The operators $W[t_0,x_0]$ may be non-monotone with respect to inputs. The following statement is true (cf., Section 13.3).

<u>Theorem 14.2.</u> Let the hypotheses of Theorem 14.1 be satisfied. Assume that $\varphi(t,x,u) \geq 0$, $\varphi_t'(t,x,u) \leq 0$, and the function $\psi_t'(t,x,u)$ is monotone and non-decreasing in u. Finally, suppose that the inequalities

$$\frac{\partial}{\partial u}\gamma_-(t,u) > \varphi[t,\gamma_-(t,u),u], \quad \frac{\partial}{\partial u}\gamma_+(t,u) > \varphi[t,\gamma_+(t,u),u]$$

are fulfilled. Then for all continuous inputs $u(t)$, $v(t)$, the relation

$$u(t_0) = v(t_0), \quad u(t) \leq v(t) \quad (t \geq t_0)$$

implies that

$$W[t_0,x_0]u(t) \leq W[t_0,x_0]v(t) \quad (t \geq t_0) \quad .$$ ∎

Now we pass to an analysis of the behaviour of vibro-solutions on an infinite interval. We shall assume that the hypotheses of Theorem 14.1 hold. If the constraints satisfy the condition

$$\sup_{-\infty < t < \infty, |u| \leq r} [|\gamma_-(t,u)| + |\gamma_+(t,u)|] = \rho(r) < \infty \quad (r > 0), \tag{14.36}$$

then for each continuous input $u(t)$, bounded on $[t_0,\infty)$, the corresponding vibro-solution $W[t_0,x_0]$ is bounded on $[t_0,\infty)$, too. In other words, the operator $W[t_0,x_0]$ acts in the space $C(t_0,\infty)$ of functions which are bounded and continuous on $[t_0,\infty)$.

Criteria ensuring the Lipschitz continuity of $W[t_0,x_0]$ in the space $C(t_0,\infty)$ are available only for some special types of the equations (14.28). Suppose that $\gamma_-(t,u)$, $\gamma_+(t,u)$ are Lipschitz continuous with respect to u and the hypothesis (14.36) holds. If $\varphi(t,x,u) \equiv \varphi(x,u)$, then it suffices that $\psi(t,x,u)$ is Lipschitz continuous with respect to u and its values are separated from zero. If $\varphi(t,x,u) \equiv \varphi(t,u)$ and is bounded, then it is sufficient that the estimate

$$\psi_x'(t,x,u) \leq -\alpha[|\varphi_t'(t,u)| + |\psi_u'(t,x,u)|] \tag{14.37}$$

holds with some $\alpha > 0$.

Let us consider asymptotically periodic inputs. We formulate the following theorem (without proof).

<u>Theorem 14.3.</u> Assume that for every $r > 0$

$$\lim_{t\to\infty} \max_{|x|,|u| \leq r} \{|\gamma_-(t,u) - \gamma_-^0(u)| + |\gamma_+(t,u) - \gamma_+^0(u)| +$$

$$+ |\varphi(t,x,u) - \varphi^0(x,u)| + |\psi(t,x,u) - \psi^0(x,u)|\} = 0 \quad , \tag{14.38}$$

where $\varphi^{o}(x,u)$ and $\psi^{o}(x,u)$ are locally Lipschitz continuous with respect to x. Suppose that neither of the functions $\varphi_t'(t,x,u)$, $\psi(t,x,u)$ can change sign if t is large enough. Then for any asympotically T-periodic input $u(t)$ $(t \geq t_o)$, the vibro-solution $W[t_o,x_o]u(t)$ is also asymptotically T-periodic. ∎

An analogous assertion remains true in the case of asymptotically almost periodic inputs.

Let the hypotheses of Theorem 14.3 hold. Assume that the range of $\psi^{o}(x,u)$ is separated from zero. Then for each T-periodic input $u_*(t)$ there is a T-periodic function $x_*(t)$ such that

$$|W[t_o,x_o]u(t) - x_*(t)| \to 0 \quad \text{as} \quad t \to \infty ,$$

provided that the continuous input $u(t)$ $(t \geq t_o)$ satisfies the following requirements:

$$\{u(t_o),x_o\} \in \Omega(W,t_o) \quad \text{and} \quad \lim_{t\to\infty} |u(t) - u_*(t)| = 0 .$$

14.5. Vibro-solutions of parametrized equations

Let us now consider the equation

$$\frac{dx}{dt} = \varphi[t,x,u(t);\theta]u'(t) + \psi[t,x,u(t);\theta] \tag{14.39}$$

with the right-hand side dependent on some parameter θ. In addition, impose the constraints

$$\gamma_-(t,u;\theta), \quad \gamma_+(t,u;\theta) \tag{14.40}$$

which depend on the same parameter θ whose value is to be chosen as an element of a certain normed space Ξ. In applications one is primarily interested in the situation when some vectors or vector-valued functions can be taken as the values of θ. We shall assume that for any fixed value of θ constraints (14.40) satisfy a standard condition in the form of the inequality

$$\gamma_-(t,u;\theta) \leq \gamma_+(t,u;\theta) . \tag{14.41}$$

In the case of equation (14.39) subject to the constraints (14.40) we are interested in properties of the vibro-solutions which satisfy the initial condition (14.6) and are defined on a finite interval $t_o \leq t \leq t_1$. Correspondingly, we shall assume that for any considered fixed value of the parameter θ the functions $\varphi(t,x,u;\theta)$ and $\psi(t,x,u;\theta)$ are defined at $t_o \leq t \leq t_1$, $-\infty < x,u < \infty$, and the constraints (14.40) are defined at $t_o \leq t \leq t_1$, $-\infty < u < \infty$. Besides, we shall postulate that all these functions satisfy the hypotheses of Theorem 14.1: all of them are jointly continuous, while the functions

$$\varphi(t,x,u;\theta),\quad \varphi_t'(t,x,u;\theta),\quad \varphi_x'(t,x,u;\theta),\quad \psi(t,x,u;\theta) \tag{14.42}$$

are jointly continuous and satisfy the local Lipschitz condition with respect to x.

We give some examples referring to the case of a scalar equation (14.39) and scalar function $\theta = \theta(t)$ taken as a parameter.

The hypotheses of Theorem 14.1 can be verified in the simplest way when

$$\varphi(t,x,u;\theta) = \xi[t,x,u;\theta(t)],\quad \psi(t,x,u;\theta) = \eta[t,x,u;\theta(t)]$$

and

$$\gamma_-(t,u;\theta) = \zeta_-[t,u;\theta(t)],\quad \gamma_+(t,u;\theta) = \zeta_+[t,u;\theta(t)] \quad ,$$

with some standard functions $\xi(t,x,u;\theta)$, $\eta(t,x,u;\theta)$ and $\zeta_-(t,u;\theta)$, $\zeta_+(t,u;\theta)$ of four and three variables, respectively. It should only be remarked that

$$\varphi_t'(t,x,u;\theta) = \xi_t'[t,x,u;\theta(t)] + \xi_\theta'[t,x,u;\theta(t)]\theta'(t) \quad ,$$

and therefore, to fulfil the hypotheses of Theorem 14.1, we must assume that the functions $\theta = \theta(t)$ which define values of the parameter are continuously differentiable.

Consider another example. Let

$$\varphi(t,x,u;\theta) = \int_{t_o}^{t} \Phi(x,u,s)\theta(s)ds \quad ,$$

$$\psi(t,x,u;\theta) = \int_{t_o}^{t} \Psi(x,u,s)\theta(s)ds \quad .$$

If the functions $\Phi(x,u,s)$, $\Phi'_x(x,u,s)$, $\Psi(x,u,s)$ are not only continuous but satisfy the Lipschitz condition with respect to x, then the corresponding functions (14.42) fulfil the hypotheses of Theorem 14.1 also in the case of continuous (not necessarily differentiable) functions $\theta(t)$ taken as the parameter θ.

Let us fix some value of θ. Suppose that the relation

$$\gamma_-[t_o,u(t_o);\theta] \leq x_o \leq \gamma_+[t_o,u(t_o);\theta] \tag{14.43}$$

holds for any x_o and any continuous input $u(t)$ $(t_o \leq t \leq t_1)$. Then, under the assumptions we have introduced, a vibro-solution

$$x(t) = W[t_o,x_o;\theta]u(t) \quad (t_o \leq t \leq t_1) \tag{14.44}$$

of the equation (14.39) subject to constraints (14.40) is well-defined.

We shall formulate a theorem that will characterize the dependence of vibro-solution (14.44) on θ.

So far, θ was treated as a fixed parameter in this section. From now on, we shall postulate that the functions (14.42) and constraints (14.40) assume the same values as some functionals which are appropriately dependent on four or three independent variables. For all functionals of this form we shall preserve the same notations (14.42) and (14.40), respectively.

We shall say that the *right-hand side of the equation* (14.39) *and the constraints* (14.40) *are Lipschitz continuous* if the following hypotheses are fulfilled:

a. The functionals (14.40) are jointly continuous and for every fixed $r > 0$ Lipschitz continuous with respect to u, θ in the domain $t \leq t \leq t_1$; $|u|, \|\theta\| \leq r$.

b. The functionals (14.42) are jointly continuous, for every fixed $r > 0$ they satisfy the Lipschitz condition with respect to x and θ in the domain $t_o \leq t \leq t_1$; $|u|, |x|, \|\theta\| \leq r$. In addition, $\psi(t,x,u;\theta)$ is Lipschitz continuous with respect to u in this domain.

We shall treat $W[t_o,x_o;\theta]$ as operators whose values are determined by two variables θ and $u(t)$ (the parameter and the input). The operator $W[t_o,x_o;\theta]$ is said to satisfy the Lipschitz condition with respect to all its variables, if for each fixed $r > 0$ there is $\ell(r) < \infty$ such that at all values θ_1, θ_2 of parameter θ and at all continuous inputs $u_1(t)$, $u_2(t)$ $(t_o \leq t \leq t_1)$ for which

$$\gamma_-[t_o,u_1(t_o);\theta_1] \leq x_o \leq \gamma_+[t_o,u_1(t_o);\theta_1] \ ,$$

$$\gamma_-[t_o,u_2(t_o);\theta_2] \leq x_o \leq \gamma_+[t_o,u_2(t_o);\theta_2]$$

and

$$\|\theta_1\|,\ \|\theta_2\|,\ \|u_1(t)\|_{t_o,t_1},\ \|u_2(t)\|_{t_o,t_1} \leq r \ ,$$

the estimate

$$\|W[t_o,x_o;\theta_1]u_1(t) - W[t_o,x_o;\theta_2]u_2(t)\|_{t_o,t_1} \leq$$

$$\leq \ell(r)\ [\|\theta_1 - \theta_2\| + \|u_1(t) - u_2(t)\|_{t_o,t_1}]$$

holds.

Theorem 14.4. Assume that the right-hand sides of the equation (14.39) and the constraints (14.40) are Lipschitz continuous. Then the operators (14.44) satisfy Lipschitz condition with respect to all arguments.

Proof. Details of the proof are unwieldy, therefore we give here only an outline of it.

At first, consider the equation (14.39) without constraints. For every fixed value of the parameter one can use the substitution (12.21). The right-hand side of equation (12.23) resulting from that substitution depends upon the parameter θ and is Lipschitz continuous with respect to it. Therefore, solutions of the equation (12.23) are Lipschitz continuous both with respect to the parameter θ and the input $u(t)$. Without any loss of generality, we can assume there that for $\|\theta\| \leq r$ and $|u(t)| \leq r$ $(t_o \leq t \leq t_1)$ all vibro-solutions of equation (14.39) are defined on $[t_o,t_1]$. Then we pass to the equation with constraints and take advantage of Lemma 14.2 to complete the proof. ∎

In many cases the right-hand sides of the equation (14.39) are well-defined for parameters θ whose values form a non-closed set in Ξ. Under the hypotheses of Theorem 14.4, one can in a natural way introduce a notion of the solution to that equation for values of θ in a larger set (to this end, it is sufficient to construct the closure of (14.44) with respect to θ).

15. Variable hysteron

15.1. Description of hysteron by differential equations

We now return to the study of hysterons. Let the functions $\Phi_\ell(u)$ and $\Phi_r(u)$ that have entered the definition of a hysteron W_* (cf., Section 3.2), be defined for all $u \in (-\infty,\infty)$. Let functions $\Pi(u;M)$ be solutions of the differential equation

$$\frac{dx}{du} = \varphi(x,u) \qquad (15.1)$$

with a jointly continuous and smooth in x function $\varphi(x,u)$. Then we shall say that the *hysteron* W_* *is smooth*.

Denote by W a transducer whose states $\{u,x\}$ are characterized by the inequalities

$$\gamma_-(u) \leq x \leq \gamma_+(u) \quad (-\infty < u < \infty) , \qquad (15.2)$$

where

$$\gamma_-(u) = \min\{\Phi_\ell(u),\Phi_r(u)\}, \quad \gamma_+(u) = \max\{\Phi_\ell(u),\Phi_r(u)\} \qquad (15.3)$$

and the operators of input-output relations

$$x(t) = W[t_0,x_0]u(t) \quad (t \geq t_0) \qquad (15.4)$$

in the case of continuous inputs $u(t)$ are defined as the vibro-solutions of the equation

$$\frac{dx}{dt} = \varphi[x,u(t)]u'(t) \tag{15.5}$$

subject to the constraints (15.3) and with the initial condition

$$x(t_o) = x_o \ . \tag{15.6}$$

It follows directly from the suitable definitions that *the hysteron* W_* *and the transducer* W *coincide.*

In the above considerations, the hysteron W_* was taken as a reference. Outputs of the hysteron were treated as the vibro-solutions of equation (15.5) with constraints (15.3). There are some vibro-correct equations with constraints which do not define any hysteron, because in general solutions of equation (15.1) subject to constraints (15.3) do not satisfy all hypotheses a1 - a5 (cf., Section 3.2) and hence do not represent the defining curves of any hysteron.

As a first example let us consider the generalized play $L(\Gamma_\ell,\Gamma_r)$ (cf., Section 2.2). For that hysteron, equation (15.5) and constraints (15.3) have the form

$$\frac{dx}{dt} = 0 \ , \quad \gamma_-(u) = \Gamma_r(u) \ , \quad \gamma_+(u) = \Gamma_\ell(u) \ . \tag{15.7}$$

Another example is offered by a hysteron $U(h,E)$ which has the form of a stop with Young's modulus E and yield limits $-h,h$. For this hysteron, equation (15.5) and constraints (15.3) assume respectively the form

$$\frac{dx}{dt} = Eu'(t) \ , \quad \gamma_-(u) = -h \ , \quad \gamma_+(u) = h \ . \tag{15.8}$$

15.2. Variable hysteron

Our next purpose is to establish a characterization for hysterons whose characteristics are variable in time. For this, it seems natural to develop a procedure which would be analogous to the construction used for defining the multiple integrals.

Let a one-parameter family $\widetilde{W} = W^t$ $(t \geq t_o)$ of the hysterons be given, with the parameter t interpreted as time.

a. At first assume that the function $\widetilde{W}$ (whose values are hysterons) is piecewise constant. Then the points $\tau_0 = t_0, \tau_1, \tau_2, \ldots$ divide the half-line $[t_0,\infty)$ into the intervals $[\tau_{i-1},\tau_i)$ where W has a constant value

$$W^t = W^{\tau_{i-1}} \quad (\tau_{i-1} \le t < \tau_i;\ i = 1, 2, \ldots) \ . \tag{15.9}$$

Let the domains $\Omega(\tau_i)$ of the feasible states of the hysterons W^{τ_i} be connected by the inclusions

$$\Omega(\tau_i) \subset \Omega(\tau_{i-1}) \quad (i = 1, 2, \ldots) \ . \tag{15.10}$$

For any continuous input $u(t)$ $(t \ge t_0)$ and each initial state $\{u(t_0),x_0\} \in \Omega(\tau_0)$ define the corresponding output signal

$$x(t) = \tilde{W}[t_0,x_0]u(t) \quad (t \ge t_0) \ , \tag{15.11}$$

by setting

$$x(t) = \begin{cases} W^{\tau_0}[\tau_0,x_0]u(t) \ , & \text{for} \quad \tau_0 \le t \le \tau_1 \ , \\ W^{\tau_1}[\tau_1,x(\tau_1)]u(t), & \text{for} \quad \tau_1 \le t \le \tau_2 \ , \\ W^{\tau_2}[\tau_2,x(\tau_2)]u(t), & \text{for} \quad \tau_2 \le t \le \tau_3 \ , \\ \cdots\cdots\cdots\cdots\cdots\cdots & \\ W^{\tau_{i-1}}[\tau_{i-1},x(\tau_{i-1})]u(t), & \text{for} \quad \tau_{i-1} \le t \le \tau_i \ , \\ \cdots\cdots\cdots\cdots\cdots\cdots & \end{cases} \tag{15.12}$$

The input-output relations (15.11) define a transducer $\tilde{W}$ with a variable domain $\Omega(\tilde{W},t)$ of the feasible states. For any t, $\Omega(\tilde{W},t)$ coincides with the domain $\Omega(\tau_{i-1})$ of the feasible states for the hysteron $W^{\tau_{i-1}}$, provided that $\tau_{i-1} \le t \le \tau$.

By the semigroup identities for operators $W^{\tau_i}[t_0,x_0]$ it follows that the analogous relations are also true for the operator $\tilde{W}[t_0,x_0]$. The continuity of operators $W^{\tau_i}[t_0,x_0]$ implies that an analogous property remains valid also for $\tilde{W}[t_0,x_0]$. If $W^{\tau_i}[t_0,x_0]$ are Lipschitz continuous, the same holds for $\tilde{W}[t_0,x_0]$, etc.

<u>b.</u> In turn, let us now drop the condition (15.10). Then, formulae (15.12)

make no more sense and in order to construct a variable hysteron $\tilde{W}$, we need a different approach.

Let $\Omega(W)$ be the set of the feasible states of the hysteron W. Denote by $P = P(W)$ a transformation of the plane $\{u,x\}$ such that all elements of $\Omega(W)$ are its fixed points. In turn, in the case of $M_o = \{u_o,x_o\} \notin \Omega(W)$, the value $P(W)M_o$ is defined as the closest to M_o point of $\Omega(W)$ which belongs to the straight line $u = u_o$ (see Figure 15.1). The ordinate of point $P(W)M_o$ will be denoted by $P(W,u_o)x_o$.

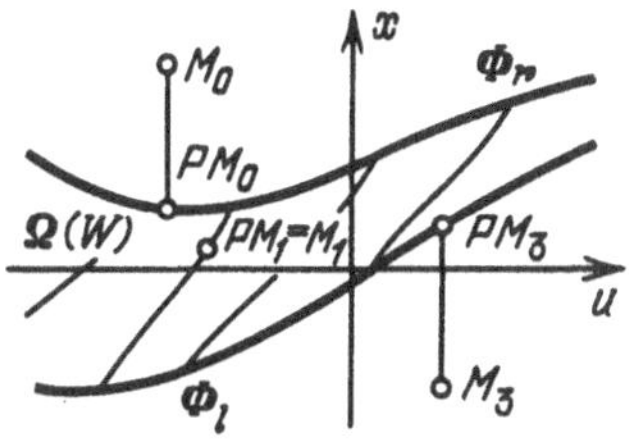

Fig. 15.1

Let us return to the construction of the variable hysteron $\tilde{W}$ with given piecewise constant function (15.9). Define now the output signal (15.11) by the following formula, to a certain degree reminiscent of (15.12),

$$x(t) = \begin{cases} W^{\tau_o}[\tau_o,x_o]u(t), & \text{for } \tau_o \leq t < \tau_1, \\ P[W^{\tau_1},u(\tau_1)]W^{\tau_o}[\tau_o,x_o]u(\tau_1), & \text{for } t = \tau_1, \\ W^{\tau_1}[\tau_1,x(\tau_1)]u(t), & \text{for } \tau_1 \leq t < \tau_2, \\ P[W^{\tau_2},u(\tau_2)]W^{\tau_1}[\tau_1,x(\tau_1)]u(\tau_2), & \text{for } t = \tau_2, \\ \cdots\cdots\cdots & \\ W^{\tau_{i-1}}[\tau_{i-1},x(\tau_{i-1})]u(t), & \text{for } \tau_{i-1} \leq t < \tau_i, \\ P[W^{\tau_i},u(\tau_i)]W^{\tau_{i-1}}[\tau_{i-1},x(\tau_{i-1})]u(\tau_i), & \text{for } t = \tau_i, \\ \cdots\cdots\cdots & \end{cases} \tag{15.13}$$

The output signal (15.13) is not necessarily continuous. It may be discontinuous at all instants $\tau_1, \tau_2, \ldots$ of the transfer from one constant hysteron to another. If the inclusion (15.10) holds at a certain fixed i, the output signal is continuous in the point τ_i.

c. Now we pass to more general situation when the function W^t does not have the form (15.9). We confine ourselves to the case of inputs defined on a finite interval $[t_0,t_1]$.

Divide the interval $[t_0,t_1]$ by the points

$$\tau_0 = t_0, \tau_1, \tau_2, \dots, \tau_n = t_1$$

into finite number of subintervals. As usually, by the *mesh size* $\delta(T)$ of the partitioning T we mean

$$\delta(T) = \max_{i=1,\dots,n} |\tau_i - \tau_{i-1}| .$$

To the partitioning T we associate the variable hysteron $\tilde{W}(T)$ defined by equalities (15.12). This hysteron will be considered as an approximation of the variable hysteron $\tilde{W}$ generated by function W^t (analogously to the approximation of an integral by its integral sums).

Take a fixed initial state

$$\{u(t_0),x_0\} \in \Omega(W^{t_0}) \tag{15.14}$$

and a fixed continuous input $u(t)$ $(t_0 \leq t \leq t_1)$. For each partitioning T of the interval $[t_0,t_1]$, the corresponding output

$$x(t;T) = \tilde{W}[t_0,x_0;T]u(t) \quad (t_0 \leq t \leq t_1) \tag{15.15}$$

is defined. Consider different partitions T and assume that, as $\delta(T)$ tends to zero, the outputs (15.15) will converge to a certain function $y(t)$. This function will be denoted by

$$y(t) = \tilde{W}[t_0,x_0]u(t) \tag{15.16}$$

and treated as an *output of the variable hysteron* $\tilde{W}$ *generated by function* W^t.

If the limits (15.16) exist for all continuous inputs and initial conditions such that (15.14) is satisfied, equality (15.16) defines the input-output relations of a variable hysteron $\tilde{W}$ with the domain $\Omega(\tilde{W}) = \Omega(W^{t_0})$ of the initial states.

d. Due to arbitrariness of the interval $[t_0,t_1]$, in "good" cases the variable hysteron $\tilde{W}$ is defined on the inputs $u(t)$ for all $t \geq t_0$. For every t , the domain $\Omega(\tilde{W},t)$ of feasible states of the variable hysteron $\tilde{W}$ coincides with $\Omega(W^t)$; for variable hysterons the semi-group identity holds, etc.

The introduced definition of a variable hysteron has a quite general character. In this definition, a character of the convergence of the outputs (15.15) towards the output (15.16) is not specified; it may be pointwise, uniform, in the metric of any function space, etc.

There is no satisfactory answer to the question what are the conditions which guarantee that the function W^t defines a variable hysteron $\tilde{W}$. There are no sufficiently general estimates which would characterize the closeness of $\tilde{W}$ and hysterons $\tilde{W}(T)$ constructed at given partitionings, etc. An exception will be given by a class of variable hysterons considered in Section 15.3.

The presented construction of variable hysterons can be generalized in various directions. For instance, to construct variable hysterons $\tilde{W}(T)$ which approximate $\tilde{W}$, one can make use of the piecewise constant functions whose values on $[\tau_{i-1},\tau_i)$ coincide with some W^{ξ_i} rather than with $W^{\tau_{i-1}}$, where ξ_i is an arbitrary point of $[\tau_{i-1},\tau_i]$ or a point chosen according to any special rule. One can admit only partitionings which satisfy some additional conditions, etc.

15.3. Variable hysteron governed by differential equations

Consider a function W^s whose values are hysterons which represent vibro-solutions of the equation

$$\frac{dx}{dt} = \varphi[s,x,u(t)]u'(t) \tag{15.17}$$

subject to the constraints

$$\gamma_-(s,u), \quad \gamma_+(s,u) \quad (-\infty < s, u < \infty) \quad . \tag{15.18}$$

Theorem 15.1. Let the functions

$$\varphi(t,x,u), \quad \varphi'_t(t,x,u), \quad \varphi'_x(t,x,u) \tag{15.19}$$

be continuous and locally Lipschitz continuous with respect to x. Assume that the constraints (15.18) are continuous. Then, the function W^t defines a variable hysteron $\tilde{W}$ with outputs (15.15) which converge to the output (15.16) (at a fixed input $u(t)$) uniformly for values t from any finite interval. The function (15.16) is a vibro-solution of the equation

$$\frac{dx}{dt} = \varphi[t,x,u(t)]u'(t) \tag{15.20}$$

with constraints

$$\gamma_-(t,u), \quad \gamma_+(t,u) \quad . \tag{15.21}$$

∎

To prove this theorem (in view of tedious details we do not present all the arguments), one can introduce a special substitution as in Chapter 12.

Under hypotheses of Theorem 15.1, the outputs (15.16) are continuous. The operator $\tilde{W}[t_0,x_0]$ is also locally Lipschitz continuous in the space $C(t_0,t_1)$. The question concerning properties of the operators $\tilde{W}[t_0,x_0]$ on classes of inputs different from the space C of continuous functions remains almost completely open.

As a first example we consider a *variable play* $\tilde{L}$ with the equation (15.17) and constraints (15.18) in the form (cf. (15.7))

$$\frac{dx}{dt} = 0, \quad \gamma_-(s,u) = \Gamma_r(s,u), \quad \gamma_+(s,u) = \Gamma_\ell(s,u) \quad .$$

The hypotheses of Theorem 15.1 hold. It is not difficult to establish the estimate

$$\|\tilde{L}[t_0,x_0]u(t) - \tilde{L}[t_0,x_0]v(t)\|_{t_0,t_1} \le \ell(r)\, \|u(t) - v(t)\|_{t_0,t_1}$$
$$(\|u(t)\|_{t_0,t_1}, \ \|v(t)\|_{t_0,t_1} \le r) \quad ,$$

provided that the constraints $\gamma_-(t,u)$ and $\gamma_+(t,u)$ are for $t \in [t_0,t_1]$ uniformly Lipschitz continuous in $u \in [-r,r]$ with Lipschitz constant $\ell(r)$:

$$|\Gamma_\ell(t,u) - \Gamma_\ell(t,v)|, \ |\Gamma_r(t,u) - \Gamma_r(t,v)| \le \ell(r)|u - v|$$

$(t_0 \leq t \leq t;\ |u|,|v| \leq r)$.

The second example concerns a *variable stop* $\tilde{U}$ with the equations (15.17) and constraints (15.18) in the form (cf.,(15.8))

$$\frac{dx}{dt} = E(s)u'(t),\quad \gamma_-(s,u) = -h(s),\quad \gamma_+(s,u) = h(s) \quad .$$

Assumptions of Theorem 15.1 are fulfilled if the function $h(t)$ is continuous, the modulus $E(t)$ continuously differentiable and the estimate

$$\|\tilde{U}[t_0,x_0]u(t) - \tilde{U}[t_0,x_0]v(t)\|_{t_0,t_1} \leq$$
$$\leq 2\left[\max_{t_0 \leq t \leq t_1} E(t) + \mathrm{Var}_{t_0}^{t_1} E(t)\right] \|u(t) - v(t)\|_{t_0,t_1}$$

holds.

15.4. Infinitesimal hysteron

Let a transducer $\tilde{V}$ with scalar inputs u and outputs x be given. We assume that the domain $\Omega(\tilde{V},t)$ of its feasible states is known for all t, all the continuous inputs $u(t)$ are admissible and the operators $\tilde{V}[t_0,x_0]$ of the input-output correspondences are defined. For any $\alpha > 0$ define

$$u_\alpha(t) = u[t_0 + \alpha(t - t_0)] \qquad (t \geq t_0)$$

and assume that for each continuous input $u(t)$

$$\lim_{\alpha\to\infty} \tilde{V}[t_0,x_0]u_\alpha\left(t_0 + \frac{t - t_0}{\alpha}\right) = V^{t_0}[t_0,x_0]u(t) \quad , \qquad (15.22)$$

where $V^{t_0}[t_0,x_0]$ are the operators which characterize a certain hysteron V^{t_0} with the domain $\Omega(V^{t_0}) = \Omega(\tilde{V};T)$ of feasible states. In such a case, we shall say that *at the time* t_0 *the transducer* $\tilde{V}$ *is a hysteron.* The hysteron V^{t_0} will be called *infinitesimal for* V *in the point* t_0 .

If the hypotheses of Theorem 15.1 hold, the variable transducer $\tilde{V} = \tilde{W}$

remains a hysteron for all t, while the hysteron V^t which is infinitesimal for $\tilde{W}$ in the point t coincides with W^t throughout. Nevertheless, a transducer which in every point t is identical with the hysteron W^t may not be a variable hysteron itself. As an example take a transducer $\tilde{V}$ whose input-output relations are defined by the vibro-solutions $x(t) = \tilde{V}[t_o,x_o]u(t)$ of the equation

$$\frac{dx}{dt} = \varphi[t,x,u(t)]u'(t) + \psi[t,x,u(t)] \quad . \tag{15.23}$$

Let us assume that the functions

$$\varphi(t,x,u), \quad \varphi_t'(t,x,u), \quad \varphi_x'(t,x,u), \quad \psi(t,x,u) \tag{15.24}$$

are jointly continuous with respect to all variables and satisfy the local Lipschitz condition with respect to x, whereas the constraints $\gamma_-(t,u)$ and $\gamma_+(t,u)$ are as previously continuous.

We now give an additional characterization of the transducer $\tilde{V}$ constructed according to the equation (15.23) and the constraints (15.21).

To this end, consider the partitioning T of the interval $[t_o,t_1]$ by the points

$$\tau_o = t_o < \tau_1 < \tau_2 < \ldots < \tau_n = t_1 \quad ;$$

to this partitioning associate a transducer $\tilde{V}(T)$ whose outputs $x(t) = \tilde{V}[t_o,x_o;T]u(t)$, which correspond to any continuous input $u(t)$, are defined by the equality

$$x(t) = \begin{cases} W^{\tau_o}[\tau_o,x_o]u(t) + (t-\tau_o)\psi[\tau_o,x_o,u(\tau_o)] & \text{for} \quad \tau_o \leqq t < \tau_1 , \\ P[W^{\tau_1},u(\tau_1)]\{W^{\tau_o}[\tau_o,x_o]u(\tau_1) + (\tau_1-\tau_o)\psi[\tau_o,x_o,u(\tau_o)]\} & \text{for} \quad t = \tau_1 , \\ W^{\tau_1}[\tau_1,x(\tau_1)]u(t) + (t-\tau_1)\psi[\tau,x(\tau_1),u(\tau_1)] & \text{for} \quad \tau_1 \leqq t < \tau_2 , \\ \cdots\cdots\cdots & \\ W^{\tau_{i-1}}[\tau_{i-1},x(\tau_{i-1})]u(t) + (t-\tau_{i-1})\psi[\tau_{i-1},x(\tau_{i-1}),u(\tau_{i-1})] & \text{for} \quad \tau_{i-1} \leqq t < \tau_j , \\ P[W^{\tau_i},u(\tau_i)]\{W^{\tau_{i-1}}[\tau_{i-1},x(\tau_{i-1})]u(\tau_i) + (\tau_i-\tau_{i-1})\psi[\tau_{i-1},x(\tau_{i-1}),u(\tau_{i-1})]\} & \text{for} \quad t = \tau_j , \\ \cdots\cdots\cdots & \end{cases} \tag{15.25}$$

analogous to (15.13).

Theorem 15.2. Let the functions (15.24) and constraints (15.21) be jointly continuous with respect to all variables and, moreover, functions (15.24) satisfy local Lipschitz condition with respect to x. Then, for every continuous input $u(t)$ $(t_0 \leq t \leq t_1)$, the functions $\mathfrak{V}[t_0,x_0;T]u(t)$, defined by the equality (15.25), uniformly on $[t_0,t_1]$ converge to $\mathfrak{V}[t_0,x_0]u(t)$, provided that the mesh size $\delta(T)$ of the partitioning T tends to zero. ∎

Any transducer $\mathfrak{V}$ which corresponds to the vibro-solutions of equation (15.23) with constraints (15.21) will be called a *variable hysteron with the drift* $\psi(t,x,u)$. The results we have presented in Section 14.4 are useful at studying variable hysterons with drift. In the further exposition we shall not develop this line.

15.5. A special class of transducers

Often, systems with time-dependent characteristics are treated as some transducers. In many cases, their analysis can be reduced to a study of some variable hysterons which are dependent upon a certain parameter (although, the transducers themselves do not coincide with those hysterons). We shall give a simple example for this.

Let continuous functions

$$\gamma_-(t,u,v),\quad \gamma_+(t,u,v),\quad \varphi(t,x,u,v),\quad \psi(t,x,u,v)\quad (t\in[t_0,t_1];\,u,v,x\in R^1)$$

be given so that

$$\gamma_-(t,u,v) \leq \gamma_+(t,u,v) \tag{15.26}$$

and the functions $\varphi(t,x,u,v)$ and $\psi(t,x,u,v)$ satisfy local Lipschitz condition with respect to x. Next, assume that operators H_-, H_φ and H_ψ, acting in the space $C(t_0,t_1)$, have been defined so that the values of operator H_φ be continuously differentiable as functions of t. We assume that all operators H_φ are of Volterra type, though our further considerations will not use this hypothesis.

In the case under consideration, for each continuously differentiable input $u(t)$ $(t_0 \leq t \leq t_1)$, the differential equation

$$\frac{dx}{dt} = \varphi[t,x,u(t), H_\varphi u(t)]u'(t) + \psi[t,x,u(t),H_\psi u(t)] \tag{15.27}$$

with constraints

$$\gamma_-[t,u(t),H_-u(t)], \quad \gamma_+[t,u(t),H_-u(t)] \tag{15.28}$$

admits a unique solution

$$x(t) = Z[t_0,x_0]u(t) \qquad (t_0 \leq t \leq t_1) \tag{15.29}$$

which satisfies the initial condition $x(t_0) = x_0$, where

$$\gamma_-[t_0,u(t_0),H_-u(t_0)] \leq x_0 \leq \gamma_+[t_0,u(t_0),H_-u(t_0)] \quad . \tag{15.30}$$

The operators (15.29) characterize a certain transducer Z defined on smooth inputs. An important question concerns the vibro-correctness of the transducer Z, i.e., a possibility of extending the operators $Z[t_0,x_0]$ by continuity (in the metric of the space $C(t_0,t_1)$) onto all continuous inputs $u(t)$, such that x_0 satisfies the condition (15.30). If the transducer Z is vibro-correct, we shall preserve the notations Z and $Z[t_0,x_0]$ for the corresponding extensions onto all continuous inputs.

The following result follows from Theorem 14.4.

Theorem 15.3. Assume that the functions

$$\varphi(t,x,u,v), \quad \varphi'_t(t,x,u,v), \quad \varphi'_x(t,x,u,v), \quad \varphi'_v(t,x,u,v)$$

are jointly continuous with respect to all variables, satisfy local Lipschitz condition with respect to x and v. Assume that the function $\psi(t,x,u,v)$ is jointly continuous with respect to all variables and satisfies local Lipschitz condition with respect to x, u and v ; the functions $\gamma_-(t,u,v)$ and $\gamma_+(t,u,v)$ are jointly continuous with respect to all variables; they satisfy local Lipschitz condition with respect to u and v . Further, suppose that H_- and H_ψ , treated as operators from $C(t_0,t_1)$ into $C(t_0,t_1)$, are Lipschitz continuous on each ball in the space $C(t_0,t_1)$, while H_φ (also considered on any ball of the space $C(t_0,t_1)$) is Lipschitz continuous as an operator from $C(t_0,t_1)$ into the space $C^1(t_0,t_1)$ of continuously differentiable functions.

Then the transducer Z is vibro-correct. Moreover, on any ball of the space $C(t_0,t_1)$ every $Z[t_0,x_0]$ is Lipschitz continuous as an operator from $C(t_0,t_1)$ into $C(t_0,t_1)$.

To introduce operators $Z[t_0,x_0]$ which are directly defined on all continuous inputs (certainly, the condition (15.30) must be fulfilled), we can use so-called "diagonalization". In such a construction, at first a continuous input $\xi = \xi(t)$ $(t_0 \leq t \leq t_1)$ must be chosen, then for the differential equation

$$\frac{dx}{dt} = \varphi[t,x,u(t),H_\varphi \xi(t)]u'(t) + \psi[t,x,u(t),H_\psi \xi(t)] \tag{15.31}$$

with constraints

$$\gamma_-[t,u,H_- \xi(t)], \quad \gamma_+[t,u,H_- \xi(t)] \tag{15.32}$$

the corresponding variable hysteron $\widetilde{W}(\xi)$ with drift is to be defined. The operator

$$x(t) = \widetilde{W}[t_0,x_0;\xi]u(t) \tag{15.33}$$

generated by hysteron $\widetilde{W}(\xi)$, is in particular defined on the input $u(t) \equiv \xi(t)$. Under natural conditions, the equality

$$Z[t_0,x_0]\xi(t) \equiv \widetilde{W}[t_0,x_0;\xi]\xi(t) \tag{15.34}$$

holds.

The hypothesis which says that the operator H_φ acts from $C(t_0,t_1)$ into $C^1(t_0,t_1)$ plays an essential role in Theorem 15.3. If $K(t,s,u)$ and $K_t'(t,s,u)$ are jointly continuous with respect to all variables, then the above hypothesis is satisfied by the integral operators

$$H_\varphi u(t) = \int_{t_0}^{t} K[t,s,u(s)]ds \quad . \tag{15.35}$$

This hypothesis is indeed essential. For example, the apparently quite simple equation

$$\frac{dx}{dt} = \varphi[t,x,u(t),\ H(\alpha)u(t)]u'(t) \quad , \tag{15.36}$$

where

$$H(\alpha)u(t) = \begin{cases} u(t_o) & \text{at} \quad t_o \leq t \leq t_o + \alpha \quad , \\ u(t-\alpha) & \text{at} \quad t_o + \alpha \leq t \leq t_1 \quad , \end{cases} \tag{15.37}$$

is, in general, not vibro-correct.

Part 4. Multidimensional hysterons

He was fresh as a young radish and unpretentious as a rake.
O' Henry

16. Multidimensional play and stop defined on smooth inputs

16.1. A simple example

Let a rectangular coil Γ with sides parallel to the coordinate axes in the plane $\{u_1,u_2\}$ be given. Consider also a pivot M which is placed in the interior of Γ and can move within the plane according to any arbitrary continuous rule. The coil will remain fixed as long as the pivot moves only in its interior. If, however, the pivot touches the coil, the latter starts moving. Assume that only translations of the coil may occur, i.e., its sides remain parallel to the coordinate axes. To describe the corresponding motion laws, a few notations are to be introduced.

Let the length of the horizontal and the vertical sides of the coil be respectively equal to $2h_1$ and $2h_2$, and the location of the coil be characterized by the coordinates $\{\xi_1,\xi_2\}$ of its center (see Figure 16.1). Denote by $\{u_1,u_2\}$ the coordinates of the pivot. Then the point $\{u_1,u_2,\xi_1,\xi_2\}$ of the four-dimensional space R^4 is to be considered as a state of the system L which consists of the coil Γ and pivot M. The prism

$$|\xi_1 - u_1| \le h_1, \quad |\xi_2 - u_2| \le h_2, \quad -\infty < u_1, u_2 < \infty \tag{16.1}$$

represents the corresponding domain $\Omega(L) \subset R^4$ of feasible states. If a state $\{u_1(t_0), u_2(t_0), \xi_1(t_0), \xi_2(t_0)\}$ of the system L at the initial time moment $t = t_0$ is given, then each continuous rule

$$u_1 = u_1(t), \quad u_2 = u_2(t) \qquad (t \ge t_0) \tag{16.2}$$

which describes movement of the pivot, uniquely determines the movement of the whole coil:

$$\xi_1 = \xi_1(t), \quad \xi_2 = \xi_2(t) \qquad (t \ge t_0) \ . \tag{16.3}$$

Therefore, the system L can be treated as a deterministic transducer with the two-dimensional input $u = \{u_1, u_2\}$ and the two-dimensional output $\xi = \{\xi_1, \xi_2\}$.

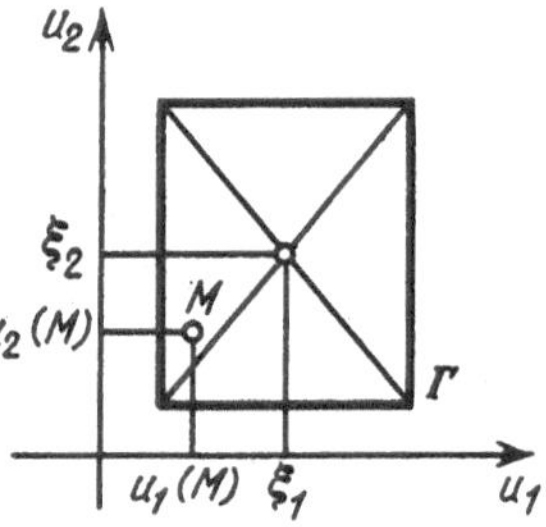

Fig. 16.1

Suppose now that the displacement of the coil in the horizontal direction is completely characterized by function $u_1(t)$ which represents variations of the first coordinate of the pivot, whereas only variations $u_2(t)$ of its second coordinate are responsible for the movement in the vertical direction. Then the output (16.3) is given by the equalities

$$\xi_1(t) = L[t_0, \xi_1(t_0); h_1]u_1(t), \quad \xi_2(t) = L[t_0, \xi_2(t_0); h_2]u_2(t), \tag{16.4}$$

where $L[t_0, \xi_0; h]$ represents an ordinary one-dimensional play (cf., Section 2.1). Therefore, the system L is referred to as an *ordinary two-dimensional play.*

Each property of a one-dimensional play in an obvious way induces the corresponding property of the two-dimensional play . Therefore, the two-dimensional play is static, controllable, vibro-correct with respect to noises at the input (moreover, its input-output and input-state relations fulfil the Lipschitz condition with the corresponding coefficients), satisfies the semigroup identities, can be described by vibro-solutions of some differential equations with discontinuous right-hand sides, etc.

The system which consists of the coil Γ and pivot M can also be considered as a transducer U with the inputs (16.2) and the outputs

$$x_1(t) = u_1(t) - \xi_1(t), \quad x_2(t) = u_2(t) - \xi_2(t) \quad (t \geq t_0) . \tag{16.5}$$

The domain $\Omega(U)$ of all feasible states of the transducer U consists of the points $\{u_1, u_2, x_1, x_2\}$ such that

$$|x_1| \leq h_1, \quad |x_2| \leq h_2, \quad -\infty < u_1,\, u_2 < \infty .$$

Due to (16.4) and (16.5), the transducer U is characterized by equalities

$$x_1(t) = U[t_0, x_1(t_0); h_1]u_1(t), \quad x_2(t) = U[t_0, x_2(t_0); h_2]u_2(t), \quad (t \geq t_0), \tag{16.6}$$

where $U[t_0, x_0; h]$ represents the corresponding one-dimensional stop. Then the transducer U is called a *two-dimensional stop*. Any two-dimensional stop is deterministic, static, controllable and vibro-correct. Operators (16.6) are Lipschitz continuous, may be characterized as vibro-solutions of some equations, etc.

A passage to higher dimensions is quite simple. For instance, by an ordinary N-dimensional play we mean a transducer whose input and output

$$u(t) = \{u_1(t), \ldots, u_N(t)\}, \quad \xi(t) = \{\xi_1(t), \ldots, \xi_N(t)\} \quad (t \geq t_0)$$

are N-dimensional, defined by

$$\xi_i(t) = L[t_0, \xi_i(t_0); h_i]u_i(t) \quad (t \geq t_0;\ i = 1, 2, \ldots, N) ,$$

and whose domain of feasible states comprises all points $\{u_1, \ldots, u_N; \xi_1, \ldots, \xi_N\}$ such that

$$|u_i - \xi_i| \leq h_i, \quad -\infty < u_1, \ldots, u_N < \infty .$$

16.2. A general notion

In two-dimensional case, we can consider coils which have a more general form than rectangle (in Figure 16.2, positions of the coil and pivot at two different time instants are depicted). Upon such an extension, the analysis of dependences between the vector inputs and vector outputs becomes much more complicated, because a decomposition of the input-output relations into system of one-dimensional relations is then impossible. In N-dimensional case, further complications occur.

We now describe some vector nonlinearities of hysteresis form that play a principal role in plasticity theory. In the mathematical models of plasticity theory due to von Mises, Tresca et al., nonlinearities of that type characterize the dependences between deformation and stress tensors.

First we assume that the inputs are smooth or at least piecewise smooth, then we shall extend the study onto arbitrary continuous inputs.

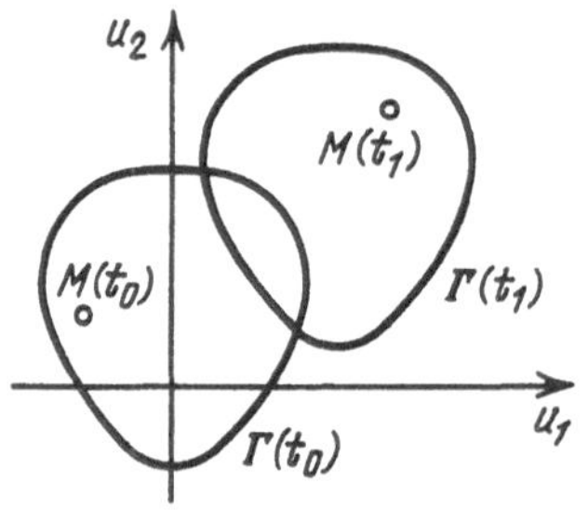

Fig. 16.2

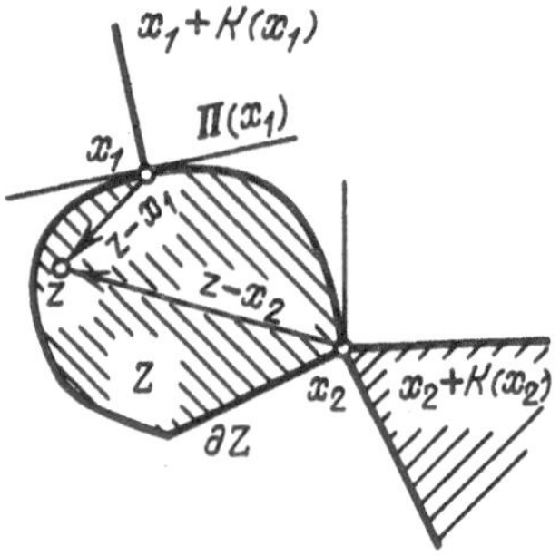

Fig. 16.3

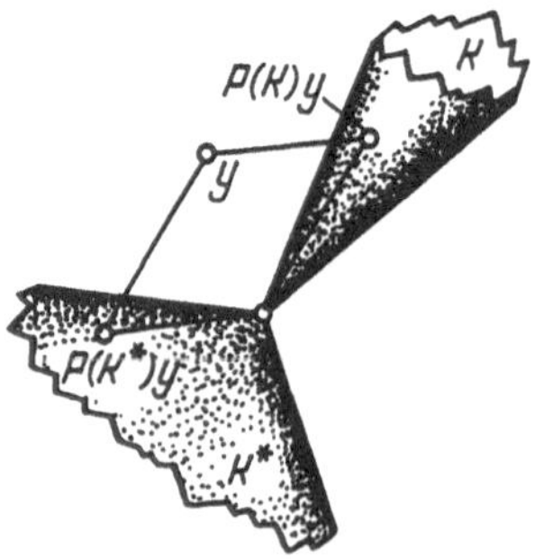

Fig. 16.4

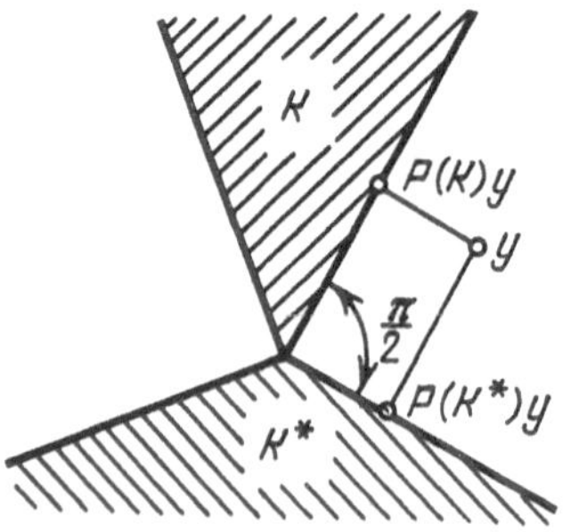

Fig. 16.5

Consider a bounded, convex and closed domain Z in the Euclidean space R^N ; let ∂Z be its boundary and int Z denote the set of its interior points. For each point $x \in Z$, we define the cones

$$K(x) = K(x;Z) = \{y \in R^N: (y,z-x) \leq 0 \quad \text{for every} \quad z \in Z\} , \tag{16.7}$$

$$K^*(x) = K^*(x;Z) = \{y \in R^N: (y,u) \leq 0 \quad \text{for every} \quad u \in K(x)\}. \tag{16.8}$$

The first of them is referred to as a *normal cone*, the second as a *tangent cone in the point* x . If $x \in \text{int } Z$, $K(x)$ contains only the point 0 and $K^*(x)$ coincides with R^N . If $x \in \partial Z$, $x + K^*(x)$ is equal to the intersection of the closed halfspaces which contain Z and include x as a boundary point. If $x \in \partial Z$ and is a regular point of the boundary (i.e., there is exactly one hyperplane supporting Z at x), then $K(x)$ is a ray and $K^*(x)$ a halfspace.

The formula

$$K(x) = \{y \in R^N: (y,u) \leq 0 \quad \text{for every} \quad u \in K^*(x)\} , \tag{16.9}$$

converse to (16.8), is true, too. In Figure 16.3, the cones $K(x_1)$ and $K(x_2)$ appropriately shifted to x_1 and x_2 are displayed.

Denote by $P(K)$ the operator which assigns to each point $x \in R^N$ the closest point $P(K)x$ to it of the cone $K \subset R^N$. Obviously,

$$|P(K)x - P(K)y| \leq |x - y| \qquad (x, y \in R^N) . \tag{16.10}$$

If $K^* = \{y: (y,u) \leq 0 \text{ for every } u \in K\}$, then (see Fig. 16.4)

$$P(K)y + P(K^*)y \equiv y \qquad (y \in R^N) . \tag{16.11}$$

In particular,

$$P[K(x)]y + P[K^*(x)]y \equiv y \qquad (x \in Z,\ y \in R^N) . \tag{16.12}$$

<u>Proof.</u> The equality (16.11) is obvious if $y \in K \cup K^*$ (in this case, one of the terms on the left-hand side is equal to zero, whereas the other equals y).

Suppose that $y \notin K \cup K^*$. Since $P(K)y$ is the closest to y point of the cone K ,

$$y - P(K)y \in K^* \quad .$$

Denote by y_1 the point of the ray $\alpha[y - P(K)y]$ $(\alpha \geq 0)$ which is closest to y. If the point y_1 were different from $P(K^*)y$, the scalar product $(P(K)y, P(K^*)y)$ would be positive and this would contradict the definition of the cone K^*. Consequently, $y_1 = P(K^*)y$ and the points $y, P(K)y, P(K^*)y$ are located (together with the point 0) in a certain two-dimensional hyperplane. Thus, it is sufficient to restrict the consideration to the case $N = 2$ (see Figure 16.5).

∎

Let us note that

$$(P(K)y, P(K^*)y) \equiv 0 \qquad (y \in R^N) \quad . \tag{16.13}$$

By definition of the operators $P[K(x)]$ it follows that the estimate

$$(P[K(x)]y, z - x) \leq 0 \qquad (y \in R^N;\ x, z \in Z)$$

holds. Therefore

$$(P[K(x)]y - P[K(z)]y, x - z) \geq 0 \qquad (y \in R^N;\ x, z \in Z) \tag{16.14}$$

and, due to (16.12),

$$(P[K^*(x)]y - P[K^*(z)]y, x - z) \geq 0 \qquad (y \in R^N;\ x, z \in Z) \quad . \tag{16.15}$$

Now we define a *multidimensional play* and *multidimensional stop* as transducers with vector inputs and outputs.

By a *play with the characteristic* Z we mean the transducer $L = L(Z)$ with the domain

$$\Omega(L) = \{\{u,\xi\}:\ u, \xi \in R^N;\ u - \xi \in Z\} \tag{16.16}$$

of the feasible states whose output

$$\xi(t) = L[t_0,\xi_0]u(t) = L[t_0,\xi_0;Z]u(t) \qquad (t \geq t_0) \tag{16.17}$$

(with values in R^N), corresponding to the piecewise smooth input

$$u = u(t) \qquad (t \geq t_o) \quad , \tag{16.18}$$

is defined as the solution of the vector differential equation

$$\frac{d\xi}{dt} = P[K(u(t) - \xi)]u'(t) \tag{16.19}$$

which satisfies the initial condition

$$\xi(t_o) = \xi_o \quad . \tag{16.20}$$

The initial state $\{u(t_o),\xi_o\}$ must be an element of $\Omega(L)$.

By a *stop with the characteristic* Z we mean the transducer $U = U(Z)$ with the domain

$$\Omega(U) = \{\{u,x\}: \; u, x \in R^N; \; x \in Z\} \tag{16.21}$$

of the feasible states, such that the output

$$x(t) = U[t_o,x_o]u(t) = U[t_o,x_o;Z]u(t) \qquad (t \geq t_o) \quad , \tag{16.22}$$

corresponding to the piecewise smooth input (16.18), is defined as the solution of the vector differential equation

$$\frac{dx}{dt} = P[K^*(x)]u'(t) \quad , \tag{16.23}$$

satisfying the initial condition

$$x(t_o) = x_o \quad . \tag{16.24}$$

The initial state must belong to $\Omega(U)$.

The solutions (16.17) and (16.22) are to be understood in the classical sense as absolutely continuous functions which satisfy the appropriate differential equations almost everywhere. Since the right-hand sides of these equations are discontinuous with respect to the state variable, the question concerning existence of the solutions is non-trivial (a similar situation has arisen in Part 3 at the study of vibro-solutions to differential equations with constraints).

If Z is a parallelepiped, $L(Z)$ and $U(Z)$ appropriately turn into the ordinary play and ordinary stop as considered in Section 16.1. To be completely precise, in Section 16.1 arbitrary continuous inputs were admissible, whereas the operators (16.17) and (16.22) are so far defined only on piecewise smooth inputs.

16.3. Correctness of the definitions of play and stop

The definitions of a play and stop with characteristic Z will be correct only if we show that for arbitrary initial conditions the solutions of equations (16.19) and (16.23) with values in the suitable domains are uniquely determined.

If the solutions of equations (16.19) and (16.23) exist, then due to (16.12) they satisfy the relation

$$\xi(t) + x(t) \equiv u(t) - u(t_0) + \xi_0 + x_0 \quad . \tag{16.25}$$

Moreover, solvability of either of those equations implies the same for the other. Therefore, it is sufficient to restrict the study to one of them.

Theorem 16.1. For each piecewise smooth input (16.18), the equation (16.23) has a unique classical solution with values in Z, such that the initial condition (16.24) is satisfied.

Proof. Since the right-hand side of the equation (16.23) is discontinuous with respect to x, we first derive an auxiliary differential inclusion whose solvability follows from standard theorems. Next we show that the solution (with values in Z) of the differential inclusion is also a classical solution of equation (16.23). The uniqueness will follow by simple estimates. It is sufficient to restrict the consideration to smooth inputs.

Let Q be the operator which assigns to each point $x \in R^N$ the closest to it point (it is uniquely defined) of the domain Z. The estimate

$$|Qx - Qy| \leq |x - y| \qquad (x, y \in R^N) \quad ,$$

analogous to (16.10), follows in an obvious way.

Consider a continuously differentiable input (16.18) and assign to each pair $\{t,x\}$ $(t \geq t_0,\ x \in R^N)$ the set

$$F(t,x) = \bigcap_{\varepsilon > 0} \overline{\{w:\ w = P[K^*(Qy)]u'(t),\ |y-x| \leq \varepsilon\}} \quad ; \tag{16.26}$$

this set consists of the limits of all convergent sequences

$$w_n = P[K^*(Qy_n)]u'(t) \qquad (n = 1,\ 2,\ \dots) \quad ,$$

where $y_n \to x$. For the sets (16.26), we introduce the following differential inclusion *)

$$\frac{dx}{dt} \in \text{co}\ F(t,x) \quad . \tag{16.27}$$

The right-hand side of (16.27) satisfies all the hypotheses of the classical Filippov's theorem on the solvability of differential inclusions. Hence, there are solutions of the inclusion (16.27) which satisfy the initial condition (16.24). By $x_0(t)$ we shall denote a solution with values in Z.

Since $x_0(t) \in Z$ $(t \geq t_0)$, $x_0'(t) \in K^*[x_0(t)]$ for all $t \geq t_0^*$ such that the derivative $x_0'(t)$ exists. On the other hand, $x_0'(t) \in \in \text{co}\ F[t,x_0(t)]$. Consequently, the vector-function $x_0(t)$ satisfies the equation (16.23) provided that the intersection $\text{co}\ F[t,x_0(t)] \cap K^*[x_0(t)]$ for almost all t contains only the vector $P[K^*(x_0(t))]u'(t)$. The last property is obvious.

It remains to prove uniqueness of the solution to equation (16.23) which satisfies the initial condition (16.24). Suppose that $x(t)$ and $z(t)$ are two such solutions, and $w(t) = x(t) - z(t)$. Then $w(t_0) = 0$ and (16.12) implies that for almost all t,

$$\frac{d}{dt}\,|w(t)|^2 = 2(w'(t),w(t)) =$$

$$= 2(P[K^*(x(t))]u'(t) - P[K^*(z(t))]u'(t),x(t) - z(t)) \leq 0$$

---

*) By co M and $\overline{\text{co}}$ M we denote the convex and the closed convex hull of the set M, respectively.

holds. Consequently, $w(t) = 0$ for all $t \geq t_0$. ∎

16.4. Properties of play and stop

a. Directly due to definitions, multidimensional plays and stops with an arbitrary characteristic Z are deterministic and static transducers. Their controllability is evident; the semigroup identities hold for both of them (because both are deterministic).

b. Questions concerning the vibro-correctness of a play and a stop, their dependence upon small perturbations of the characteristic, etc., are more difficult. In the subsequent sections these questions will be discussed for some special classes of characteristics.

c. A transducer W is called *convergent* if the scalar function

$$\alpha(t) = W[t_0,x_0]u(t) - W[t_0,y_0]u(t) \qquad (t \geq t_0) \tag{16.28}$$

is non-increasing for all initial states $\{u(t_0),x_0\}$ and $\{u(t_0),y_0\}$. *Multidimensional plays and stops are convergent.* Hence, their outputs, corresponding to a periodic input, are asymptotically periodic. If the characteristic Z is polyhedral, then at any T-periodic input $u(t)$ and any initial state x_0 the following estimate, recently obtained by A.A. Vladimirov, holds:

$$|W[t_0,x_0]u(t) - x_*(t)| < C \exp[-\gamma(t-t_0)] \qquad (t \geq t_0) ,$$

where $C, \gamma > 0$, and $x_*(t)$ is some T-periodic function.

d. The definitions of a multidimensional play and stop comprehend the classical solutions to some differential equations with discontinuous right-hand sides (i.e., absolutely continuous functions which fulfil the equations for almost all t). It turns out that functions (16.17) and (16.22) have both unilateral derivatives for all t, with the right derivatives equal to the right-hand sides of the equations for all t (not only almost everwhere). We mention this subtle property without giving any proof.

16.5. On the classical solutions of equations with discontinuous right-hand sides

The equations (16.19) and (16.23), in particular, have discontinuous right-hand sides but still admit classical solutions (absolutely continuous functions satisfying the equation for almost all t). It would be of interest for numerous applications to develop a similar analysis for some larger classes of such equations.

Theorem 16.1 , on the existence and uniqueness of the classical solution, without any problem can be applied to equations of the form

$$\frac{dx}{dt} = P[K^*(x;Z)]f(t,x) \quad , \tag{16.29}$$

where Z again denotes a bounded, closed and convex set, and the vector-function $f(t,x)$ is Lipschitz continuous with respect to x and integrable in t over any finite interval for every fixed x . Due to this assertion, outputs of the play and stop can be defined for any absolutely continuous input (not necessarily piecewise smooth).

Equations of the form (16.29) arise in many optimization problems. Similar equations play an essential role in the theory of sliding systems, etc.

Conditions which guarantee that any classical solution of a differential equation with discontinuous right-hand side has for every t the right derivative $D_r x(t)$ and this derivative is equal to the right-hand side of the equation are known only in very particular cases. As it has been already mentioned, the solutions (16.17) and (16.22) of equations (16.19) and (16.23) have such a property. This property is true for a special class of equations (16.29), namely

$$\frac{dx}{dt} = P[K^*(x;Z)]\text{grad } V(x) \quad , \tag{16.30}$$

where $V(x)$ $(x \in R^N)$ is a twice continuously differentiable function. For each initial value $x(t_0) = x_0 \in Z$,there exists a unique absolutely continuous solution $x_0(t)$ of equation (16.30) with values in Z , such that for all $t \geq t_0$ the equality

$$D_r\, x_0(t) = P[K^*(x_0(t);Z]\text{grad } V[x_0(t)] \tag{16.31}$$

holds. Equations of the form (16.30) arise from the construction and study of various gradient algorithms for minimization of smooth functions on convex domains, as well as from minimization algorithms for non-smooth convex functions.

If $V(x)$ is convex, the right-sided differentiability of the solutions to (16.30) for all t may be shown by using the well-known Komura's theorems on the properties of solutions to the Cauchy problem for differential inclusions with the right-hand side monotone in sense of Minty. Some special constructions must be applied if $V(x)$ is non-convex.

17. Strictly convex characteristics

17.1. Vibro-correctness modulus

In the sequel, we shall study properties of a play and a stop which have strictly convex characteristic Z (i.e., for any segment contained in Z, all its interior points belong to $\operatorname{int} Z$).

For each $r \geq 0$ let us define

$$\alpha(r;Z) = \inf_{x \in \partial Z} \inf_{y \notin K^*(x;Z), |y| = r} \rho(x+y, Z)$$

or, equivalently,

$$\alpha(r;Z) = \inf_{x \in \partial Z} \inf_{y \in \partial K^*(x;Z), |y| = r} \rho(x+y, Z) \quad . \tag{17.2}$$

In (17.1) and (17.2), we have used the notations: $K^*(x;Z)$ - the tangent cone to Z at the point x, $\rho(w,S)$ - the Euclidean distance between the point w and the set S. By the strict convexity of Z, function (17.1) is strictly increasing with respect to r. This function is continuous and $\alpha(0;Z) = 0$. Therefore, the inverse function

$$\lambda(\varepsilon;Z) = \alpha^{-1}(\varepsilon;Z) \qquad (0 \leq \varepsilon < \infty) \tag{17.3}$$

is strictly increasing and continuous, too, as well as $\lambda(0;Z) = 0$.

We shall call the non-decreasing continuous function $\lambda(\varepsilon)$ $(\varepsilon \geq 0)$ a *vibro-correctness modulus of transducer* W, if for all admissible inputs $u(t)$, $v(t)$ $(t \geq t_0)$ and for

$$\{u(t_0), x_0\}, \quad \{v(t_0), x_0\} \in \Omega(W) \ , \tag{17.4}$$

the estimate

$$\|W[t_o,x_o]u(t) - W[t_o,x_o]v(t)\|_{t_o,t_1} \leq \lambda[\|u(t) - v(t)\|_{t_o,t_1}]$$
$$(t_1 > t_o) \tag{17.5}$$

takes place.

If $\lambda(\varepsilon)$ is the vibro-correctness modulus of a transducer W and $\lambda(0) = 0$, then the *transducer* W *is vibro-correct* (sometimes it is more convenient to say that W is then *uniformly vibro-correct*). Clearly, as it follows from its definition, the vibro-correctness modulus is non-unique.

Theorem 17.1. The plays and stops with a strictly convex characteristic are vibro-correct. The function (17.3) represents a vibro-correctness modulus of the play while the function *)

$$\lambda_1(\varepsilon;Z) = \min\{\varepsilon + \lambda(\varepsilon;Z),\ \operatorname{diam} Z\} \qquad (\varepsilon \geq 0) \tag{17.6}$$

is the vibro-correctness modulus for the corresponding stop.

Proof. For the play and stop, there are simple relations which connect their defining operators (cf., Chapter 16). Thus, we shall confine ourselves to showing that function (17.3) is a vibro-correctness modulus of the play.

Consider piecewise smooth inputs $u(t)$ and $v(t)$ on a fixed time interval $t_o \leq t \leq t_1$. Let $u(t_o) - x_o \in Z$ and $v(t_o) - x_o \in Z$, then the outputs

$$x(t) = L[t_o;x_o;Z]u(t), \quad y(t) = L[t_o,x_o;Z]v(t) \quad (t_o \leq t \leq t_1) \tag{17.7}$$

are well-defined. Because of

$$\frac{d}{dt}|x(t) - y(t)|^2 = 2(x'(t),x(t) - y(t)) - 2(y'(t),x(t) - y(t)), \tag{17.8}$$

for proving the theorem it is enough to show that the inequalities

$$(x'(\tau),x(\tau) - y(\tau)) \leq 0, \quad (y'(\tau),x(\tau) - y(\tau)) \geq 0 \tag{17.9}$$

*) diam Z denotes the diameter of the set Z, i.e., $\operatorname{diam} Z = \sup\{|x - y| : x, y \in Z\}$.

are satisfied in each point $\tau \in (t_o,t_1)$ where both functions (17.7) are differentiable and

$$|x(\tau) - y(\tau)| \geq \lambda[\|u(t) - v(t)\|_{t_o,t_1};Z] \tag{17.10}$$

holds.

The proofs of both relations (17.9) are identical. We shall prove the first of them. Suppose the converse. Let

$$(x'(\tau),x(\tau) - y(\tau)) > 0 \quad , \tag{17.11}$$

then the equality

$$x'(\tau) = P[K(u(\tau) - x(\tau);Z)]u'(\tau)$$

implies that the vector $x(\tau) - y(\tau)$ does not belong to the cone $K^*[u(\tau) - x(\tau);Z]$. Assume that, in addition to (17.11), the inequality (17.10) is fulfilled. Then, by the strict convexity of Z,

$$\rho\{[u(\tau) - x(\tau)] + [x(\tau) - y(\tau)],Z\} > \|u(t) - v(t)\|_{t_o,t_1} .$$

The last estimate may be rewritten in the form

$$\rho\{[u(\tau)-v(\tau)] + [v(\tau) - y(\tau)],Z\} > \|u(t) - v(t)\|_{t_o,t_1} . \tag{17.12}$$

However, $v(\tau) - y(\tau) \in Z$, whereas by (17.12) it follows that

$$|u(\tau) - v(\tau)| > \|u(t) - v(t)\|_{t_o,t_1} .$$

We have arrived at contradiction. ∎

17.2. Hölder condition

In many cases, one may give simple estimates of the function (17.3) from above.

A characteristic Z is said to be *strictly convex with* γ *representing the index of strict convexity*, if for any $x, y \in Z$ the ball

$$\left\{w: \left|w - \frac{x+y}{2}\right| \leq \gamma\, |x-y|^2\right\}$$

is contained in Z . For such characteristics, the estimate

$$\lambda(\varepsilon;Z) \leq \varepsilon + \sqrt{\frac{\varepsilon}{2\gamma}} \tag{17.13}$$

holds and, because $2\gamma \operatorname{diam} Z \leq 1$,

$$\lambda_1(\varepsilon;Z) \leq \min\left\{2\varepsilon + \sqrt{\frac{\varepsilon}{2\gamma}}\, , \frac{1}{2\gamma}\right\} . \tag{17.14}$$

By Theorem 17.1 and the estimates (17.13), (17.14), *for any ball* $\|u(t)\|_{t_0,t_1} \leq r$ *the operators which are defining for a play and stop with strictly convex characteristic* Z *satisfy the Hölder condition with index* 1/2. In the case of a play, the coefficient in Hölder condition is dependent on r and γ. For the stop, it only depends upon γ :

$$\|L[t_0,x_0;Z]u(t) - L[t_0,x_0;Z]v(t)\|_{t_0,t_1} \leq$$

$$\leq \ell(r,\gamma)\,\|u(t) - v(t)\|^{1/2}_{t_0,t_1} \quad (\|u(t)\|_{t_0,t_1},\ \|v(t)\|_{t_0,t_1} \leq r) \tag{17.15}$$

and

$$\|U[t_0,x_0;Z]u(t) - U[t_0,x_0;Z]v(t)\|_{t_0,t_1} \leq$$

$$\leq \frac{3}{\sqrt{2\gamma}}\,\|u(t) - v(t)\|^{1/2}_{t_0,t_1} \quad (\|u(t)\|_{t_0,t_1},\ \|v(t)\|_{t_0,t_1} \leq r). \tag{17.16}$$

Balls and ellipsoids represent strictly convex characteristics. For the ball Z_0 with radius ρ_0 , the function (17.1) is defined by

$$\alpha(r;Z_0) = \sqrt{\rho_0^2 + r^2} - \rho_0 , \tag{17.17}$$

hence

$$\lambda(\varepsilon;Z_0) = \sqrt{\varepsilon^2 + 2\rho_0\varepsilon}, \quad \lambda_1(\varepsilon;Z_0) = \varepsilon + \sqrt{\varepsilon^2 + 2\rho_0\varepsilon} . \tag{17.18}$$

The corresponding index of strict convexity is equal to $(4\rho_0)^{-1}$. Let Z_1 denote the ellipsoid with half-axes $\rho_1 \leq \rho_2 \leq \ldots \leq \rho_N$; then

$$\lambda(\varepsilon;Z_1) \leq \sqrt{\varepsilon^2 + 2\rho_N \varepsilon}, \quad \lambda_1(\varepsilon;Z_1) \leq \varepsilon + \sqrt{\varepsilon^2 + 2\rho_N \varepsilon} \tag{17.19}$$

and the index of strict convexity is equal to $(4\rho_N)^{-1}$. For the stop with the ball Z_0 as characteristic, the Hölder condition (17.16) is satisfied with coefficient $2(\sqrt{5} - 1)\sqrt{\rho_0}$.

Suppose that for each point x which belongs to the boundary ∂Z of a characteristic Z there is a ball Z_0 with radius ρ_0 , such that $Z \subset Z_0$ and $x \in \partial Z_0$. (Then, certainly, the characteristic Z is strictly convex; however, this property does not take place for every strictly convex characteristic). *Under the above conditions, functions* (17.18) *are vibro-correctness moduli of the play* $L(Z)$ *and stop* $U(Z)$ *with characteristic* Z .

A continuous and monotone function $\delta(r)$ $(r \geq 0)$, positive at $r > 0$, is called *modulus of uniform convexity* of the set Z if for $x, y \in Z$ it follows that the ball

$$\left\{w\colon \left|w - \frac{x + y}{2}\right| \leq \delta(|x - y|)\right\}$$

is contained in Z . For each strictly convex set $Z \subset R^N$, one may construct its modulus of uniform convexity. By the definitions,

$$\lambda(\varepsilon;Z) \leq \varepsilon + \delta^{-1}\left(\frac{\varepsilon}{2}\right) . \tag{17.20}$$

This inequality appears also useful in applications of Theorem 17.1.

17.3. Passage to continuous inputs

So far we have only admitted piecewise smooth inputs $u(t)$ $(t \geq t_0)$ at the study of the play and stop. Theorem 17.1 makes it possible to pass to an arbitrary continuous input $u(t)$ in the case of a strictly convex characteristic (as usual, the initial state is to be traced). To this end, let us construct a sequence of smooth inputs $u_n(t)$ which converges to $u(t)$ uniformly on each finite interval $[t_0,t_1]$ and satisfies the condition $u_n(t_0) = u(t_0)$. At a fixed initial state, the input $u_n(t)$ generates an output $x_n(t)$ such that, by Theorem 17.1, the sequence

$x_n(t)$ converges to some continuous vector-function $x(t)$ uniformly on each finite interval. This function $x(t)$ will be treated as the output corresponding to the input $u(t)$. By Theorem 17.1, the output $x(t)$ does not depend upon the choice of the approximating sequence $u_n(t)$.

The above considerations are equally applicable to plays and stops. We shall preserve the notations $L[t_o,x_o;Z]$ and $U[t_o,x_o;Z]$ for the suitable defining operators, extended onto continuous inputs.

The properties of the play and stop (see Chapter 16) do not change upon passing to arbitrary continuous inputs. They remain deterministic, static, controllable and vibro-correct transducers. Theorem 17.1 on the vibro-correctness moduli (17.3) and (17.6) of the play and stop remains true also for the extended operators.

17.4. Strong convergence

As it has been shown in Chapter 16, the play and stop, generated by an arbitrary bounded, convex and closed characteristic Z, are convergent on piecewise smooth inputs.

A convergent transducer W will be called *strongly convergent* if for each $\varepsilon > 0$ there exists a $\nu = \nu(\varepsilon) > 0$ such that for any admissible input $u(t)$ $(t \geq t_o)$ and for all initial states $\{u(t_o),x_o\}$, $\{u(t_o),y_o\} \in \Omega(W)$, the inequality

$$|u(t_1) - u(t_o)| < \nu , \tag{17.21}$$

holding at some $t_1 > t_o$, implies the estimate

$$|W[t_o,x_o]u(t_1) - W[t_o,y_o]u(t_1)| < \varepsilon . \tag{17.22}$$

Theorem 17.2. If the characteristic Z is strictly convex, then the play $L(Z)$ and the stop $U(Z)$ are strongly convergent for all continuous inputs.

17.5. Perturbation of characteristics

It is natural to expect that "close" characteristics Z and Z_1 define "close" describing operators of the corresponding play and stop.

What we need is to give detailed and complete descriptions of the class of inputs to be considered, characterizations of the closeness for characteristics and operators; of interest are quantitative estimates. We may relatively easily deduce similar statements for the case when both characteristics Z and Z_1 (or one of them) are strictly convex. We confine ourselves to the case of a play.

Theorem 17.3. Let the characteristic Z be strictly convex and $\delta(r)$ denote its modulus of uniform convexity. Then for each piecewise smooth input $u(t)$ $(t \geq t_0)$ and for $u(t_0) - x_0 \in Z$, $u(t_0) - y_0 \in Z_1$ the estimate

$$|L[t_0,x_0;Z]u(t) - L[t_0,y_0;Z_1]u(t)| \leq$$

$$\leq \max\left\{|x_0 - y_0|,\ \rho_X(Z,Z_1) + \delta^{-1}\left[\frac{3}{2}\rho_X(Z,Z_1)\right]\right\} \quad (t \geq t_0) \tag{17.23}$$

holds, where $\rho_X(Z,Z_1)$ is the Hausdorff distance between the characteristic Z_1 (which is non-necessarily strictly convex) and Z .

Proof. Define

$$x(t) = L[t_0,x_0;Z]u(t), \quad y(t) = L[t_0,y_0;Z]u(t) \quad (t \geq t_0) \quad . \tag{17.24}$$

By equality (17.8), for proving the estimate (17.23) it is enough to show that the relations (17.10) hold at all those values $t > t_0$ for which both functions (17.24) are differentiable and

$$|x(\tau) - y(\tau)| \geq \rho_X(Z,Z_1) + \delta^{-1}\left[\frac{3}{2}\rho_X(Z,Z_1)\right] \quad . \tag{17.25}$$

In the proof of Theorem 17.1, both relations (17.9) had the same structure and admitted the same proof. Now the situation is different, those relations differ in their meaning. Nevertheless, we only give a proof of the inequality

$$(y'(\tau),y(\tau) - x(\tau)) \leq 0 \quad . \tag{17.26}$$

The second relation is simpler and admits an analogous proof.

Suppose that (17.26) is false. Then

$$(y'(\tau), y(\tau) - x(\tau)) > 0 \tag{17.27}$$

and

$$|y(\tau) - x(\tau)| \geq \rho_X(Z, Z_1) + \delta^{-1}\left[\frac{3}{2}\rho_X(Z, Z_1)\right] . \tag{17.28}$$

$y'(\tau) \neq 0$ due to (17.27), thus $u(\tau) - y(\tau) \in \partial Z_1$. By definition of the play, the vector $y'(\tau)$ is an element of the normal cone $K[u(\tau) - y(\tau); Z_1]$, hence again in view of (17.27) the difference $y(\tau) - x(\tau)$ does not belong to the cone $K^*[u(\tau) - y(\tau); Z_1]$.

By the definition of the Hausdorff distance between sets, there is a point $\xi \in Z$ such that

$$|[u(\tau) - y(\tau)] - \xi| \leq \rho_X(Z, Z_1) . \tag{17.29}$$

(17.28) and (17.29) imply the estimate

$$|[u(\tau) - x(\tau)] - \xi| \geq \delta^{-1}\left[\frac{3}{2}\rho_X(Z, Z_1)\right] . \tag{17.30}$$

At the same time, $u(\tau) - x(\tau)$ and ξ belong to Z. Therefore, by (17.30) and definition of the uniform convexity modulus it follows that the ball T_0 with center in $w_0 = \frac{1}{2}[u(\tau) - x(\tau) + \xi]$ and radius $\frac{3}{2}\rho_X(Z, Z_1)$ is contained in Z. Take $w_1 = w_0 + \frac{1}{2}[u(\tau) - y(\tau) - \xi]$. The ball T_1 with center at w_1 and radius $\rho_X(Z, Z_1)$ is contained in the ball T_0, hence $T_1 \subset Z$.

As it has been already shown, $u(\tau) - y(\tau) \in \partial Z_1$ and $y(\tau) - x(\tau) \notin K^*[u(\tau) - y(\tau); Z_1]$. Consequently, all points of the segment with the ends $u(\tau) - y(\tau)$ and $u(\tau) - x(\tau)$, which are different from $u(\tau) - y(\tau)$, do not belong to Z_1. In particular, $w_1 \notin Z_1$. But then there exists a point η of the ball T_1, such that $\rho(\eta, Z_1)$ is greater than the radius $\rho_X(Z, Z_1)$. On the other hand,

$$\rho(\eta, Z_1) \leq \rho_X(Z, Z_1)$$

because $\eta \in Z$. We have arrived at contradiction. ∎

Under the hypotheses of Theorem 17.3, the following estimate, analogous to (17.23) (and admitting an analogous proof),

$$|L[t_0,x_0;Z]u(t) - L[t_0,y_0;Z_1]u(t)| \le$$

$$\le \max\{|x_0 - y_0|,\ 3\lambda[\rho_X(Z,Z_1);Z]\} \quad (t \ge t_0) \tag{17.31}$$

holds with $\lambda(\varepsilon;Z)$ taken in the form (17.3).

If both characteristics Z and Z_1 are strictly convex, then Theorem 17.1 implies that the estimate (17.23) remains true for any continuous input $u(t)$. The estimate (17.31) admits then a sharpening: for every continuous input $u(t)$, the inequality

$$|L[t_0,x_0;Z]u(t) - L[t_0,y_0;Z_1]u(t)| \le$$

$$\le \max\{|x_0 - y_0|,\ \lambda[\rho_X(Z,Z_1);Z],\ \lambda[\rho_X(Z,Z_1);Z_1]\} \quad (t \ge t_0) \tag{17.32}$$

is satisfied.

17.6. Vibro-correctness modulus and differential inclusions

Let us again consider the play $L = L(Z)$ with characteristic Z which may be not strictly convex.

Denote by $B(\varepsilon) = B(\varepsilon;Z)$ the set of points $\xi \in R^N$ which admit the representation

$$\xi = L[t_0,x_0;Z]u(t_1) - L[t_0,x_0;Z]v(t_1) \ , \tag{17.33}$$

where $x_0 \in R^N$ and continuous, piecewise smooth inputs $u(t)$, $v(t)$ $(t \ge t_0)$ satisfy the conditions

$$u(t_0) - x_0 \in Z,\quad v(t_0) - x_0 \in Z \tag{17.34}$$

and

$$\|u(t) - v(t)\|_{t_0,t_1} \le \varepsilon \ . \tag{17.35}$$

Using estimates on the lengths of vectors (17.33), we can analyze properties of the vibro-correctness moduli of the play $L(Z)$. For each $x \in R^N$ and $\varepsilon > 0$ define the closed set

$$A(x,\varepsilon) = \begin{cases} \bigcup\limits_{y\in Z;\rho(y-x,Z)\le\varepsilon} \{z:\ z\in K(y;Z), |z|\le 1\} , & \text{if } \rho(x,Z-Z)\le\varepsilon , \\ \text{singleton } \{0\} , & \text{if } \rho(x,Z-Z)>\varepsilon . \end{cases} \tag{17.36}$$

Henceforth, $Z_1 - Z_2$ will denote the set which contains all points $u - v$ such that $u \in Z_1$ and $v \in Z_2$. Let us define

$$R(x;\varepsilon) = \text{co } A(-x;\varepsilon) - \text{co } A(x;\varepsilon) \tag{17.37}$$

and consider the differential inclusion

$$\frac{dx}{dt} \in R(x;\varepsilon) . \tag{17.38}$$

All sets (17.37) are bounded, convex and closed. The right-hand side of equation (17.38) is upper semicontinuous with respect to x. Hence, by virtue of Filippov's theorem, there exist solutions to equation (17.38) for any initial condition. By $B_1(\varepsilon) = B_1(\varepsilon;Z)$ we denote the set which comprises all values at $t \ge t_0$ of the solutions to equation (17.38) with the homogeneous initial condition $x(t_0) = 0$.

For any given characteristic Z, one can easily characterize the sets (17.37). Hence, one can also easily establish some estimates for the solutions of equation (17.38). As it turns out, there is a close connection between the sets $B_1(\varepsilon)$ and the sets $B(\varepsilon)$ of vectors (17.33).

Theorem 17.4. The sets $B_1(\varepsilon)$ and $B(\varepsilon)$ have the same closure, i.e.,

$$\overline{B(\varepsilon,Z)} = \overline{B_1(\varepsilon;Z)} . \tag{17.39}$$

Proof. First we prove the inclusion

$$B(\varepsilon;Z) \subset B_1(\varepsilon;Z) . \tag{17.40}$$

To this end, consider the outputs

$$x(t) = L[t_0,x_0;Z]u(t), \quad y(t) = L[t_0,x_0;Z]v(t)$$

which correspond to piecewise smooth inputs $u(t)$ and $v(t)$ such that the conditions (17.34) and (17.35) are fulfilled. We are going to prove that

$$x'(\tau) - y'(\tau) \in A[-x(\tau) + y(\tau);\varepsilon] - A[x(\tau) - y(\tau);\varepsilon] \tag{17.41}$$

in each point τ where the inputs $u(t)$, $v(t)$ and the outputs $x(t)$, $y(t)$ are differentiable. Since any play is a static hysteron, without any loss of generality we can assume that the inequalities

$$|u'(t)| \leq 1, \quad |v'(t)| \leq 1 \quad (t_0 \leq t \leq t_1) \tag{17.42}$$

are satisfied.

For proving (17.41) it is enough to show that

$$x'(\tau) \in A[-x(\tau) + y(\tau);\varepsilon] , \quad y'(\tau) \in A[x(\tau) - y(\tau);\varepsilon] . \tag{17.43}$$

Proofs of both of these relations proceed similarly. Let us give here the proof for the first relation.

Since $u(\tau) - x(\tau) \in Z$ and $v(\tau) - y(\tau) \in Z$, by (17.35) it follows that

$$u(\tau) - x(\tau) \in Z \cap Z[y(\tau) - x(\tau);\varepsilon] ,$$

where

$$Z(\xi;\varepsilon) = \{w: \rho(w - \xi, Z) \leq \varepsilon\} .$$

Then the first of inequalities (17.42) implies the first of relations (17.43). Thus, inclusion (17.40) has been proved and it only remains to show that

$$B_1(\varepsilon;Z) \subset \overline{B(\varepsilon;Z)} \tag{17.44}$$

in order to complete the proof of the theorem.

Consider a continuous function $x(t)$ $(t \geq t_0)$ for which $x(t_0) = 0$ and each interval $[t_0,t_1]$ can be divided by points $t_0 = \tau_0 < \tau_1 < \dots < \tau_n = t_1$ into a finite number of parts so that on any fixed subinterval

$[\tau_{i-1},\tau_i]$ the function $x(t)$ is linear, and for all $t \in (\tau_{i-1},\tau_i)$ its derivative $x'(t)$ either belongs to $A[-x(t);\varepsilon]$ or to $-A[x(t);\varepsilon]$ (the derivative $x'(t)$ remains constant for all $t \in (\tau_{i-1},\tau_i)$). By $B_2(\varepsilon;Z)$ we shall denote the set which contains the values of all such functions at $t \geq t_o$. Clearly, $B_2(\varepsilon;Z) \subset B(\varepsilon;Z)$, and for proving (17.44) it is sufficient to show that

$$B_1(\varepsilon;Z) \subset \overline{B_2(\varepsilon;Z)} \quad .$$

To this end, in turn, it suffices to show that

$$B_1(\varepsilon;Z) \subset \overline{\bigcup_{0<\delta<\varepsilon} B_1(\delta;Z)} \tag{17.45}$$

and

$$B_1(\delta;Z) \subset B_2(\varepsilon;Z) \qquad (0 < \delta < \varepsilon) \quad . \tag{17.46}$$

Observe that $Z(x;\varepsilon) \subset Z(\alpha x;\alpha\varepsilon)$ at $0 \leq \alpha \leq 1$, thus $\alpha B_1(\varepsilon;Z) \subset$ $\subset B_1(\alpha\varepsilon;Z)$ at $0 \leq \alpha \leq 1$. This implies (17.45).

Now take some $\delta \in (0,\varepsilon)$ and an interval $[\sigma_1,\sigma_2]$ such that

$$\sigma_2 - \sigma_1 \leq \frac{1}{4} (\varepsilon - \delta) \quad .$$

Denote by $\xi(t)$ $(\sigma_1 \leq t \leq \sigma_2)$ a solution of the equation

$$\frac{dx}{dt} \in R(x;\delta) \quad . \tag{17.47}$$

For $|x - y| < \varepsilon - \delta$ the inclusion $Z(y,\delta) \subset Z(x,\varepsilon)$ holds, hence

$$\frac{d\xi(t)}{dt} \in R[\xi(\sigma_1);\tfrac{1}{2}(\varepsilon + \delta)] \qquad (\sigma_1 \leq t \leq \sigma_2)$$

and, furthermore,

$$\frac{\xi(\sigma_2) - \xi(\sigma_1)}{\sigma_2 - \sigma_1} \in R[\xi(\sigma_1);\tfrac{1}{2}(\varepsilon + \delta)] =$$

$$= \text{co } A[-\xi(\sigma_1),\tfrac{1}{2}(\varepsilon + \delta)] - \text{co } A[\xi(\sigma_1),\tfrac{1}{2}(\varepsilon + \delta)] \quad . \tag{17.48}$$

Every point of the convex hull of any set in R^N is (by the Carathéodory theorem) a convex combination of at most $N + 1$ elements of that set. Therefore, in view of (17.48), the equality

$$\frac{\xi(\sigma_2) - \xi(\sigma_1)}{\sigma_2 - \sigma_1} = \sum_{i=1}^{N+1} \alpha_i\eta_i + \sum_{i=1}^{N+1} \beta_i\zeta_i \tag{17.49}$$

follows, where

$$\alpha_i,\ \beta_i \in [0,1],\quad \alpha_1 + \ldots + \alpha_{N+1} = 1,\quad \beta_1 + \ldots + \beta_{N+1} = 1$$

and

$$\eta_i \in A[-\xi(\sigma_1),\tfrac{1}{2}(\varepsilon + \delta)],\ \zeta_i \in -A[\xi(\sigma_1),\tfrac{1}{2}(\varepsilon + \delta)] \ . \tag{17.50}$$

Split the interval $[\sigma_1,\sigma_2]$ into $2N + 2$ parts $[\tau_{i-1},\tau_i]$ so that

$$\tau_i - \tau_{i-1} = \begin{cases} \frac{1}{2}(\sigma_2 - \sigma_1)\alpha_i\ , & \text{for} \quad i = 1, 2, \ldots, N + 1 \\ \frac{1}{2}(\sigma_2 - \sigma_1)\beta_{i-N-1}, & \text{for} \quad i = N+2, N+3, \ldots, 2N+2 \ . \end{cases}$$

By

$$\xi_1(t) = F(\sigma_1,\sigma_2)\xi(t) \quad (\sigma_1 \le t \le \sigma_2)$$

denote a continuous, piecewise linear function whose derivative in the interval (τ_{i-1},τ_i) equals η_i at $i = 1,2, \ldots ,N+1$ and coincides with ξ_{i-N-1} at $i = N+2,N+3, \ldots , 2N+2$. Moreover, let us assume that function $\xi_1(t)$ satisfies the condition $\xi_1(\sigma_1) = \xi(\sigma_1)$. Then, by virtue of (17.49), the equality $\xi_1(\sigma_2) = \xi(\sigma_2)$ holds, too. Furthermore, due to (17.50) it follows that

$$\eta_i \in A[-\xi_1(t),\varepsilon]\ ,\quad \zeta_i \in -A[\xi_1(t),\varepsilon] \quad (\sigma_1 < t < \sigma_2) \ .$$

Now pass to the interval $[t_o,t_1]$. Consider some decomposition $t_o = \sigma_o < \sigma_1 < \ldots < \sigma_s = t_1$ into parts such that the length of each $[\sigma_{i-1},\sigma_i]$ does not exceed $\frac{1}{4}(\varepsilon - \delta)$. Take

$$x(t) = F(\sigma_{i-1},\sigma_i)\xi(t) \quad \text{for} \quad t \in [\sigma_{i-1},\sigma_i] \;,$$

where $\xi(t)$ is a solution of the equation (17.47), satisfying the initial condition $\xi(t_o) = 0$. By construction,

$$x(t) \in B_2(\varepsilon;Z) \quad \text{and} \quad x(t_1) = \xi(t_1) \;.$$

∎

17.7. Lower bound for vibro-correctness moduli

As already shown, the vibro-correctness modulus is non-unique. This leads to a natural question of how "accurate" are the vibro-correctness moduli of the play and stop with a strictly convex characteristic introduced in Theorem 17.1. In particular, how accurate are the estimates (17.18) for characteristics which have the form of a ball or ellipsoid? Do the operators $L[t_o,x_o;Z]$ and $U[t_o,x_o;Z]$ satisfy a standard Lipschitz condition in the case of such characteristics?

Assume that the characteristic Z is strictly convex. Define

$$\kappa(x;r) = \inf_{w\in K(x;z),\, w\neq 0} \; \sup_{(y,w)=0;\,|y|=r} \rho(x+y,Z) \quad (x \in \partial Z) \tag{17.51}$$

and

$$\kappa(r;Z) = \inf_{x\in \partial Z} \kappa(x;r) \quad (r \geq 0) \;. \tag{17.52}$$

The function (17.52) is strictly increasing, continuous and $\kappa(0;Z) = 0$. Therefore, the inverse function

$$\beta(\varepsilon;Z) = \kappa^{-1}(\varepsilon;Z) \quad (\varepsilon \geq 0) \tag{17.53}$$

is also strictly increasing, continuous and $\beta(0;Z) = 0$.

In the case of the ball Z_o with radius ρ_o, the function (17.52) coincides with (17.17) and the function (17.53) with the first of functions (17.18).

Now consider an arbitrary transducer W, with a domain of feasible states $\Omega(W) \subset R^{2N}$ and a certain class of admissible inputs $u(t)$ $(t \geq t_o)$ with values in R^N. Define a strictly increasing, continuous function

$\mu(\varepsilon)$ $(\varepsilon \geq 0,\ \mu(0) = 0)$, such that for all $\varepsilon > \varepsilon_1 > 0$ there exist $x_0, y_0 \in R^N$ and admissible inputs $u(t)$, $v(t)$ $(t \geq t_0)$ satisfying the conditions

$$\{u(t_0), x_0\}\ ,\ \{v(t_0), y_0\} \in \Omega(W),\quad \|u(t) - v(t)\|_{t_0, t_1} \leq \varepsilon$$

with

$$|W[t_0, x_0]u(t_1) - W[t_0, y_0]v(t_1)| > \mu(\varepsilon - \varepsilon_1)\ . \qquad (17.54)$$

If the transducer W is vibro-correct and has vibro-correctness moduli, then the function $\mu(\varepsilon)$ represents a lower bound for all those moduli. Thus, also in the general situation (without assumed vibro-correctness of the transducer W) we shall call the function $\mu(\varepsilon)$ $(\varepsilon \geq 0)$ a *lower bound of vibro-correctness moduli.*

Theorem 17.5. Assume that a characteristic Z is strictly convex. Then the function (17.53) is a lower bound for the vibro-correctness moduli of the play $L(Z)$.

Proof. To prove this theorem, one can use Theorem 17.4. Complementary constructions are left to the reader. ∎

Let us extend the study of the strictly convex characteristic Z . Assume that $u_0 \in \operatorname{int} Z$ and x_0 is the closest to u_0 point of the boundary ∂Z of the set Z . Then the ball $T = \{x: |x - u_0| \leq |x_0 - u_0|\}$ is contained in Z . Therefore, $K(x_0; Z)$ coincides with the ray $\alpha(x_0 - u_0)$ $(\alpha \geq 0)$ and (cf., (17.51))

$$\sup_{(y, x_0 - u_0) = 0,\ |y| = r} \rho(x_0 + y, Z) \leq$$

$$\leq \sup_{(y, x_0 - u_0) = 0,\ |y| = r} \rho(x_0 + y, T) = \sqrt{|x_0 - u_0|^2 + r^2} - |x_0 - u_0|\ .$$

Hence, as an addition to Theorem 17.5 it follows that the estimate

$$\beta(\varepsilon; Z) \geq \sqrt{2\varepsilon\ |x_0 - u_0| + \varepsilon^2} \qquad (17.55)$$

holds for the function (17.53).

By (17.55), one can easily conclude the following important (and unexpected) property: *for any strictly convex characteristic* Z *every vibro-correctness modulus of the play* $L(Z)$ *is at least of order* $\sqrt{\varepsilon}$ *for small* ε. This implies that in the case of any strictly convex characteristic Z the operators $L[t_0,x_0;Z]$ can satisfy neither the Lipschitz condition nor any Hölder condition with an index greater than $1/2$! All these observations should be once more thought over after Chapter 18 has been read.

For each unit vector $a \in R^N$ choose now points $x_+(a), x_-(a) \in \partial Z$ in which the supporting hyperplanes $\Pi_+(a)$ and $\Pi_-(a)$ are orthogonal to a. Define

$$\kappa_1(a;r) = \sup_{(y,a)=0,\,|y|=r} \min\{\rho[y + x_+(a),Z],\ \rho[y + x_-(a),Z]\} \qquad (17.56)$$

and

$$\kappa_1(r;Z) = \inf_{a\in R^N,\,|a|=1} \kappa_1(a;r) \ . \qquad (17.57)$$

By $\beta_1(\varepsilon;Z)$ denote the inverse function to $\kappa_1(r;Z)$. Clearly,

$$\beta_1(\varepsilon;Z) \geq \beta(\varepsilon;Z) \qquad (\varepsilon \geq 0) \ . \qquad (17.58)$$

The function $\beta_1(\varepsilon;Z)$ (as well as the function $\beta(\varepsilon;Z)$) is a lower bound for the vibro-correctness moduli of the play $L(Z)$ with the strictly convex characteristic Z. In view of (17.58), this bound is not weaker than the estimate given by Theorem 17.5.

In conclusion let us consider the characteristics $Z \subset R^N$ which are symmetric with respect to some subspace Π. For such characteristics, a lower bound for the vibro-correctness modulus of the N-dimensional play $L(Z)$ will be given by every lower bound for the vibro-correctness modulus of the N_1-dimensional play $L(Z \cap \Pi)$ where $N_1 = \dim \Pi$.

18. Polyhedral characteristics

18.1. Basic theorems

In this chapter, we study properties of the play $L(Z)$ and the stop

$U(Z)$ whose characteristic $Z \subset R^N$ is a polyhedron. Properties of such plays and stops are in many respects "better" than analogous properties of the play and stop with strictly convex characteristic .

Lemma 18.1. Assume that Z is a polyhedral characteristic. Then the operators $L[t_0,x_0;Z]$ satisfy Lipschitz condition with some coefficient $\gamma(Z)$ on the set of piecewise smooth inputs. If $u(t_0)-x_0 \in Z$ and $v(t_0) - x_0 \in Z$, then

$$|L[t_0,x_0;Z]u(t) - L[t_0,x_0;Z]v(t)| \leq$$

$$\leq \gamma(Z) \|u(s) - v(s)\|_{t_0,t_1} \quad (t \geq t_0) \quad . \tag{18.1}$$

This lemma will be proved in Section 18.3. It permits to extend the operators $L[t_0,x_0;Z]$ onto arbitrary continuous inputs by using standard constructions. Assume that an input $u(t)$ $(t \geq t_0)$ is continuous, then construct a sequence of smooth inputs $u_n(t)$ $(t \geq t_0)$ which converge to $u(t)$ uniformly on each finite interval and satisfy the condition $u_n(t_0) = u(t_0)$ $(n = 1, 2, \ldots)$. If $u(t_0) - x_0 \in Z$, then the functions

$$x_n(t) = L[t_0,x_0;Z]u_n(t) \quad (t \geq t_0;\ n = 1, 2, \ldots) \tag{18.2}$$

are well-defined and, by virtue of Lemma 18.2, the sequence (18.2) is uniformly on each finite interval convergent to a certain function $x(t)$. It remains to put

$$x(t) = L[t_0,x_0;Z]u(t) \quad (t \geq t_0) \quad . \tag{18.3}$$

The above definition of the outputs is obviously correct. As a consequence of Lemma 18.1, we can conclude the following property.

Theorem 18.1. Let the characteristic Z have the form of a polyhedron. Then the operators $L[t_0,x_0;Z]$ of a play satisfy the Lipschitz condition with a certain coefficient $\gamma(Z)$ on the set of continuous inputs. If continuous inputs $u(t)$, $v(t)$ $(t \geq t_0)$ satisfy the condition $u(t_0) - x_0$, $v(t_0) - x_0 \in Z$, then the estimate (18.1) holds.

∎

By definitions of the play and the stop (cf., Chapter 16), it follows for any convex characteristic Z that if $x_0 \in Z$ and $u(t_0) - \xi_0 = x_0$, then for each piecewise smooth input $u(t)$ $(t \geq t_0)$ the identity

$$L[t_0,\xi_0;Z]u(t) + U[t_0,x_0;Z]u(t) \equiv u(t) \quad (t \geq t_0) \tag{18.4}$$

is satisfied. Therefore, under the hypotheses of Lemma 18.1, operators $U[t_0,x_0;Z]$ are also Lipschitz continuous (in particular, with Lipschitz constant $1 + \gamma(Z)$). Consequently, the operators $U[t_0,x_0;Z]$ with polyhedral characteristics can be extended onto arbitrary continuous inputs. Identities (18.4) remain true upon such an extension. As a consequence of Theorem 18.1, the following result holds.

<u>Theorem 18.2.</u> Assume that Z is a polyhedral characteristic. Then the operators $U[t_0,x_0;Z]$ of a stop are Lipschitz continuous on the set of continuous inputs.

∎

Obviously, upon extension to the set of all continuous inputs, plays and stops with polyhedral characteristics remain vibro-correct transducers which are still static, deterministic and controllable. For the play, a vibro-correctness modulus is given by the function $\lambda(\varepsilon;Z) = \gamma(Z)\varepsilon$; for the stop it can be taken as $\lambda_1(\varepsilon;Z) = \varepsilon + \gamma(Z)\varepsilon$. In case of polyhedral characteristics, neither plays nor stops are strongly convergent hysterons.

Questions concerning the stability of plays and stops with polyhedral characteristics subject to perturbations of the characteristic (within the class of polyhedrons) were hardly studied. For a polyhedron $Z \subset R^N$ with arbitrarily located (N-1)-dimensional edges, such correctness has been established only in the case of small perturbations with preserved number of vertices. For $N = 2$, the last requirement is equivalent to the lack of parallel sides in a polygon.

Further, up to the end of this chapter we shall restrict our considerations to plays.

18.2. <u>Estimates of the Lipschitz constant</u>

An optimal value of the Lipschitz constant $\gamma(Z)$ in (18.1) is known only at $N = 2$. The general estimate we establish below is nearly optimal.

First let us consider the case $N = 2$. In that case, Z is a polygon with edges $a_1, \ldots, a_M$. By $\ell_1, \ldots, \ell_M$ we denote the straight lines which pass through the origin and are parallel to $a_1, \ldots, a_M$, respectively. Some of those lines may coincide (if some of the edges of the polygon Z are parallel). By $\varphi(Z)$ we denote the least non-zero angle between two lines ℓ_i and ℓ_j. We put

$$\gamma(Z) = \frac{1}{\sin \frac{1}{2} \varphi(Z)} . \tag{18.5}$$

Theorem 18.3. Let $\Omega \subset R^2$ be a polygon. Then (18.5) is the optimal Lipschitz constant in condition (18.1).

A proof of this theorem is given in Section 18.4. Theorem 18.3 yields a number of unexpected consequences. Consider, for example, a play $L(Z)$ with the characteristic Z in the form of a regular triangle ABC. Let D be the center of segment AC and D_1 denote a point close to D located outside Z. A quadrilateral characteristic Z_1 with vertices $ABCD_1$ is the closer to Z the smaller is distance between D_1 and D. At the same time, on account of Theorem 18.3, the optimal constant $\gamma(Z_1)$ in the Lipschitz condition for the play $L(Z_1)$ grows to infinity as $D_1 \to D$, although $\gamma(Z) = 2/\sqrt{3}$.

At the passage to an arbitrary N, we shall use functions $\beta_m(x_1, \ldots, x_m)$ $(m = 1, \ldots, N)$ defined in an inductive way. Set $\beta_1(x) = 1$ for $x \in R^N \setminus \{0\}$. If the function $\beta_{m-1}(x_1, \ldots, x_{m-1})$ has been introduced, we define the function $\beta_m(x_1, \ldots, x_m)$ on ordered sets of linearly independent $x_1, \ldots, x_m \in R^N$ by the equality

$$\beta_m(x_1, \ldots, x_m) =$$

$$= \sup\{|y| : y \in E, |(y, x_m)| \le 1, \rho(y, \ell) \le \beta_{m-1}(x_1, \ldots, x_{m-1})\}, \tag{18.6}$$

where E is the linear span of the vectors $x_1, \ldots, x_m$, and ℓ is the straight line contained in E and orthogonal to the vectors $x_1, \ldots, x_{m-1}$.

Theorem 18.4. Assume that $Z \subset R^N$ is a convex M-hedron with non-empty interior. Let $M \ge N - 1$ and $b_1, \ldots, b_M$ denote unit outward vector normal to (N-1)-dimensional edges of Z. Then the play $L(Z)$

is Lipschitz continuous and satisfies (18.1) with the constant

$$\gamma(Z) = \max \beta_N(b_{i_1}, \ldots, b_{i_N}) \quad , \tag{18.7}$$

where the maximum is taken over all ordered sets of linearly independent normal vectors $b_{i_1},\ldots,b_{i_N}$ $(1 \leq i_1, \ldots, i_N \leq M)$.

18.3. Proofs of Lemma 18.1 and Theorem 18.4

In order to prove Lemma 18.1 and Theorem 18.4, it is sufficient to consider the outputs

$$\xi(t) = L[t_o,\xi_o;Z]u(t), \quad \eta(t) = L[t_o,\xi_o;Z]v(t) \quad (t \geq t_o) \tag{18.8}$$

which correspond to continuously differentiable inputs $u(t)$, $v(t)$ $(t \geq t_o)$ such that

$$u(t_o) - \xi_o, \; v(t_o) - \xi_o \in Z; \quad |u(t) - v(t)| \leq \varepsilon \quad (t \geq t_o) \tag{18.9}$$

(ε is an arbitrarily fixed number) and then to prove the inequality

$$|\xi(t) - \eta(t)| \leq \gamma(Z)\varepsilon \quad (t \geq t_o) \quad , \tag{18.10}$$

where $\gamma(Z)$ is given by (18.7).

Take a number κ and set

$$E_i(\kappa) = \{y: y \in R^N, (y,b_i) = \kappa\} \quad (i = 1, \ldots, M) \quad .$$

By $P_i(\kappa)x$ $(x \in R^N)$ denote the point of the plane $E_i(\kappa)$ which is closest to x . Further, $A(\kappa)$ will stay for the set of all those operators A which admit the representation

$$Ax = (1 - \alpha)x + \alpha P_i(\nu)x \quad ,$$

where $\alpha \in [0,1]$; $i = 1, \ldots, M$; $\nu = \pm\kappa$. We shall call the finite set $\{z_o, z_1, \ldots, z_s\}$ κ-*proper* , if $z_j = A_j z_{j-1}$ $(j = 1, \ldots, s)$ with $A_j \in A(\kappa)$.

Lemma 18.2. For each $T \geq t_o$ and $\delta > 0$ there exists an $(\varepsilon + \delta)$--proper sequence $z_o, z_1, \ldots, z_s$ such that $z_o = 0$ and $z_s = \xi(T) - \eta(T)$.

Proof. Take

$$\zeta_i^+(x;\varepsilon) = \begin{cases} b_i & , \text{ if } (x,b_i) \leq \varepsilon , \\ 0 & , \text{ if } (x,b_i) > \varepsilon , \end{cases} \tag{18.11}$$

and

$$\xi_i^-(x;\varepsilon) = \begin{cases} 0 & , \text{ if } (x,b_i) < -\varepsilon, \\ -b_i & , \text{ if } (x,b_i) \geq -\varepsilon. \end{cases} \tag{18.12}$$

For proving the lemma, it is sufficient to take advantage of the fact that the set $R(x,\varepsilon)$, defined by equality (17.37), is contained in the intersection of the ball $|y| \leq 2$ with the set of all linear combinations of vectors $\zeta_i^+(x;\varepsilon)$, $\zeta_i^-(x;\varepsilon)$ $(i = 1, \ldots, M)$ with non--negative coefficients.

∎

By virtue of Lemma 18.2, for proving inequality (18.10) it is sufficient to show the estimate

$$|z_s| \leq \gamma(Z)\kappa \tag{18.13}$$

for every κ-proper sequence $z_o = 0, z_1, \ldots, z_s$.

We shall say that a κ-proper sequence $z_o^*, z_1^*, \ldots, z_s^*$ is *extremal*, if for any κ-proper sequence $z_o^*, z_1, \ldots, z_s$ with the same first element and the same number of all elements the estimate $|z_s| \leq |z_s^*|$ holds, whereas for every κ-proper sequence $z_o^*, z_1, \ldots, z_r$, where $r < s$, the strong inequality $|z_r| < |z_s^*|$ takes place. By a compactness argument it follows that the sequence $z_o = 0, z_1, \ldots, z_s$ in the proof of (18.13) can be considered as extremal.

Assume that $\bar{z}_o, \bar{z}_1, \ldots, \bar{z}_s$ is a certain fixed sequence which is extremal and κ-proper. Let

$$\bar{z}_j = (1 - \alpha_j)\bar{z}_{j-1} + \alpha_j P_{i(j)}(\nu_j)\bar{z}_{j-1} , \tag{18.14}$$

where $\alpha_j \in [0,1]$, $|\nu_j| = \kappa$. Take $c_1 = b_{i(s)}$, then denote by c_2 the first vector in the sequence $b_{i(s)}, b_{i(s-1)}, \ldots, b_{i(1)}$ which is non-parallel to c_1 , etc. After the vectors $c_1, \ldots, c_k$ have been constructed, by c_{k+1} denote the first vector of the same sequence $b_{i(s)}, b_{i(s-1)}, \ldots, b_{i(1)}$ which does not belong to the linear span of vectors $c_1, \ldots, c_k$. Extending this process, we construct a set of linearly independent vectors $c_1, \ldots, c_d$ where d is the dimension of the linear span of vectors $b_{i(s)}, b_{i(s-1)}, \ldots, b_{i(1)}$. The estimate (18.13) for extremal κ-proper sequences $z_0 = 0, z_1, \ldots, z_s$ (as well as Lemma 18.1 and Theorem 18.4) will be proved when the following statement is established.

Lemma 18.3. If $s > 0$, then the estimate

$$|Q\bar{z}_s| \leq \beta_d(c_1, \ldots, c_d)\kappa \tag{18.15}$$

holds for the projection $Q\bar{z}_s$ of vector $\bar{z}_s$ onto the linear span Π of vectors $c_1, \ldots, c_d$.

Proof. Take $f_0(x) = |x|^2$ and define

$$f_j(x) = \sup_{A \in A} |f_{j-1}(Ax)| \quad (x \in R^N;\ j = 1, 2, \ldots) .$$

Each of the above-constructed functions is convex. Consequently, the function $f_{s-j}(x)$, considered on the segment which connects the points $\bar{z}_{j-1}$ and $P_{i(j)}(\nu_j)\bar{z}_{j-1}$, assumes its maximal value in one of these points. Since the sequence $\bar{z}_0, \bar{z}_1, \ldots, \bar{z}_s$ is extremal, its maximal value is achieved in the point

$$\bar{z}_j = P_{i(j)}(\nu_j)\bar{z}_{j-1} .$$

This means, all the coefficients α_j in the formula (18.14) are just equal to 1 .

At $N = 1$, the assertion of Lemma 18.3 is evident. Suppose it is true for all $N_1 < N$. Then the estimate (18.15) holds if $d < N$. Let us now consider the case $d = N$.

Let $c_N = b_{i(j_0)}$. Then, according to this assumption, the estimate

$$|Q_1\bar{z}_s| \le \beta_{N-1}(c_1, \ldots, c_{N-1})\kappa \tag{18.16}$$

is satisfied, where Q_1 represents the operator of projection onto the linear span Π_1 of vectors $c_1, \ldots, c_{N-1}$.

Further, as it has been already shown,

$$\bar{z}_{j_o} = P_{i(j_o)}(\nu_{j_o})\bar{z}_{j_o-1} ,$$

hence

$$|(\bar{z}_{j_o}, c_N)| = \kappa$$

and the estimate (18.15) follows as a consequence of (18.16). ∎

18.4. Proof of Theorem 18.3

The Lipschitz condition (18.1) with Lipschitz constant (18.5) can be concluded from Theorem 18.4. To this end, it is sufficient to rewrite (18.6) in the form

$$\beta_m(x_1, \ldots, x_m) = \Big\{[\beta_{m-1}(x_1, \ldots, x_{m-1})]^2 + \\ + \Big[\frac{1}{\sin \mu(x_1,\ldots,x_m)} + \beta_{m-1}(x_1, \ldots, x_{m-1}) \operatorname{ctg} \mu(x_1, \ldots, x_m)\Big]^2\Big\}^{1/2} ,$$

where $\mu(x_1, \ldots, x_m)$ denotes the angle between the vector x_m and the linear span $\Pi(x_1, \ldots, x_{m-1})$ of vectors $x_1, \ldots, x_{m-1}$. We give here a simple proof for this.

By Theorem 17.4, the Lipschitz condition (18.1) with Lipschitz constant (18.5) will be shown if the estimate

$$|x(t_1)| \le \frac{\varepsilon}{\sin \frac{1}{2} \varphi(Z)} = \varepsilon_1 \tag{18.17}$$

is established for all solutions $x(t)$ $(t_o \le t \le t_1)$ of the differential inclusion (17.38), such that $x(t_o) = 0$.

By $R_1(x;\varepsilon)$ we denote the set of all linear combinations of the vectors $\zeta_i^+(x;\varepsilon)$, $\zeta_i^-(x;\varepsilon)$ $(i = 1, \ldots, M)$ with nonnegative coefficients, which are situated in the circle $|x| \leq 2$ (cf., (18.11) and (18.12)). Since $R(x;\varepsilon) \subset R_1(x;\varepsilon)$, for proving the Lipschitz condition we need to establish the estimate (18.17) for the solutions of the differential inclusion

$$\frac{dx}{dt} \in R_1(x;\varepsilon) \tag{18.18}$$

which satisfy the initial condition $x(t_o) = 0$. To derive this estimate, we can use the Lyapunov functions. Define

$$V(x) = \max\{V_o(x), V_1(x), \ldots, V_M(x)\} ,$$

where $V_o(x) = |x|^2$ and

$$V_i(x) = |x|^2 - (x,b_i)^2 + \varepsilon^2 \quad (i = 1, \ldots, M) .$$

The function $V(x)$ is convex. According to (18.18), its right derivative is non-positive in the domain $|x| > \varepsilon$. Since $V(x) \geq |x|^2$, the condition $x(t_o) = 0$ implies that $|x(t)| \geq \varepsilon_1$ for all $t \geq t_o$. Therefore, the estimate (18.17) has been proved.

It remains to prove that (18.5) is the optimal constant in Lipschitz condition (18.1). To this purpose, by virtue of Theorem 17.4, it suffices to take any $\varepsilon > 0$ and construct a solution $x(t)$ $(t_o \leq t \leq t_1)$ of equation (17.38), such that $x(t_o) = 0$ and

$$|x(t_1)| = \frac{\varepsilon}{\sin \frac{1}{2} \varphi(Z)} = \varepsilon_1 . \tag{18.19}$$

Denote by $R_2(x;\varepsilon)$ the convex hull of the vectors

$$\eta_i^+(x;\varepsilon) = \begin{cases} b_i , & \text{if} \quad (x,b_i) \leq \varepsilon \quad \text{and} \quad |x| \leq a , \\ 0 , & \text{otherwise} , \end{cases} \tag{18.20}$$

$$\eta_i^-(x;\varepsilon) = \begin{cases} -b_i , & \text{if} \quad (x,b_i) \geq -\varepsilon \quad \text{and} \quad |x| \leq a , \\ 0 , & \text{otherwise} \end{cases} \tag{18.21}$$

(cf., (18.11) and (18.12)), where the constant a represents a positive lower bound for the lengths of the sides of polygon Z. For sufficiently small positive ε, the inclusions $R_2(x;\varepsilon) \subset R(x;\varepsilon)$ take place. Thus it is then enough to find a solution $x(t)$ $(t_0 \leq t \leq t_1)$ of the equation

$$\frac{dx}{dt} \in R_2(x;\varepsilon) \quad , \tag{18.22}$$

such that $x(t_0) = 0$ and the equality (18.18) holds.

Let $\varphi(Z)$ be the angle between the straight lines ℓ_{i_1} and ℓ_{i_2}. Denote by x_1 such a point in the bisectrix of this angle that its distance from the origin equals ε_1. According to the definitions (18.20) and (18.21) it follows that for all $\alpha \in [0,1]$

$$\eta^+_{i_1}(\alpha x_1;\varepsilon) = b_{i_1}, \quad \eta^-_{i_1}(\alpha x_1;\varepsilon) = -b_{i_1} \quad ,$$

$$\eta^+_{i_2}(\alpha x_1;\varepsilon) = b_{i_2}, \quad \eta^-_{i_2}(\alpha x_1;\varepsilon) = -b_{i_2} \quad ,$$

and therefore each of the vectors

$$\tfrac{1}{2}(b_{i_1} + b_{i_2}), \quad \tfrac{1}{2}(b_{i_1} - b_{i_2}), \quad \tfrac{1}{2}(-b_{i_1} + b_{i_2}), \quad \tfrac{1}{2}(-b_{i_1} - b_{i_2})$$

belongs to all the sets $R_2(\alpha x_1;\varepsilon)$ at $0 \leq \alpha \leq 1$. However, one of the above four vectors is parallel to the vector x_1. Thus the solution $x(t)$ for which the equality (18.19) is fulfilled at some $t = t_1$ can be defined as $x(t) = c(t - t_0)x_1$ (here c denotes some constant). The proof of Theorem 18.3 has been completed. ∎

18.5. Remarks

a. As already mentioned, the authors were unable to find an optimal constant in the Lipschitz condition (18.1) if $N \geq 3$. Nevertheless, not only the upper bound (18.7) but also various lower bounds for that constant are available.

Let $\mathcal{B}$ denote the set of nonsingular square matrices of order N, such that their rows coincide with the components of one of the vectors $b_1, \ldots, b_M$. The coefficient in the Lipschitz condition (18.1) is bounded from below by

$$\gamma_1(Z) = \max_{B \in \mathcal{B}} \|B^{-1}\| \quad , \tag{18.23}$$

where $\|B^{-1}\|$ denotes the norm of matrix B^{-1} treated as a linear operator from the space R^N, with the norm defined as maximum of the absolute values of components, into the space R^N equipped with the Euclidean norm.

<u>b.</u> An input $u(t)$ $(t \geq t_0)$ with values in R^N will be called *absorbing* if the relations

$$\overline{\lim_{t\to\infty}}\, (u(t),y) = \infty \;, \qquad \underline{\lim_{t\to\infty}}\, (u(t),y) = -\infty \tag{18.24}$$

are satisfied for every nonzero $y \in R^N$.

As already mentioned, the play $L(Z)$ with a polyhedral characteristic is not strongly convergent. Nevertheless, *if a continuous input* $u(t)$ $(t \geq t_0)$ *is absorbing and*

$$u(t_0) - x_0, \quad u(t_0) - y_0 \in Z \quad ,$$

then

$$\lim_{t\to\infty} |L[t_0,x_0;Z]u(t) - L[t_0,y_0;Z]u(t)| = 0 \quad .$$

19. Arbitrary convex characteristics

19.1. Vibro-correctness of play and stop

In Chapter 17 we studied properties of plays and stops with strictly convex characteristics. In Chapter 18, polyhedral characteristics were taken into consideration. In both cases the uniform continuity of plays and stops viewed as operators in the space of continuous vector--functions has been shown. Estimates for the vibro-correctness moduli of those operators have been derived and some other properties established. A passage to arbitrary convex characteristics has turned out complicated although also in that case some facts are evident.

Let $N = 2$. Then the operators of a play and stop, considered on piecewise smooth inputs, are uniformly continuous (in norm of the space C). It would be of interest to establish various estimates for the vibro-correctness modulus in terms of geometric properties of the suitable characteristic.

For $N \geq 3$, as the examples constructed by A. F. Klepcyn [45, 46] show, the operators corresponding to a play and a stop may be no longer uniformly continuous. Nevertheless, those operators admit extension by continuity from the set of smooth or piecewise smooth inputs onto arbitrary inputs. This follows by Theorem 19.1 whose formulation and proof are given in the present chapter.

Consider the space $C = C(t_0,t_1)$ of continuous vector-functions with values in R^N , defined on a fixed interval $[t_0,t_1]$. If $u_0 = u_0(t) \in C$ and $\rho > 0$, by $B[u_0,\rho]$ we denote the set of piecewise smooth functions $v(t) \in C$ such that $\|v(t) - u_0(t)\|_{t_0,t_1} \leq \rho$.

Theorem 19.1. Let $Z \subset R^N$ be an arbitrary bounded, convex and closed characteristic with interior points. Then there exists such a number $\rho = \rho(Z) > 0$ that the operators of the play $L(Z)$ and the stop $U(Z)$ satisfy on each set $B[u_0,\rho(Z)]$ the Hölder condition with exponent $1/2$, i.e., for each fixed continuous input $u_0 = u_0(t) \in C$, from

$$v_1(t),\ v_2(t) \in B[u_0,\rho(Z)];\quad v_1(t_0) - \xi_0,\ v_2(t_0) - \xi_0 \in Z \tag{19.1}$$

it follows that the estimate

$$\|L[t_0,\xi_0;Z]v_1(t) - L[t_0,\xi_0;Z]v_2(t)\|_{t_0,t_1} \leq$$
$$\leq \gamma(u_0,\rho,Z)\sqrt{\|v_1(t) - v_2(t)\|_{t_0,t_1}} \tag{19.2}$$

holds . Moreover, if

$$v_1(t),\ v_2(t) \in B[u_0,\rho(Z)],\quad x_0 \in Z \tag{19.3}$$

then

$$\|U[t_0,x_0;Z]v_1(t) - U[t_0,x_0;Z]v_2(t)\|_{t_0,t_1} \leq$$
$$\leq \gamma_1(u_0,\rho,Z)\sqrt{\|v_1(t) - v_2(t)\|_{t_0,t_1}}\ . \tag{19.4}$$

By virtue of (16.12), it is sufficient to prove the theorem in the case of a play. A proof for that case will be given in the following two sections.

19.2. Estimate for the variation of output

The following statement will be useful in the sequel.

Lemma 19.1. Let $x^* \in \text{int } Z$, the ball $|x - x^*| \leq m$ $(m > 0)$ be contained in the characteristic Z which, in turn, is contained in the ball $|x - x^*| \leq M$. Assume that the input $v(t)$ is piecewise smooth, $v(\tau_0) - \xi_0 \in Z$ and

$$|v(t) - v(\tau_0)| \leq \varepsilon < m \qquad (\tau_0 \leq t \leq \tau_1) .$$

Then the estimate

$$\int_{\tau_0}^{\tau_1} |\dot{\xi}(t)|dt \leq \frac{M^2}{2(m-\varepsilon)} \tag{19.5}$$

holds with

$$\xi(t) = L[\tau_0,\xi_0;Z]v(t) \qquad (\tau_0 \leq t \leq \tau_1) . \tag{19.6}$$

Proof. In view of (16.7) and (16.19), for almost all $t \in [\tau_0,\tau_1]$ and all $z \in Z$ the inequality

$$(\dot{\xi}(t), v(t) - \xi(t) - z) \geq 0 \tag{19.7}$$

is satisfied. In particular, for almost all $t \in [\tau_0,\tau_1]$ the estimate

$$(\dot{\xi}(t), v(t) - \xi(t) - x^* - y) \geq 0 \qquad (y \in R^N, \ |y| \leq m)$$

holds. It is convenient to rewrite the latter estimate in the form

$$(\dot{\xi}(t),\xi(t) - v(\tau_0) + x^*) \leq (\dot{\xi}(t),v(t) - v(\tau_0)) - (\dot{\xi}(t),y)$$
$$(y \in R^N, \ |x| \leq m) .$$

Therefore,

$$(\dot{\xi}(t),\xi(t) - v(\tau_o) + x^*) \leq -(m - \varepsilon)|\dot{\xi}(t)| \quad ,$$

i.e., for almost all $t \in [\tau_o,\tau_1]$

$$2(m - \varepsilon)|\dot{\xi}(t)| \leq - \frac{d}{dt} |\xi(t) - v(\tau_o) + x^*|^2. \tag{19.8}$$

Consequently,

$$2(m - \xi) \int_{\tau_o}^{\tau_1} |\dot{\xi}(t)|dt \leq |\xi(\tau_o) - v(\tau_o) + x^*|^2 \leq M^2$$

and this implies (19.5).

∎

We can easily obtain various refinements and modifications of the bound (19.5).

By using arguments similar to those which have yielded (19.8), for almost all $t \in [\tau_o,\tau_1]$ the estimate

$$\frac{d}{dt} |\xi(t) - v(\tau_o) + x^{**}|^2 \leq 0 \tag{19.9}$$

with

$$x^{**} = \frac{\varepsilon}{m} x^* + \frac{m - \varepsilon}{m} [v(\tau_o) - \xi_o]$$

can be shown. (19.9) implies (after integration) that

$$|\xi(\tau_1) - v(\tau_o) + x^{**}| \leq |\xi(\tau_o) - v(\tau_o) + x^{**}|$$

and, because of the equality

$$x^{**} - [v(\tau_o) - \xi_o] = \frac{\varepsilon}{m} \{x^* - [v(\tau_o) - \xi_o]\} \quad ,$$

that

$$|\xi(\tau_1) - v(\tau_o) + x^{**}| \leq \frac{M\varepsilon}{m} \quad .$$

Therefore,

$$|\xi(\tau_1) - \xi(\tau_0)| \le |\xi(\tau_1) - v(\tau_0) + x^{**}| + |\xi(\tau_0) - v(\tau_0) + x^{**}| \le \frac{2M\varepsilon}{m} ,$$

and, consequently,

$$|2x^* - 2v(\tau_0) + \xi(\tau_0) + \xi(\tau_1)| \le 2M + \frac{2M\varepsilon}{m} .$$

Write now the estimate which follows from (19.8) (after integration) in the form

$$2(m-\varepsilon) \int_{\tau_0}^{\tau_1} |\dot{\xi}(t)|dt \le (\xi(\tau_0) - \xi(\tau_1), 2x^* - 2v(\tau_0) + \xi(\tau_0) + \xi(\tau_1)) .$$

Hence

$$2(m - \varepsilon) \int_{\tau_0}^{\tau_1} |\dot{\xi}(t)|dt \le \frac{2M\varepsilon}{m}\left(2M + \frac{2M\varepsilon}{m}\right) ,$$

i.e., eventually

$$\int_{\tau_0}^{\tau_1} |\dot{\xi}(t)|dt \le \frac{2M^2(m+\varepsilon)\varepsilon}{m^2(m-\varepsilon)} . \tag{19.10}$$

The estimate (19.10) means that for small ε the variation of the output $\xi(t)$ on $[\tau_0,\tau_1]$ is also small. Estimates of the form (19.10) are convenient at an analysis of a play treated as operator from the space of continuous inputs into the space of outputs with bounded variation (and into other analogous spaces).

19.3. Proof of Theorem 19.1

Let x^*, m and M be the point and numbers as in the formulation of Lemma 19.1. Let $\rho = \rho(Z)$ be any fixed number from the interval $(0,m/3)$.

Consider a fixed input $u_0 = u_0(t) \in C$ and a point ξ_0 which satisfies the condition $u_0(t_0) - \xi_0 \in Z$. By k denote such a positive integer that $|u_0(s) - u_0(t)| < \rho$ for $|s - t| \leq 1/k$ and $s, t \in [t_0,t_1]$. Then divide the interval $[t_0,t_1]$ into k equal parts by the points

$$\tau_0 = t_0 < \tau_1 < \tau_2 < \dots < \tau_k = t_1 .$$

Take

$$\xi_1(t) = L[t_0,\xi_0;Z]v_1(t), \quad \xi_2(t) = L[t_0,\xi_0;Z]v_2(t) ,$$

where $v_1(t)$ and $v_2(t)$ are piecewise smooth inputs which satisfy the relations (19.1). Since

$$|v_1(t) - v_1(\tau_{j-1})| \leq 2\|v_1(s) - u_0(s)\|_{t_0,t_1} + $$
$$+ |u_0(t) - u_0(\tau_{j-1})| < 3\rho \quad (\tau_{j-1} \leq t \leq \tau_j)$$

and analogously

$$|v_2(t) - v_2(\tau_{j-1})| < 3\rho \quad (\tau_{j-1} \leq t \leq \tau_j) ,$$

by Lemma 19.1 for all $j = 1, \dots, k$ the estimates

$$\int_{\tau_{j-1}}^{\tau_j} |\dot{\xi}_1(t)|dt \leq \frac{M^2}{2(m-3\rho)} , \quad \int_{\tau_{j-1}}^{\tau_j} |\dot{\xi}_2(t)|dt \leq \frac{M^2}{2(m-3\rho)}$$

hold. Therefore,

$$\int_{t_0}^{t_1} |\dot{\xi}_1(t)|dt \leq \frac{kM^2}{2(m-3\rho)} , \quad \int_{t_0}^{t_1} |\dot{\xi}_2(t)|dt \leq \frac{kM^2}{2(m-3\rho)} . \qquad (19.11)$$

Let us again use the inequality (19.7), at first setting

$$v(t) = v_1(t), \quad \xi(t) = \xi_1(t), \quad z = v_2(t) - \xi_2(t)$$

and then

$$v(t) = v_2(t), \quad \xi(t) = \xi_2(t), \quad z = v_1(t) - \xi_1(t) \quad .$$

This yields the relations

$$(\dot{\xi}_1(t), v_1(t) - \xi_1(t) - v_2(t) + \xi_2(t)) \geq 0 \quad ,$$

$$(\dot{\xi}_2(t), v_2(t) - \xi_2(t) - v_1(t) + \xi_1(t)) \geq 0$$

which in turn imply the estimate

$$(\dot{\xi}_1(t) - \dot{\xi}_2(t), \xi_1(t) - \xi_2(t) + v_2(t) - v_1(t)) \leq 0 \quad .$$

Consequently, for almost all $t \in [t_o, t_1]$,

$$\frac{1}{2}\frac{d}{dt} |\xi_1(t) - \xi_2(t)|^2 \leq \|v_1(s) - v_2(s)\|_{t_o,t_1} |\dot{\xi}_1(t) - \dot{\xi}_2(t)| \leq$$

$$\leq \|v_1(s) - v_2(s)\|_{t_o,t_1} [|\dot{\xi}_1(t)| + |\dot{\xi}_2(t)|] \quad .$$

Hence (this time for all $t \in [t_o, t_1]$),

$$\frac{1}{2} |\xi_1(t) - \xi_2(t)|^2 \leq \|v_1(s) - v_2(s)\|_{t_o,t_1} \int_{t_o}^{t_1} [|\dot{\xi}_1(s)| + |\dot{\xi}_2(s)|]ds \quad ,$$

and the estimate (19.2) with

$$\gamma(u_o,\rho,Z) = M \sqrt{\frac{2k}{m - 3\rho}} \tag{19.12}$$

follows from (19.11). Theorem 19.1 has been completely proved. ∎

20. Inputs with summable derivatives

20.1. Statement of the problem

In this section, we consider again the play $L(Z)$ with a strictly convex characteristic $Z \subset R^N$. As it has been shown in Chapter 17, the operators $L[t_o,\xi_o;Z]$ of the play treated as defined on the space C of continuous functions (endowed with the uniform norm), do not satisfy the Lipschitz condition on the whole space. Moreover, they do not satisfy any Hölder condition with the index larger than $1/2$. In many cases, this feature complicates the study of closed-loop systems containing plays.

As it turns out, a transfer from the space of all continuous inputs to some smaller classes permits the treatment of the play as an operator which fulfils the standard Lipschitz condition in the suitable norms (different from the norm of the space C). This important property has been proved in the case where the boundary ∂Z of the characteristic Z is smooth. It indicates the importance of a suitable choice of the class of admissible inputs and the introduction of appropriately adjusted norms in the spaces of inputs and outputs. It would be interesting to develop an analysis of the properties of the play and stop operators with arbitrary convex characteristics in various function spaces.

Below we shall use the space $S = S(t_0,t_1)$ of absolutely continuous vector-functions $u(t)$ $(t_0 \leq t \leq t_1)$, equipped with the norm

$$\|u(t)\|_{S(t_0,t_1)} = |u(t_0)| + \int_{t_0}^{t_1} |u'(t)|dt \quad . \tag{20.1}$$

We formulate (without proofs) a few theorems on the properties of the play operators defined in the space S.

20.2. Lipschitz condition

Throughout this section, we assume that the boundary ∂Z of the characteristic Z is smooth (in each point of ∂Z there exists a unique external normal vector and it is continuously dependent upon the point). By virtue of Theorem 16.1 it follows that the operator $L[t_0,\xi_0;Z]$ acts from S into S; it is defined (cf., Chapter 16) on the inputs $u(t)$ $(t \geq t_0)$ for which $u(t_0) - \xi_0 \in Z$.

Theorem 20.1. For the play $L[t_0,\xi_0;Z]$, treated as an operator from S into S, the following Lipschitz condition holds on every ball in the space S: for

$$\|u(t)\|_{S(t_0,t_1)}, \ \|v(t)\|_{S(t_0,t_1)} \leq r; \quad u(t_0) - \xi_0, v(t_0) - \xi_0 \in Z \ , \tag{20.2}$$

the inequality

$$||L[t_o,\xi_o,Z]u(t) - L[t_o,\xi_o;Z]v(t)||_{S(t_o,t_1)} \leq$$

$$\leq \gamma(r)\ ||u(t) - v(t)||_{S(t_o,t_1)}$$

is satisfied.

The global Lipschitz condition in the whole space S does not hold in the case of a play (treated as an operator from S into S). Neither are the operators $L[t_o,\xi_o;Z]$ uniformly continuous on the whole S (as operators from S into S). Therefore, upon passing from the space C to the space S both for inputs and outputs, the uniform continuity on the whole space is lost but a local Lipschitz condition still holds as formulated in the following theorem.

Theorem 20.2. The play $L[t_o,\xi_o;Z]$, considered as an operator from the space $S = S(t_o,t_1)$ into the space $C = C(t_o,t_1)$, satisfies the following global Lipschitz condition: for all $u(t), v(t) \in S$ such that

$$u(t_o) - \xi_o,\ v(t_o) - \xi_o \in Z\ , \tag{20.4}$$

the inequality

$$||L[t_o,\xi_o;Z]u(t) - L[t_o,\xi_o;Z]v(t)||_{t_o,t_1} \leq$$

$$\leq \gamma_o||u(t) - v(t)||_{S(t_o,t_1)} \tag{20.5}$$

is fulfilled. ∎

It would be of interest to find sets in the space C on which the play operators are Lipschitz continuous (as operators from C into C). No satisfactory answer to that problem is known.

The following fact has been established. If $L[t_o,\xi_o;Z]$ is considered as an operator from C into C , then on each ball of the space S it satisfies the Hölder condition with exponent 2/3 , i.e., (20.2) implies the estimate

$$||L[t_o,\xi_o;Z]u(t) - L[t_o,\xi_o;Z]v(t)||_{t_o,t_1} \leq$$

$$\leq \gamma(r)\,\|u(t) - v(t)\|^{2/3}_{t_0,t_1} \quad . \tag{20.6}$$

20.3. Remarks

a. The play $L[t_0,\xi_0;Z]$ with polyhedral characteristic satisfies the global Lipschitz condition as an operator from S into S .

b. The scheme exposed in Chapter 6 (from Section 6.6 on) permits an extension of the play operators onto the space Ξ of line inputs. In particular, those operators can be considered as defined on the space of all (possibly discontinuous) inputs with bounded variation. The assertions of Section 20.2 remain valid upon replacing the space S by the space V of functions with bounded variation.

c. It seems prospective to consider the play and stop in Hölder and Wiener spaces. Recall that the Wiener space V_β , with $\beta > 1$, is defined as the space of functions $u(t)$ which are defined on a certain interval $[t_0,t_1]$ and have the finite norm

$$\|u(t)\|_{V_\beta} = |u(t_0)| + \sup \left\{ \sum_{i=1}^{n} |u(\tau_i) - u(\tau_{i-1})|^\beta \right\}^{1/\beta}, \tag{20.7}$$

where the upper bound on the right-hand side is to be taken over all partitionings

$$\tau_0 = t_0 < \tau_1 < \tau_2 < \ldots < \tau_n = t_1$$

of the interval $[t_0,t_1]$.

21. Vibro-correct equations with vector input

21.1. Statement of the problem

This section is adherent to Part 3. Here we examine properties of solutions to the differential equation

$$\frac{dx}{dt} = f[t,x,u(t),u'(t)] \quad , \tag{21.1}$$

satisfying the initial condition

$$x(t_o) = x_o \quad . \tag{21.2}$$

The inputs $u(t)$ $(t \geq t_o)$ with values in some R^{N_1} are taken into consideration. (21.1) is treated as an equation in R^N, its solutions

$$x(t) = V[t_o,x_o]u(t) \tag{21.3}$$

(if they exist) are considered as outputs of a certain transducer V.

Assume that the vector-function $f(t,x,u,v)$ is jointly continuous. If it satisfies with respect to x the local Lipschitz condition, then the solutions (21.3) exist and are unique in the case of smooth and piecewise smooth inputs $u(t)$. We are interested in specifying conditions which would ensure the vibro-correctness of equation (21.1). As it turns out, the theory of vibro-correct equations and the corresponding vibro-solutions, developed in Part 3, is no longer true in the case of vector inputs. The vibro-correctness property is preserved only by equations which satisfy some special algebraic hypotheses. Fortunately, those hypotheses are fulfilled in many situations of practical importance.

The notions of vibro-correctness for differential equations, vibro-correctness on constant inputs, vibro-solutions, etc. (cf., Chapters 11 - 13) do not change upon passing to vector inputs.

Theorem 21.1. Assume that equation (21.1) is vibro-correct on constant inputs. Then it admits the representation

$$\frac{dx}{dt} = \varphi[t,x,u(t)]u'(t) + \psi[t,x,u(t)] \quad . \tag{21.4}$$

Proof. It proceeds along the lines of the proof of Theorem 11.1. ∎

Smoothness of the matrix-function $\varphi(t,x,u)$ and the vector-function $\psi(t,x,u)$ do not guarantee the vibro-correctness of equation (21.4). Let, for instance, the input $u(t) = \{u^1(t), u^2(t)\}$ be two-dimensional and equation (21.4) have the scalar form

$$\frac{dx}{dt} = u^1(t) \frac{d}{dt} u^1(t) + u^1(t) \frac{d}{dt} u^2(t) \quad .$$

For the initial condition $x(0) = 0$ and input $u_*(t) = \{0,0\}$, the solution (21.3) is identically equal to zero in the case of the same initial condition, and for the inputs

$$u_n(t) = \left\{\frac{1}{\sqrt{n}} \sin nt, \frac{1}{\sqrt{n}} (\cos nt - 1)\right\} \quad (n = 1, 2, \ldots) .$$

the corresponding solutions are

$$x_n(t) = -\frac{t}{2} + \frac{1}{4n} (1 - \cos 2nt + \sin 2nt) \quad .$$

The sequence $x_n(t)$ does not converge to zero as $n \to \infty$, i.e., the considered equation is not vibro-correct.

21.2. Frobenius condition

Below we shall use so-called Frobenius condition, important in various problems of differential geometry, complete integrability, etc.

Consider a smooth function $\varphi(x,u)$ $(x \in R^N,\ u \in R^{N_1})$ with values in the space of linear operators acting from R^{N_1} into R^N, i.e., a function with values in the form of $(N \times N_1)$-vectors. The function $\varphi(x,u)$ is said to satisfy the *Frobenius condition* if for all x, u the equality

$$\left\{\left\{\frac{\partial}{\partial u} \varphi(x,u) + \left[\frac{\partial}{\partial x} \varphi(x,u)\right] \varphi(x,u)\right\}\xi\right\}\eta =$$

$$= \left\{\left\{\frac{\partial}{\partial u} \varphi(x,u) + \left[\frac{\partial}{\partial x} \varphi(x,u)\right]\varphi(x,u)\right\}\eta\right\}\xi \quad (\xi, \eta \in R^{N_1}) \tag{21.5}$$

holds. The derivatives with respect to vector arguments in (21.5) are understood in the classical sense.

Let us consider the equation

$$\frac{dx}{du} = \varphi(x,u) \tag{21.6}$$

which comprehends the derivative taken with respect to vector u. *If the Frobenius condition is satisfied, then for any initial condition equation* (21.6) *is locally solvable.* This well-known property will be of principal importance in our considerations. (21.6) represents a system of partial differential equations; the above statement is referred to as a *theorem on complete integrability* for that system.

If $N = 1$ and the input $u = \mathrm{col}\{u^1, \ldots, u^{N_1}\}$ is treated as a column vector, then $\varphi(x,u) = \{\varphi^1(x,u), \ldots, \varphi^{N_1}(x,u)\}$ represents a functional on R^{N_1}. It is convenient to consider $\varphi(x,u)$ as a vector. The derivative $\frac{\partial}{\partial u}\varphi(x,u)$ is to be considered as a mapping of the space R^{N_1} of column vectors ξ into the space R^{N_1} of row vectors, described by the equality

$$\left[\frac{\partial}{\partial u}\varphi(x,u)\right]\xi = \left\{\left[\frac{\partial}{\partial u^i}\varphi^j(x,u)\right]_{i,j=1,\ldots,N^1} \xi\right\}^T .$$

The derivative $\frac{\partial}{\partial x}\varphi(x,u)$ is a mapping

$$\left[\frac{\partial}{\partial x}\varphi(x,u)\right]\alpha = \left\{\alpha\,\frac{\partial}{\partial x}\varphi^1(x,u), \ldots, \alpha\,\frac{\partial}{\partial x}\varphi^{N_1}(x,u)\right\}$$

defined on one-dimensional space R^1 of real numbers α. Hence, the Frobenius condition is equivalent to postulating symmetry of the matrix

$$\frac{\partial}{\partial u}\varphi(x,u) + \left[\frac{\partial}{\partial x}\varphi(x,u)\right]\varphi(x,u) .$$

This symmetry condition can be expressed in the form of the following system of scalar equalities

$$\frac{\partial}{\partial u^i}\varphi^j + \left(\frac{\partial}{\partial x}\varphi^j\right)\varphi^i = \frac{\partial}{\partial u^j}\varphi^i + \left(\frac{\partial}{\partial x}\varphi^i\right)\varphi^j \quad (i, j = 1, \ldots, N_1;\ i \neq j). \tag{21.7}$$

If $N = 1$ and $N = 2$, the Frobenius condition (21.7) reduces to the single scalar equality

$$\frac{\partial}{\partial u^1}\varphi^2 + \left(\frac{\partial}{\partial x}\varphi^2\right)\varphi^1 = \frac{\partial}{\partial u^2}\varphi^1 + \left(\frac{\partial}{\partial x}\varphi^1\right)\varphi^2 . \tag{21.8}$$

In the case of $N > 1$, a coordinate form of the Frobenius conditions becomes more complicate. They are represented by the system of equalities

$$\frac{\partial}{\partial u_j}\varphi_{\nu i} + \sum_{\mu=1}^{N}\left(\frac{\partial}{\partial x_\mu}\varphi_{\nu i}\right)\varphi_{\mu j} =$$

$$= \frac{\partial}{\partial u_i} \varphi_{\nu j} + \sum_{\mu=1}^{N} \left(\frac{\partial}{\partial x_\mu} \varphi_{\mu j} \right) \varphi_{\mu i} \quad (\nu = 1,\dots,N \; ; \quad i,j = 1,\dots,N_1 \;) \; . \tag{21.9}$$

For $N = N_1 = 2$, system (21.9) reduces to the following two scalar equalities

$$\frac{\partial}{\partial u^2} \varphi_{11} + \left(\frac{\partial}{\partial x^1} \varphi_{11}\right) \varphi_{12} + \left(\frac{\partial}{\partial x^2} \varphi_{11}\right) \varphi_{22} =$$

$$= \frac{\partial}{\partial u^1} \varphi_{12} + \left(\frac{\partial}{\partial x^1} \varphi_{12}\right) \varphi_{11} + \left(\frac{\partial}{\partial x^2} \varphi_{12}\right) \varphi_{21} \; , \tag{21.10}$$

$$\frac{\partial}{\partial u^2} \varphi_{21} + \left(\frac{\partial}{\partial x_1} \varphi_{21}\right) \varphi_{12} + \left(\frac{\partial}{\partial x^2} \varphi_{21}\right) \varphi_{22} =$$

$$= \frac{\partial}{\partial u^1} \varphi_{22} + \left(\frac{\partial}{\partial x^1} \varphi_{22}\right) \varphi_{11} + \left(\frac{\partial}{\partial x^2} \varphi_{22}\right) \varphi_{21} \; .$$

21.3. Necessary condition of vibro-correctness

Let us continue the study of equation (21.4). For simplicity, in the sequel we shall assume that the matrix function $\varphi(t,x,u)$ is twice continuously differentiable.

Theorem 21.2. Assume that the equation (21.4) is vibro-correct on constant inputs. Then for each fixed $t = t_*$ the matrix-function $\varphi(t_*,x,u)$ fulfils the Frobenius condition.

Proof. Suppose that the Frobenius condition does not hold in some point t_*, x_*, u_* . Then there exist a linear functional ℓ defined on R^N , vectors ξ, $\eta \in R^{N_1}$ and positive reals c, δ , such that for t, $s \in [t_* - \delta, t_* + \delta]$; $|x - x_*|$, $|y - y_*| < \delta$; $|u - u_*|$, $|w - u_*| < \delta$, the inequality

$$\ell\{\{[\varphi_u'(t,x,u) + \varphi_x'(t,x,u)\, \varphi(t,x,u)]\xi\}\eta\} \geq$$
$$\geq \ell\{\{[\varphi_u'(s,y,w) + \varphi_x'(s,y,w)\, \varphi(s,y,w)]\eta\}\xi\} + c \tag{21.11}$$

is satisfied. Fix a positive integer n and divide the interval

$[t_*,t_*+\delta]$ by the points

$$\sigma_o = t_* < \sigma_1 < \sigma_2 < \ldots < \sigma_{4n} = t_* + \delta \tag{21.12}$$

into $4n$ equal parts.

Construct a continuous input $u_n(t)$ $(t \geq t_*)$ which is linear on each of the subintervals $[\sigma_{i-1},\sigma_i]$, equal to $u_* + \frac{1}{\sqrt{n}}(\xi+\eta)$ for $t \geq t_o + \delta$, and defined by the equalities

$$u_n(\sigma_{4k}) = u_* + \frac{1}{\sqrt{n}}(\xi+\eta) \quad ,$$

$$u_n(\sigma_{4k+1}) = u_* - \frac{1}{\sqrt{n}}(\xi-\eta) \quad ,$$

$$u_n(\sigma_{4k+2}) = u_* - \frac{1}{\sqrt{n}}(\xi+\eta) \quad ,$$

$$u_n(\sigma_{4k+3}) = u_* + \frac{1}{\sqrt{n}}(\xi-\eta)$$

in the points (21.12). Since

$$\|u_n(t) - u_*\|_{t_*,t_*+\delta} \to 0 \quad \text{for every} \quad \delta > 0 \quad ,$$

due to the vibro-correctness of equation (21.4) on constant inputs, for some $t_1 > t_*$ the equality

$$\lim_{n\to\infty} \|x_n(t) - x_*(t)\|_{t_*,t_1} = 0 \tag{21.13}$$

is satisfied with

$$x_n(t) = V[t_*,x_*]u_n(t), \quad x_*(t) = V[t_*,x_*]u_*(t), \quad u_*(t) \equiv u_* \quad .$$

Without any loss of generality we can assume $t_1 = t_* + \delta$. Since

$$x_n(t_1) - x_*(t_1) = \sum_{i=1}^{4n} \int_{\sigma_{i-1}}^{\sigma_i} \varphi[s,x_n(s),u_n(s)]u_n'(s)ds +$$

$$+\int_{t_*}^{t_1}\{\psi[s,x_n(s),u_n(s)] - \psi[s,x_*(s),u_*]\}ds \quad ,$$

we have

$$\ell[x_n(t_1) - x_*(t_1)] = \sum_{i=1}^{4n} F_i + \sigma(1) \quad , \tag{21.14}$$

where

$$F_i = \int_{\sigma_{i-1}}^{\sigma_i} \ell\{\varphi[s,x_n(s),u_n(s)]\}ds$$

and the vectors v_i represent values of the derivative $u_n'(t)$ on (σ_{i-1},σ_i) . By virtue of (21.14), the proof will be completed if we succeed to show that

$$\overline{\lim_{n\to\infty}} \sum_{i=1}^{4n} F_i > 0 \quad . \tag{21.15}$$

Consider the sum $F_{4k+1} + F_{4k+3}$. By the equality

$$F_{4k+1} + F_{4k+3} =$$

$$= -\int_{\sigma_{4k}}^{\sigma_{4k+1}} \ell\left\{\varphi[s,x_n(s),u_* + \frac{\xi + \eta}{\sqrt{n}} - \frac{8\sqrt{n}}{\delta}(s-\sigma_{4k})\xi]\ \frac{8\sqrt{n}}{\delta}\ \xi\right\}ds +$$

$$+ \int_{\sigma_{4k+2}}^{\sigma_{4k+3}} \ell\left\{\varphi[s,x_n(s),u_* - \frac{\xi + \eta}{\sqrt{n}} + \frac{8\sqrt{n}}{\delta}(s-\sigma_{4k+2})\xi]\ \frac{8\sqrt{n}}{\delta}\ \xi\right\}ds \quad ,$$

in view of the relations (21.14) and by the differentiability of the matrix function $\varphi(t,x,u)$, the relations

$$F_{4k+1} + F_{4k+3} =$$

$$= \int_{\sigma_{4k}}^{\sigma_{4k+1}} \ell\left\{\varphi_x'[s,x_n(s),u_n(s)]\left[x_n\left(s + \frac{\delta}{2n}\right) - x_n(s)\right]\frac{8\sqrt{n}}{\delta}\ \xi\right\}ds +$$

$$+ \int_{\sigma_{4k}}^{\sigma_{4k+1}} \ell\Big\{\varphi_u'[s,x_n(s),u_n(s)]\Big[\frac{16\sqrt{n}}{\delta}(s - \sigma_{4k})\xi -$$

$$- \frac{2}{\sqrt{n}}(\xi + \eta)\Big]\frac{8\sqrt{n}}{\delta}\xi\Big\}ds + o\Big(\frac{1}{n}\Big)$$

are true. But by virtue of (21.4), the equality

$$x_n\Big(s + \frac{\delta}{2n}\Big) - x_n(s) =$$

$$= \varphi[\sigma_{4k},x_n(\sigma_{4k}),u_*]\Big[\frac{16\sqrt{n}}{\delta}(s - \sigma_{4k})\xi - \frac{2}{\sqrt{n}}(\xi + \eta)\Big] + o\Big(\frac{1}{\sqrt{n}}\Big)$$

is satisfied for $s \in [\sigma_{4k}, \sigma_{4k+1}]$. Thus,

$$F_{4k+1} + F_{4k+3} =$$

$$= \int_{\sigma_{4k}}^{\sigma_{4k+1}} \ell\Big\{\chi_k\Big[\frac{16\sqrt{n}}{\delta}(s - \sigma_{4k})\xi - \frac{2}{\sqrt{n}}(\xi + \eta)\Big]\frac{8\sqrt{n}}{\delta}\xi\Big\}ds + o\Big(\frac{1}{n}\Big) ,$$

where

$$\chi_k = \varphi_x'[\sigma_{4k},x_n(\sigma_{4k}),u_*]\varphi[\sigma_{4k},x_n(\sigma_{4k}),u_*] + \varphi_u'[\sigma_{4k},x_n(\sigma_{4k}),u_*] ,$$

and, after an integration,

$$F_{4k+1} + F_{4k+3} = \frac{4}{n}\ell[(\chi_k\ \eta)\xi] + o\Big(\frac{1}{n}\Big) .$$

Analogously,

$$F_{4k+2} + F_{4k+4} = -\frac{4}{n}\ell[(\chi_k'\ \xi)\eta] + o\Big(\frac{1}{n}\Big) ,$$

where

$$\chi_k' = \varphi_x'[\sigma_{4k+1},x_n(\sigma_{4k+1}),u_*]\varphi[\sigma_{4k+1},x_n(\sigma_{4k+1}),u_*] +$$

$$+ \varphi_u'[\sigma_{4k+1},x_n(\sigma_{4k+1}),u_*] .$$

Therefore, (21.11) implies the estimate

$$F_{4k+1} + F_{4k+2} + F_{4k+3} + F_{4k+4} \geq \frac{4c}{n} + o\left(\frac{1}{n}\right) ,$$

and, consequently,

$$\sum_{i=1}^{4n} F_i \geq 4c + o(1) .$$

This means that the relation (21.15) (therefore also Theorem 21.2) has been proved.

∎

21.4. Sufficient condition of vibro-correctness

For equations (21.4) with vector input, an analogue of Theorem 12.1 can be formulated as follows.

Theorem 21.3. Let the matrix-function $\varphi(t,x,u)$ be twice continuously differentiable and the vector-function $\psi(t,x,u)$ be once continuously differentiable. Assume that for each $t = t_*$ the function $\varphi(t_*,x,u)$ satisfies the Frobenius condition. Then equation (21.4) is vibro-correct.

Proof. As in the proof of Theorem 12.1, let us consider the auxiliary equation

$$\frac{dx}{du} = \varphi(t,x,u) , \tag{21.16}$$

with t treated as a parameter. By the Frobenius condition, for each initial condition

$$x(u_0) = x_0 , \tag{21.17}$$

equation (21.16) admits a unique solution

$$x(u) = Q(u,u_0,x_0,t) .$$

If solutions of equation (21.4) are postulated to have the form

$$x(t) = Q[u_0, u(t), z(t), t] \quad , \tag{21.18}$$

then in order to determine the function $z(t)$, an equation

$$\frac{dz}{dt} = F[t, z, u(t)] \quad , \tag{21.19}$$

whose right-hand side does not depend explicitly upon $u'(t)$, is to be solved. Hence, the solutions of equation (21.19) represent values of an operator acting from C into C, defined on continuous inputs. Substituting these solutions in (21.18), we get an operator $V[t_0, x_0]$ which transforms continuous inputs $u(t)$ into continuous outputs $x(t)$. This operator is a continuous extension of the operator (21.3) defined on smooth inputs. A possibility of extending the operator (21.3) by continuity onto the space of continuous inputs is equivalent to the vibro-correctness of equation (21.4).

Details of the proof are identical with the corresponding parts of the proof of Theorem 12.1.

∎

21.5. Remarks

a. Analogously to the construction developed in Chapter 13 for vibro-correct equations with scalar inputs, one can introduce (e.g., under the hypotheses of Theorem 21.3) a notion of vibro-solutions to the equation (21.4) with vector inputs. The notation (21.3) will be preserved.

b. If every vibro-solution (21.3) is defined for all $t \geq t_0$, the equation (21.4) will be called globally vibro-correct. To ensure the global vibro-correctness, it is sufficient that for t, u on any fixed ball the vector-function $\psi(t,x,u)$ satisfies global Lipschitz condition with respect to x and the derivative $\varphi'_x(t,x,u)$ is uniformly bounded (certainly, provided that the hypotheses of Theorem 21.3 are fulfilled). An analogue of Theorem 13.2 is then true, as well. It would be of interest to establish criteria of the global vibro-correctness in terms of various unilateral estimates.

c. A detailed analysis of the operators (21.3) would be of interest. One can easily see that (21.3), treated as operators from C into C, under natural assumptions satisfy the local Lipschitz condition.

d. The reader has certainly noticed that in the case of vibro-correct equations with vector inputs the assumptions on smoothness of the right-hand sides are stronger than those used in Part 3 at the study of equations with scalar inputs. The sharpening of assumptions does not follow from the intrinsic structure of the problem.

22. Equations with vector inputs and smooth constraints

22.1. Constraints

As it has been shown, in Part 3 (Chapters 14, 15) vibro-correct equations with constraints can be used for describing ordinary and variable hysterons. For hysterons with vector inputs and outputs, analogous constructions have been successfully developed only in some special cases.

Let a closed bounded domain $Z \subset R^N$ with the smooth boundary ∂Z be given. Denote by $n(x)$ $(x \in \partial Z)$ the unit outward normal to Z. By a *constraint*

$$\Pi = \{\partial Z, b(x)\} \tag{22.1}$$

we shall mean the pair consisting of ∂Z and a smooth vector-function $b(x)$ $(x \in \partial Z)$ such that

$$(b(x), n(x)) > 0 \quad (x \in \partial Z) \quad . \tag{22.2}$$

$P(x)$ $(x \in \partial Z)$ will denote the linear operator of projection along the vector $b(x)$ onto the $(N-1)$-dimensional subspace orthogonal to the vector $n(x)$. Obviously,

$$P(x)h = h - \frac{(h, n(x))}{(b(x), n(x))} b(x) \quad (h \in R^N) \quad . \tag{22.3}$$

Consider the equation

$$\frac{dx}{dt} = g(t,x) \tag{22.4}$$

with the right-hand side $g(x,t)$ jointly continuous in $t \in R^1$, $x \in R^N$, and satisfying the local Lipschitz condition with respect to x. By the *solutions of equation* (22.4) *with constraint* (22.1) we shall mean solutions $x(t)$ $(t \geq t_0)$ of the equation

$$\frac{dx}{dt} = f(t,x) \tag{22.5}$$

with the initial condition

$$x(t_o) = x_o \in Z \quad , \tag{22.6}$$

where

$$f(t,x) = \begin{cases} g(t,x), & \text{if } x \in \text{int } Z \quad , \quad \text{or} \\ & \text{if } x \in \partial Z \text{ and } (g(t,x),n(x)) \leq 0 \quad ; \\ P(x)g(t,x), & \text{if } x \in \partial Z \text{ and } (g(t,x),n(x)) \geq 0 \quad . \end{cases}$$

These solutions exist for all $t \geq t_o$. They are uniquely determined by the initial condition (22.6).

Let us pass to the vibro-correct equations

$$\frac{dx}{dt} = \varphi[t,x,u(t)]u'(t) + \psi[t,x,u(t)] \tag{22.7}$$

with vector inputs $u(t)$. If the functions $\varphi(t,x,u)$, $\psi(t,x,u)$ are sufficiently smooth, for each smooth input $u(t)$ $(t \geq t_o)$ there exists a unique solution

$$x(t) = W[t_o,x_o;\Pi]u(t) \tag{22.8}$$

of equation (22.7) with constraint (22.1), satisfying the initial condition (22.6). In Part 3, the solutions of vibro-correct equations with constraints have been constructed for arbitrary continuous scalar inputs. Analogous constructions are not possible in the general vector case.

A relatively simple treatment is possible in the important case where the equation (22.7) admits the form

$$\frac{dx}{dt} = A(x)u'(t) \quad , \tag{22.9}$$

with a non-singular matrix function $A(x)$ $(x \in R^N)$ that satisfies Frobenius condition and with $b(x) = A(x)A^T(x)n(x)$ $(x \in \partial Z)$. In this case, the possibility of passing to arbitrary continuous inputs (vibro-correctness) is assured, in particular, by postulating that for all different $x, y \in \partial Z$, vectors $A^T(x)n(x)$ and $A^T(y)n(y)$ have different directions.

In the sequel, the two-dimensional case is considered in a more detail.

22.2. Planar motion

a. First, let us consider the elementary equation

$$\frac{dx}{dt} = u'(t) \quad . \tag{22.10}$$

Fix a point x_0 of the curve ∂Z. Denote by $\theta_1(x)$, $\theta_2(x)$ the angles from $n(x_0)$ and $b(x)$ to $n(x)$, oriented counter-clockwise. Let Γ denote the part of curve ∂Z which is contained in the boundary of the convex hull of Z. As it turns out, the transducer W which associates the outputs defined by equalities (22.8) to smooth inputs is uniformly vibro-correct if and only if the following four hypotheses are satisfied:

(a1) $n(x) = n(y)$ implies that $b(x) = b(y)$, while from $n(x) = -n(y)$ it follows that $b(x) = -b(y)$.

(a2) Function $b(x)$ is constant on each connected component of the set $\partial Z \setminus \Gamma$.

(a3) Function $\arg b(x)$ is non-decreasing when x runs counter-clockwise along ∂Z.

(a4) The equality

$$\int_{\partial Z} \operatorname{tg} \theta_2(x)\, d\theta_1(x) = 0 \tag{22.11}$$

holds.

b. In the previous statement, the set Z could be non-convex. It would be of interest to set up some additional properties of the transducer W which would imply convexity of Z.

We shall call the transducer W *weakly convergent* if for each $\varepsilon > 0$ there exists $\gamma > 0$ for which by $x_0, y_0 \in Z$ and $|u(t_1) - u(t_0)| < \gamma$ it follows that

$$|W[t_0,x_0]u(t) - W[t_0,y_0]u(t)| < \varepsilon \quad (t > t_1) \quad .$$

It turns out that the transducer W with outputs (22.8) corresponding to smooth inputs is uniformly vibro-correct and weakly convergent if and only if the following two hypotheses are satisfied:

(b1) If $x \neq y$ $(x, y \in \partial Z)$ then $b(x) \neq b(y)$, and $n(x) = -n(y)$ implies that $b(x) = -b(y)$.

(b2) Equality (22.11) is true.

c. Choose a point x_o of the curve ∂Z . Let the curvilinear integral

$$h(x) = \int_{x_o}^{x} A^{-1}(\xi)d\xi \qquad (x \in \partial Z)$$

(by the Frobenius condition, its value is independent of the integration path) determine a mapping of the curve ∂Z into the curve which is the boundary of a certain domain Z^* . The transducer W , defined by the solutions of equation (22.9) with constraint (22.1), is uniformly vibro--correct if and only if the transducer W^* , appropriately defined by the solutions of equation (22.10) with constraint $\{\partial Z^*, b^*(y)\}$, where

$$b^*[h(x)] = A^{-1}(x)b(x) \qquad (x \in \partial Z) ,$$

has the same property. This implies that natural analogues of the assertions formulated above for equation (22.10) remain true also for equation (22.9).

In particular, the transducer W defined by solutions of equation (22.9) with constraint (22.1) is uniformly vibro-correct and weakly convergent if and only if the following hypotheses are satisfied:

(c1) If $x \neq y$ $(x, y \in \partial Z)$, then $\arg A^T(x)n(x) \neq \arg A^T(y)n(y)$ and $\arg A^{-1}(x)b(x) \neq \arg A^{-1}(y)b(y)$. In turn, if $\arg A^T(x)n(x) = -\arg A^T(y)n(y)$, then $\arg A^{-1}(x)b(x) = -\arg A^{-1}(y)b(y)$.

(c2) The equality

$$\int_{\partial Z} \operatorname{tg} \theta_2^*(x)d\theta_1^*(x) = 0$$

holds with $\theta_1^*(x)$, $\theta_2^*(x)$ representing oriented counter-clockwise angles from $n(x_o)$ and $A^{-1}(x)b(x)$ to $A^T(x)n(x)$, respectively.

22.3. Other descriptions

In Parts 1-3, four different descriptions of hysterons were used:

- by a system of defining curves,
- by canonical representations,
- by vibro-correct equations with constraints,
- by specifying some general properties of the hysteron.

Relations between various descriptions were established. The most important of them are given by the identification theorems.

We now shall develop a similar study also in the case of hysterons with vector inputs. As an example, let us formulate an analogue of the identification theorem.

Assume that $Z = A_1A_2 \dots A_M$ is a convex polygon which has no parallel edges. Denote by n_i the unit outward normal to the edge $A_{i-1}A_i$ and by K_i the supporting cone for Z in vertex A_i. Let unit vectors $b_1,\dots,b_M$ be given such that $(b_i,n_i) > 0$ and the angles between b_i and b_1 increase with i. By P_i denote the projector in direction b_i onto the straight line which contains the origin and is parallel to the edge $A_{i-1}A_i$. Define

$$f(x,v) = \begin{cases} v, & \text{if } x \in \operatorname{int} Z, \\ & \text{if } x \text{ is an interior point of the edge } A_{i-1}A_i \\ & \text{and } (v,n_i) \leq 0, \\ & \text{if } x = A_i \text{ and } v \in K_i; \\ P_i v, & \text{if } x \text{ is an interior point of the edge } A_{i-1}A_i \\ & \text{and } (v,n_i) \geq 0, \\ & \text{if } x = A_i \text{ and } v \notin K_i \text{ but } P_i v \in K_i; \\ P_{i-1} v, & \text{if } x = A_i \text{ and } v \notin K_i \text{ but } P_{i-1} v \in K_i; \\ 0, & \text{if } x = A_i \text{ and } v, P_{i-1} v, P_i v \notin K_i. \end{cases}$$

For any smooth input $u(t)$ $(t \geq t_0)$, the Cauchy problem

$$\frac{dx}{dt} = f[x,u'(t)], \quad x(t_0) = x_0 \in Z$$

has a unique solution

$$x(t) = V[t_0,x_0]u(t) \quad (t \geq t_0). \tag{22.12}$$

The transducer V defined by operators (22.12), is uniformly vibro-correct and, moreover, it satisfies global Lipschitz condition if $|P_MP_{M-1} \dots P_1h| < |h|$ where h is a non-trivial vector parallel to

A_1A_M . Let *M* denote the class of the constructed vibro-correct transducers.

Assume that W_0 is a static, uniformly vibro-correct transducer with two-dimensional inputs $u(t)$, two-dimensional outputs $x(t)$ and the domain $Z = A_1A_2 \dots A_M$ of feasible states characterized by values of the output. For the output the following is true: if $u(t) - u(t_0) + x_0 \in Z$ $(t_0 \leq t \leq t_1)$, then the equality $W_0[t_0,x_0]u(t) = u(t) - u(t_0) + x_0$ holds. In turn, $u(t) - v(t) = u(t_0) - v(t_0)$ $(t_0 \leq t \leq t_1)$ implies that $W_0[t_0,x_0(t)]u(t) = W_0[t_0,x_0]v(t)$ $(t_0 \leq t \leq t_1,\ x_0 \in Z)$. Then the transducer W_0 belongs to the class *M* .

Clearly, the above statement does not replace the general identification theorem which has been formulated in Part 2 for transducers with scalar inputs and scalar outputs. It only refers to the transducers with two-dimensional inputs and outputs, such that their properties for the outputs which belong to the interior of the domain of feasible states are characterized by a simple linear operator (a priori known).

The authors do not know any non-trivial form of the identification theorem for transducers with hysteresis nonlinearity if either inputs or outputs are more than two-dimensional. A number of hypothetical statements have been recently disproved by some sophisticated counter-examples due to A.F. Klepcyn.

Part 5. Discontinuous nonlinearities

Ah, cruel Three! In such an hour,
Beneath such dreamy weather,
To beg a tale of breath too weak
To stir the tiniest feather!
Yet what can one poor voice avail
Against three tongues together?
L. Carroll

23. Static elements

23.1. Continuous characteristics

An element W with inputs u(t) and outputs x(t) is called a *static element* if its output $x(t_*)$ is for any fixed $t = t_*$ uniquely determined by the corresponding value $u(t_*)$ of the input. In other words, for a static element W its input-output relation can be reduced to the form

$$x(t) = f[t,u(t)] \quad , \qquad (23.1)$$

where f(t,u) is a function of two variables. The function f(t,u) is then referred to as a *characteristic of element* W . If the characteristic is independent of t , the element W is called *stationary*; otherwise, it will be *non-stationary*.

In the above definition of the static element, there are no restrictions concerning the nature of variables t, u, x . In the most important situations that originate from system theory and some other applications, t represents continuous or discrete time. In some applications, t may also be an element of R^{N_0} with $N_0 > 1$. The values u , x of the inputs and outputs, respectively, are some scalars or real vectors.

Throughout this chapter we shall assume that t (variable) belongs to a given closed and bounded set Ω in some finite-dimensional space; $u \in R^N$, $x \in R^M$ (correspondingly, the characteristic f(t,u) is defined for $t \in \Omega$, $u \in R^N$, and its values belong to R^M).

For the static element W, its input - output relations are given by the *superposition operator*

$$\mathfrak{f} u(t) = f[t,u(t)] . \qquad (23.2)$$

We shall now discuss some properties of the operator (23.2), of importance in the study of systems composed of static elements. For this, by analogy to the scalar case we introduce the notions $C(\Omega)$ and $L_p = L_p(\Omega)$ $(1 \leq p \leq \infty)$ of the spaces of vector functions defined on Ω, with values in R^N or R^M.

The characteristic $f(t,u)$ will be called *continuous* if it is a *Carathéodory function*, i.e., continuous with respect to u for almost all t and measurable with respect to t for every u.

Theorem 23.1. Suppose that the characteristic $f(t,u)$ is continuous. Then the operator (23.2) maps L_p into L_q, where $1 \leq p,q < \infty$, if and only if for almost all $t \in \Omega$

$$|f(t,u)| \leq a(t) + b|u|^{p/q} \quad (u \in R^N) , \qquad (23.3)$$

where $a(t) \in L_q$. If the operator (23.2) maps L_p into L_q $(1 \leq p,q < \infty)$, then it is continuous from L_p into L_q.

Theorem 23.2. Let the characteristic $f(t,u)$ be continuous and

$$|f(t,u)| \leq a(t) + \varphi(u) \quad (t \in \Omega,\ u \in R^N) , \qquad (23.4)$$

where $a(t) \in L_q$ $(1 \leq q < \infty)$ and $\varphi(u)$ is continuous. Then the operator (23.2) acts from L_∞ (hence, also from C) into L_q and is continuous.

If $f(t,u)$ is continuous with respect to the pair (t,u), then (23.2) is a continuous operator in the space C.

The continuity of operator (23.2) is equivalent to the correctness of the corresponding static element with respect to small perturbations of the inputs, i.e., small noises at the input contribute to small variations of the output, both measured in norms of the appropriate spaces.

In the preceding chapters, the input-output operators of hysterons had been first defined over some small classes of inputs (with only piecewise monotone functions admissible as inputs) and only afterwards we discussed the possibility of extending them to larger classes.

A similar analysis is of interest also in the case of static elements. We recall here the following result.

Theorem 23.3. Let the characteristic $f(t,u)$ be continuous. Assume $\mathfrak{f}\, C \subset L_q$ and the convergence of functions $u_k(t) \in C$ to a function $u_*(t) \in C$ in L_p imply the convergence of the sequence $\mathfrak{f}\, u_k(t)$ to $\mathfrak{f}\, u_*(t)$ in the norm of the space L_q, where $1 \leq p,q < \infty$. Then the operator (23.2) maps L_p into L_q and is continuous there. ∎

Since the spaces C and L_p are infinite-dimensional, operators which are continuous there are not necessarily uniformly continuous on bounded sets. We refer to [70] for supplementary conditions ensuring the uniform continuity of operator (23.2). If the function $f(t,u)$ is continuous with respect to (t,u) and $\mathfrak{f}$ is treated as an operator from L_∞ into L_∞ (or, alternatively, from C into C), then this operator is uniformly continuous on every ball.

23.2. Elements with discontinuous characteristics

An analysis of a static element turns out much more difficult if its characteristic $f(t,u)$ is discontinuous as a function of u. As a rule, such an element can be regarded as a mathematical model of real system only if a certain strong requirement is imposed in addition; for instance, for any measurable input the corresponding output of that element is postulated to be measurable, too.

We begin with a discussion of this requirement.

The characteristic $f(t,u)$ will be called a *superpositionally measurable function* (or *SM-function*), if $f[t,u(t)]$ is measurable at every measurable $u(t)$. To ensure the superpositional measurability, it is enough to assume the measurability of the function $\mathfrak{f}\, u(t)$ at continuous $u(t)$. Clearly, Carathéodory functions (continuous characteristics) are superpositionally measurable. Sums and products of SM-functions are again SM-functions. The pointwise limit $f(t,u)$ of every sequence $f_k(t,u)$ of SM-functions is an SM-function, too.

The last property often simplifies an analysis of concrete discontinuous characteristics. For example, the characteristic $f(t,u) = \text{sign}\, u$ $(u \in R^1)$ is an SM-function, since it is the limit of the sequence of continuous functions $f_k(t,u) = \frac{2}{\pi} \text{arc tg}\,(ku)$ $(k=1,2,\dots)$.

Borel measurable functions belong to the class of SM-functions. Recall that the vector-function

$$f(t,u) = \{f_1(t,u),\dots,f_M(t,u)\} \quad (t \in \Omega,\ u \in R^N) \tag{23.5}$$

is called *Borel measurable (measurable in the Borel sense)*, if every Lebesgue set $\{\{t,u\}: f_j(t,u) \leq \mu\}$ (for all j and μ) is measurable in the Borel sense. The class of Borel measurable functions is quite large, though not all Carathéodory functions are Borel measurable, in particular.

Before studying singularities which are typical for static elements with discontinuous characteristics, it is useful to get acquainted with so-called monsters.

A function $\chi(t,u)$ $(t \in \Omega;\ u \in R^N)$ is called a *monster*, if it assumes only values 0 and 1, and, at the same time, for every measurable $u(t)$, the function $\chi[t,u(t)]$ is almost everywhere equal to zero, while on the other hand, for every fixed $t_* \in \Omega$ the equality $\chi(t_*,u) = 0$ holds at most on a countable set of $u \in R^N$.

Theorem 23.4. If the continuum hypothesis is true, then monsters do exist.

Proof. Let us introduce a complete ordering of the set Ω and space C (both are continua), with an index α of the complete ordering, so that each element has at most a countable number of predecessors. For each $t_\alpha \in \Omega$, let

$$\chi(t_\alpha,u) = \begin{cases} 0, & \text{if } u = u_\beta(t_\alpha) \text{ for } \beta < \alpha \\ 1, & \text{for other } u \in R^N. \end{cases} \tag{23.6}$$

Function (23.6) represents a monster, because for any continuous $u(t)$ $\chi[t,u(t)]$ is different from 0 at most on a countable set. ∎

Were the continuum hypothesis false, one only could state that the existence of monsters does not contradict the Zermelo-Fraenkel axioms of set theory.

Every monster is superpositionally measurable, but is not measurable as a function of two variables. On the other hand, a jointly measurable function $f(t,u)$ is not SM-measurable, when $f(t,u) = 0$ for $u \neq 0$ and $f(t,0)$ is non-measurable. Consequently, neither the class of SM-functions nor the class of jointly measurable functions are contained each in the other.

Consider a static element W with the characteristic

$$f(t,u) = \chi(t,u)\psi(t,u) \quad (t \in \Omega;\ u \in R^N) \quad , \tag{23.7}$$

where $\chi(t,u)$ is the monster (23.6) and $\psi(t,u)$ represents any function with values in R^M. For every measurable input $u(t)$, the output $\mathfrak{f}\, u(t)$ is almost everywhere zero (by definition of the monster), i.e., the operator $\mathfrak{f}$ transforms each L_p into each L_q. At the same time (again by definition of the monster and since $\psi(t,u)$ is arbitrary), estimates of the form (23.3) do not hold for the function (23.7). Therefore, after passing to elements with discontinuous characteristics which are superpositionally measurable, estimates (23.3) and (23.4) become only sufficient conditions for (23.2) to be an operator in the suitable spaces.

Nevertheless, also in the case of static elements with discontinuous characteristics one can establish several important properties which just reflect the fact that (23.2) is an operator from L_p into L_q. The subsequent section is devoted to this problem.

23.3. Estimates of the outputs

Denote by $||x(t)||_p$ the L_p-norm of function $x(t)$.

Lemma 23.1. If the operator (23.2) acts from L_p into L_q, where $1 \leq p,q < \infty$, then

$$||\mathfrak{f}\, u(t)||_q^q \leq a + b\, ||u(t)||_p^p \quad (u(t) \in L_p) \tag{23.8}$$

holds with some positive constant a and

$$b = \sup\{\beta : \beta = \|\mathfrak{f}\, u(t)\|_q^q,\ \|u(t)\|_p \le 1\} < \infty . \tag{23.9}$$

Proof. Suppose that $b = \infty$, then there exists a sequence $u_k(t) \in L_p$ such that $\|u_k(t)\|_p \le 1$ and $\|\mathfrak{f}\, u_k(t)\|_q > k^2$. Since the norm in L_p is absolutely continuous, given the sequence $u_k(t)$, one can construct a new sequence $v_k(t) \in L_p$ of functions with disjoint supports, such that $\|v_k(t)\|_p^p < 2^{-2k}$ and $\|\mathfrak{f}\, v_k(t)\|_q \ge k$. Then the function $v(t) = v_1(t) + \dots + v_k(t) + \dots$ would belong to L_p , although $\mathfrak{f}\, v(t) \notin L_q$, because

$$\|\mathfrak{f}\, v(t) - f(t,0)\|_q^q = \sum_{k=1}^{\infty} \|\mathfrak{f}\, v_k(t) - f(t,0)\|_q^q = \infty .$$

This contradiction implies that b must be finite.

Assume that $u(t) \in L_p$ and r is the integer part of the $\|u(t)\|_p^p$. Next, divide the set Ω into non-intersecting parts $\Omega_1, \dots, \Omega_{r+1}$ so that each of the functions $u_j(t) = \kappa_j(t)u(t)$, where $\kappa_j(t) = 1$ for each $t \in \Omega_j$ and $\kappa_j(t) = 0$ for $t \notin \Omega_j$, belong to the unit ball in the space L_p (this can be done due to the absolute continuity of the norm in L_p). Then,

$$\|\mathfrak{f}\, u(t)\|_q^q = \sum_{j=1}^{r+1} \int_{\Omega_j} |f[t,u(t)]|^q\, dt \le \sum_{j=1}^{r+1} \|\mathfrak{f}\, u_j(t)\|_q^q \le b(r+1) ,$$

hence (23.8) follows.

∎

The lemma we have just proved enables us to establish analogues of the estimates (23.3).

Theorem 23.5. Suppose that the operator (23.2) acts from L_p into L_q , where $1 \le p, q < \infty$. Then there exist $a(t) \in L_q(\Omega; R^1)$ and $b > 0$, such that for any given measurable function $u(t)$ the estimate

$$|f[t,u(t)]| \le a(t) + b|u(t)|^{p/q} \tag{23.10}$$

holds at almost all $t \in \Omega$.

Proof. We shall restrict ourselves to the case $p = q = 1$ and scalar-valued functions $f(t,u)$. Define the constant b by (23.9) and set

$$g(t,u) = \begin{cases} |f(t,u)| - b|u| & , \text{ if } |f(t,u)| \geq b|u| , \\ 0 & , \text{ if } |f(t,u)| \leq b|u| . \end{cases} \tag{23.11}$$

The superposition mapping g, generated by (23.11), also acts in the space L_1. To prove the theorem, one should construct a function $a(t) \in L_1$ such that for every measurable function $u(t)$ at almost all $t \in \Omega$,

$$g[t,u(t)] \leq a(t) . \tag{23.12}$$

Since the space L_1 is separable, one can construct functions $u_j(t) \in L_1$ $(j = 1,2,\dots)$ so that the sequence $x_j(t) = g\, u_j(t)$ is dense in $g\, L_1$. By (23.11), each of the functions

$$z_k(t) = \max\{g\, u_1(t),\dots, g\, u_k(t)\} \quad , \quad k = 1,2,\dots$$

belongs to the range of the operator g. The sequence $z_k(t)$ is monotone in view of its construction and, due to Lemma 23.1, the estimate $\|z_k(t)\|_1 \leq a$ is satisfied with some positive constant a. By the classic Lebesgue theorem, the sequence $z_k(t)$ is convergent to some function $a(t) \in L_1$.

The function $a(t)$ is then an upper bound for all the functions $g\, u_j(t)$ which form a dense set in $g\, L_1$. This implies that the estimate (23.12) is true for all $u(t) \in L_1$.

Also in the case of any measurable function $v(t)$, the estimate (23.12) holds at almost all $t \in \Omega$, since in view of the above proof, it is true for every function

$$v_k(t) = \min\{|v(t)|,k\}\text{sign } v(t) .$$

∎

In the next section we give a characterization of some classes of discontinuous characteristics which satisfy the estimates (23.3).

23.4. Proper characteristics

A scalar function $g(u)$ $(u \in R^1)$ is called *proper* if the pre-image $g^{-1}(I)$ of each interval $I = (a,\infty)$ is either empty or has a non-zero inner measure. The class of proper functions includes all continuous and all non-decreasing functions, as well as functions which in every point of their domain are either left lower semicontinuous or right upper semicontinuous, etc.

A function $f(t,u)$ $(t \in \Omega,\ u \in R^1)$ will be called *proper* if $f(t_*,u)$ is a proper function in u for almost every fixed $t_* \in \Omega$.

If a proper function $f(t,u)$ is jointly measurable and bounded for almost all t, then also the function

$$h(t) = \sup_{u \in R} f(t,u) \quad (t \in \Omega) \tag{23.13}$$

is measurable and for each $\varepsilon > 0$ there exists a measurable function $u_\varepsilon(t)$ such that for almost all $t \in \Omega$,

$$h(t) - \varepsilon \leq f[t,u_\varepsilon(t)] \leq h(t) \quad . \tag{23.14}$$

Due to this property and by repeating the arguments used in the proof of Theorem 23.5, one can easily show the following.

Theorem 23.6. Let the operator (23.2), generated by the scalar-valued, superpositionally and jointly measurable function $f(t,u)$ $(t \in \Omega,\ -\infty < u < \infty)$, transform L_p into L_q, where $1 \leq p,q < \infty$. For each $\mu > 0$ assume that the function $|f(t,u)| - \mu|u|$ is proper. Then for almost all $t \in \Omega$ the estimate

$$|f(t,u)| \leq a(t) + b|u|^{p/q} \quad (-\infty < u < \infty) \quad , \tag{23.15}$$

holds with $a(t) \in L_q$ and b defined by (23.9).

Clearly, the characteristic $f(t,u)$ need not necessarily be proper to guarantee that (23.15) is true. The properness hypothesis might be replaced, in particular, by postulating the measurability of $f(t,u)$ in Borel sense.

23.5. Continuity on a fixed input

For any static element W with continuous characteristic, the input-output operator (23.2) is continuous from L_p into L_q, provided that it transforms each input from L_p into an output from L_q (cf., Theorem 23.1). For discontinuous characteristics, the input--output operator (23.2) does not necessarily have an analogous property.

It may now happen that the operator $\mathfrak{f}$ still remains continuous despite of the discontinuity of its characteristic with respect to u in each point. As an example we can consider the operator $\mathfrak{f}$ generated by a monster.

If $\mathfrak{f}$ acts from L_p into L_q, where $1 \le p,q < \infty$, and is discontinuous at least in one point of the space L_p, then the set of its discontinuity points is dense in L_p. In Chapter 24, we shall construct static elements with discontinuous characteristics, such that discontinuity points of the operator $\mathfrak{f}$ form a "principal" part of the space L_p. Here we shall restrict ourselves to just one general statement.

Theorem 23.7. Suppose that the operator (23.2) acts from L_p into L_q, where $1 \le p,q < \infty$. Let $u_0(t) \in L_p$ and the function

$$\varphi(t,v) = f[t,u_0(t) + v] \quad (t \in \Omega,\ v \in R^N) \tag{23.16}$$

for almost all $t \in \Omega$ be continuous with respect to v in the point $v = 0$. Then the operator (23.2) is continuous in the point $u_0(t)$, i.e.,

$$\lim_{\|u(t)-u_0(t)\|_p \to 0} \|\mathfrak{f}\, u(t) - \mathfrak{f}\, u_0(t)\|_q = 0 . \tag{23.17}$$

Proof. Let $u_k(t) \in L_p$ and $\|u_k(t) - u_0(t)\|_p \to 0$. Then the sequence $u_k(t)$ converges in measure to $u_0(t)$ and, by the continuity of function (23.16) with respect to v, the sequence $\mathfrak{f}\, u_k(t)$ is convergent in measure to $\mathfrak{f}\, u_0(t)$.

In view of $\|u_k(t) - u_0(t)\|_p \to 0$, integrals of the functions $|u_k(t)|^p$ are uniformly absolutely continuous. Therefore, Theorem 23.5 (estimate (23.10)) implies the uniform absolute continuity for integrals of the functions $|\mathfrak{f}\, u_k(t)|^q$ and, consequently, for integrals of the functions $|\mathfrak{f}\, u_k(t) - \mathfrak{f}\, u_o(t)|^q$.

Hence, we can pass to the limit under integrals in the equality

$$\|\mathfrak{f}\, u_k(t) - \mathfrak{f}\, u_o(t)\|_q^q = \int_\Omega |\mathfrak{f}\, u_k(t) - \mathfrak{f}\, u_o(t)|^q \, dt \quad ,$$

with the integrands converging in measure to zero. Eventually, this yields the convergence

$$\|\mathfrak{f}\, u_k(t) - \mathfrak{f}\, u_o(t)\|_q \to 0 \quad .$$

∎

Suppose, for instance, that a static element W is stationary and its characteristic $f(u)$ has exactly one discontinuity point $u_o \in R^N$. According to Theorem 23.7, only those inputs $u_*(t) \in L_p$ that assume the value u_o on a certain set of positive measure, will then be the discontinuity points of operator (23.2).

23.6. Additional remarks

a. By an *L-characteristic of operator (23.2)* we shall understand the set of points $\{\alpha, \beta\}$ in the plane such that $\mathfrak{f}$ acts from $L_{1/\alpha}$ into $L_{1/\beta}$. Various general properties of L-characteristics follow as a consequence of Theorem 23.5 .

A *set* G located in the quadrant $\alpha, \beta \geq 0$ *is* 0*-concave* , if $\{\alpha_o, \beta_o\} \in G$ implies $\{\lambda\alpha_o, \lambda\beta_o\} \in G$ for $\lambda \geq 1$, and $\{\lambda\alpha_o, \beta_o\} \in G$ for $0 \leq \lambda \leq 1$. If $\xi(\alpha)$ is a non-decreasing, concave and non-negative function, then both the set G_1 of the points $\{\alpha, \beta\}$ such that $\beta > \xi(\alpha)$ and the set G_2 of the points $\{\alpha, \beta\}$ for which $\beta \geq \xi(\alpha)$ will be 0-concave. L-characteristics of all operators (23.2) are also 0-concave. If the function $\xi(\alpha)$ is non-decreasing and non-negative, moreover $\xi(\alpha)\,\alpha^{-1}$ is non-increasing, then each of the sets $\{\{\alpha,\beta\} : \beta > \xi(\alpha)\}$, $\{\{\alpha,\beta\} : \beta \geq \xi(\alpha)\}$ is the L-characteristic of some operator (23.2). If, for example,

$$f(t,u) = a_1(t)\,|u|^{\gamma_1} + a_2(t)\,|u|^{\gamma_2} + \dots + a_k(t)\,|u|^{\gamma_k} \quad (t \in \Omega,\ u \in R^1), \tag{23.18}$$

where $0 \leq \gamma_1 < \gamma_2 < \dots < \gamma_k$, $a_j(t) \in L_{p_j}$, $p_j = r_j^{-1}$, then the L-characteristic of operator (23.2) has the form of the polyhedron

$G = \{\{\alpha,\beta\} : \beta \geq \alpha\, \gamma_j + r_j,\ j = 1, \ldots, k\}$.

b. If the inputs and outputs are observed (measured, computed, etc.) up to the values on a set of measure zero, the problem of reconstructing the characteristic of a static element from those observations turns out to be hopeless in the general situation. Thus, for reconstructing a characteristic, an additional qualitative information on that characteristic must be available (in particular, concerning its continuity or monotonicity).

An interesting question concerns the possibility of determining the values of operator (23.2) over a certain class of inputs if its values are known for inputs $u(t)$ from some other class U . For example, if V is the set of all continuous inputs, then by virtue of Lusin's theorem the outputs are determined (up to values on a set of measure zero) for every measurable input. What would happen if, instead, V were chosen as the set of all analytic inputs or the set of all continuously differentiable inputs ? The answer is rather surprising ! For these smaller classes of inputs, the corresponding input-output relations uniquely specify the outputs solely in the case of very special continuous inputs.

Proof. For the proof, it is sufficient to construct a continuous function $\varphi(t)$ $(t \in \Omega = [0,1])$ such that its graph intersects the graph of any continuously differentiable function in points with abscissae from a set of measure zero. Then let $f(t,u) = 0$ for $u \neq \varphi(t)$ and $f(t,u) = 1$ for $u = \varphi(t)$.

∎

c. So far we have only considered operators (23.2) acting from L_p into L_q , where $1 \leq p,q < \infty$. The cases, where the spaces L_∞ or C are taken as the classes of inputs or outputs, cause no new difficulties. In particular, the following holds.

Theorem 23.8. Suppose that the operator (23.2) acts from C into L_p , where $1 \leq p \leq \infty$. Then for any $\rho > 0$ there is a function $a(t,\rho) \in L_p$, such that for each $u(t) \in C$ from the ball $\|u(t)\|_C \leq \rho$, at almost all $t \in \Omega$ the estimate

$$|f[t,u(t)| \leq a(t,\rho) \tag{23.19}$$

holds.

d. At the end, let us note an additional feature which may equally be pleasant or disappointing in dependence upon a specific problem.

Under the hypotheses of Theorem 23.5, there exists a function $g(t,u)$ satisfying the estimate

$$|g(t,u)| \leq a(t) + b|u|^{p/q}$$

such that for each measurable $u(t)$, at almost all $t \in \Omega$ the equality

$$f[t,u(t)] = g[t,u(t)]$$

holds.

24. Elements with monotone characteristics

24.1. Cones

A convex and closed set K in the Banach space E is called a *cone* if from $x \in K$ it follows that $\lambda x \in K$ for $\lambda \geq 0$, and $x, -x \in K$ imply $x = 0$. Each cone K induces a *partial order in* E; we shall write $x \prec y$ if $y - x \in K$. Clearly, $x \prec y$ and $y \prec z$ together imply $x \prec z$. In turn, from $x \prec y$ and $y \prec x$ it follows that $x = y$. These inequalities can be multiplied by non-negative numbers. Multiplied by a negative number, the inequality changes its direction. Inequalities having the same direction can be added by sides. One can pass to the limit in the inequalities, etc.

There are three cones in R^1: the half-line $u \geq 0$, half-line $u \leq 0$ and point $u = 0$. In R^2, besides the cone consisting of the origin, there are cones which have the form of rays with the origin as starting point or the form of closed sets contained between pairs of such rays, that define an angle less than π. Each cone K in R^N (except of the singleton containing only the origin) can be described by the equality

$$K = \{x : x = \lambda z, \lambda \geq 0, z \in F_0\} , \tag{24.1}$$

where F_0 is a bounded, convex and closed set which does not contain the origin.

As an important example of a cone in R^N equipped with a certain base, we can take the set K_+ of vectors with non-negative components. Clearly, the cone K_+ is dependent upon the choice of base. For the partial order generated by K_+, the inequality $u \prec v$ denotes that each component of vector u is majorized by the appropriate component of vector v.

In infinite-dimensional spaces, the structure of a cone may be quite complex. In particular, (24.1) will not describe then all cones of practical importance. That characterization only applies to so-called *cones admitting plastering*. In the spaces C and L_p of scalar functions, the cones K_+ of non-negative functions play a special role, however K_+ admits plastering only in the space L_1.

Suppose we intend to construct a cone K in some space E of vector functions with values in R^N. To this end, as a rule we choose a certain cone T in R^N and then construct $K = K(T)$ as a sequence of vector functions $u(t) \in E$ with values in T. In particular, if $T = K_+ \subset R^N$, then the cone $K(T) \subset E$ can also be denoted by K_+. The relation $x(t) \prec y(t)$ means then that the appropriate scalar inequalities are true for all components of the vector functions $x(t)$ and $y(t)$.

A cone K is called *solid* if it contains at least one interior point. A cone K is called *reproducing* if each element $x \in E$ can be represented in the form $x = u - v$ where $u, v \in K$. Every reproducing cone is *non-flattened* in the following sense: there exists minimal constant $\alpha(K)$ (*non-flattening constant*) such that for each $x \in E$ the representation $x = u - v$ holds with $u, v \in K$ and $\|u\|, \|v\| \leq \alpha(K)\|x\|$.

Every solid cone is reproducing. Let us note that the cones K_+ of non-negative functions are solid in C and L_∞, whereas in all spaces L_p with $1 \leq p < \infty$ those cones are only reproducing.

24.2. Special classes of cones

Any partial order in the space E with cone K in a natural way gives rise to the notions of monotone (non-decreasing and non-increasing) sequences, upper and lower bounded sets (in contrast to the boundedness in norm), upper and lower limits of the sets. If there is a minimal element u in the set of the upper limits of a set S , then it is called a *supremum* of the set S and is denoted as $u = \sup S$. Analogously we introduce an *infimum* $\inf S$. For the conical segment $\langle u,v\rangle = \{x : u \prec x \prec v\}$ we have $u = \inf \langle u,v\rangle$ and $v = \sup \langle u,v\rangle$.

If every finite set contains its supremum (hence also infimum), the cone K is called *minihedral*. If every upper (lower) bounded set admits a supremum (infimum), the cone K is called *strongly minihedral*. Every minihedral cone is reproducing, but not necessarily solid.

In R^N, the class of strongly minihedral cones coincides with the class of minihedral cones, as well as it coincides with the class of all cones K_+ (i.e., the cones which contain all vectors with components non-negative in a certain base). The cones K_+ are minihedral in the space C , without being strongly minihedral; K_+ are strongly minihedral in L_p $(1 \le p < \infty)$.

A norm in the space E is called *semi-monotone* if there exists a minimal constant $\beta(K)$ (*normality constant of cone* K) such that $0 \prec x \prec y$ implies $\|x\| \le \beta(K)\|y\|$. If a norm is semi-monotone, then the corresponding cone is called *normal*. A cone K is normal if and only if $x_n \prec z_n \prec y_n$ together with the convergence of both x_n and y_n to the same limit ξ imply $\|z_n - \xi\| \to 0$. Any set bounded in norm is bounded in sense of partial order if and only if the cone K is solid. In turn, any set bounded in sense of partial order is also bounded in norm if and only if the cone K is normal. The cones K_+ in the spaces C and L_p are normal.

If any monotone sequence bounded with respect to a partial order is convergent in norm, the cone K is called *regular*. Every regular cone is normal. The cone K_+ is regular in L_p $(1 \le p < \infty)$, whereas this property does not hold in the spaces C and L_∞ .

If every monotone sequence, bounded in a norm, is convergent in the same norm, the cone K is called *completely regular*. Every completely regular cone is *regular*, too. Every cone that admits plastering is completely regular (the converse is not true!), hence any cone in R^N is completely regular. The cone K_+ in L_p $(1 \le p < \infty)$ is completely regular. Let us also note that some regular cones are not completely regular.

Elements $x,y \in E$ are *comparable* if either $x \prec y$ or $y \prec x$. A set $W \subset E$ is called *linearly ordered* , if all pairs of its elements are comparable. A cone K is referred to as *semi-regular* if it is normal and if there are elements of K arbitrarily close each to the other in every linearly ordered continuum $W \subset E$. Every regular cone is also semi-regular, thus any normal cone in a separable Banach space (e.g., the cone K_+ in C) is semi-regular.

24.3. Monotone characteristics

A scalar characteristic $f(t,u)$ $(t \in \Omega,\ u \in R^1)$ is called *monotone* (or *non-decreasing*), if for every (for almost every) $t \in \Omega$ it is non--decreasing as a function of u . Clearly, every stationary monotone characteristic $f(u)$ is superpositionally measurable. The superpositional measurability of a non-stationary characteristic $f(t,u)$ is equivalent to the *Borel property* (mod 0) of $f(t,u)$ (i. e., $f(t,u)$ differs from a certain Borel function only at points $\{t,u\}$ with t from a set of measure zero).

We shall now consider static elements which have both vector inputs and outputs.

Let T_1 and T_2 be some cones in R^N and R^M , respectively. A characteristic $f(t,u)$ $(t \in \Omega,\ u \in R^N)$ with values in R^M will be called *$\{T_1,T_2\}$-monotone* (or *$\{T_1,T_2\}$-non-decreasing*) if $u_1 \prec u_2$ implies $f(t,u_1) \prec f(t,u_2)$. If T_1 and T_2 are the cones of vectors $u = \{u_1, \ldots, u_N\}$ with non-negative components (in some bases), we shall simply speak of the *monotonicity of the characteristic* , without specifying the cones. In this case, the monotonicity of the characteristic

$$f(t,u) = \{f_1(t,u_1, \ldots, u_N), \ldots, f_M(t,u_1, \ldots, u_N)\} \tag{24.2}$$

is equivalent to the monotonicity of every function $f_j(t,u_1, \ldots, u_N)$ with respect to each variable u_i $(i = 1, \ldots, N)$.

Let us now consider spaces E_1 and E_2 of functions with values correspondingly in R^N and R^M . We shall order the spaces E_1 and E_2 by introducing cones $K_1 = K(T_1)$ and $K_2 = K(T_2)$ of functions with values in T_1 and T_2 , respectively. If the characteristic $f(t,u)$ of a static element is $\{T_1,T_2\}$-monotone and the operator (23.2)

$$\mathfrak{f}\, u(t) = f[t,u(t)] \tag{24.3}$$

acts from E_1 into E_2 , then this operator is $\{K_1,K_2\}$-monotone.

Scalar monotone functions have, as it is well known, a remarkable property: the set of their discontinuity points is at most countable, i.e., the "principal part" of the domain of a monotone function consists of its continuity points. Our purpose consists in setting an analogue of this statement for the operator (24.3).

The last question is by no means trivial, as in the general case monotone operators may be discontinuous in every point. As the first example, we can consider the operator $Au = \kappa(t;u)$, acting from the space R^1, equipped with natural order, into the space $L_\infty(-\pi,\pi)$ ordered by the cone K_+, where $\kappa(t;u)$ is the characteristic function of the set $G(u) = \{s : -\pi \leq s \leq \pi, \text{ tg } s < u\}$. As the second example we take the operator

$$Ax(t) = \lim_{\varepsilon\to 0} \underset{0\leq t\leq\varepsilon}{\text{vrai max}} \text{ arc tg } x(t)$$

acting from $L_1(-\infty,\infty)$ into R^1.

At a first glance, these examples might seem depressing.

A set is called *meagre* if it is union of a countable family of nowhere dense sets. Complete spaces are not meagre, therefore the points of any complete space E that do not belong to a fixed meagre set form a "principal part" of the space E.

Let an operator A with the domain $\mathcal{D}(A) \subset E_1$ act from E_1 into E_2. The point $x_0 \in \mathcal{D}(A)$ (not necessarily belonging to $\mathcal{D}(A)$) is called an *α-discontinuity point of the operator A* if in each neighbourhood of x_0 there are points $y, z \in \mathcal{D}(A)$ such that $\|Ay - Az\| \geq \alpha$. For every $\alpha > 0$, the set $F(\alpha)$ of the α-discontinuity points is closed. Any discontinuity point of the operator A is also its α-discontinuity point at some $\alpha > 0$.

Theorem 24.1. Suppose that the characteristic $f(t,u)$ ($t \in \Omega$, $u \in R^N$) is superpositionally measurable and $\{T_1,T_2\}$-monotone, where T_1 is solid in R^N. Further, assume that the set $\mathcal{D} = \mathcal{D}(\mathfrak{f})$ of the continuous functions $u(t)$ satisfying $\mathfrak{f}\, u(t) \in L_p$ $(1 \leq p < \infty)$ is dense in C (for example, $\mathfrak{f}\, C \subset L_p$). Then the set of the discontinuity points of $\mathfrak{f}$, treated as an operator from C into L_p, is meagre in C.

Theorem 24.2. Assume that the $\{T_1,T_2\}$-monotone (with T_1 solid in R^N) characteristic $f(t,u)$ defines the operator $\mathfrak{f}$ acting from L_p into L_q , where $1 \leq p,q < \infty$. Then the set of the discontinuity points of $\mathfrak{f}$, treated as an operator from L_p into L_q , is meagre in L_p .

Theorems 24.1 and 24.2 will be proved in the subsequent sections. Both of them can be extended onto the elements with characteristics $g(t,u) = f_1(t,u) - f_2(t,u)$, provided $f_1(t,u)$ and $f_2(t,u)$ are monotone. We only should then remember that each of the operators $\mathfrak{f}_1$ and $\mathfrak{f}_2$ has to be defined in the suitable space.

24.4. Proof of Theorem 24.1

Theorem 24.1 is a consequence of the following

Lemma 24.1. Let the space E_1 be partially ordered by a solid cone K_1 , and the space K_2 by a regular cone K_2 . Let the operator A , acting from E_1 into E_2 and defined on a set $\mathcal{D}(A) \subset E_1$, be monotone. Then the set of discontinuity points of the operator A is meagre.

Proof. Suppose the converse, then some set $F(\alpha)$ of the α-discontinuity points of the operator A would contain a ball $K = \{x : x \in E_1 , \|x - x_0\| < r\}$ with non-zero radius. Let u be a fixed, normalized $(\|u\| = 1)$, interior element of the cone K_1 .

Construct the sequence of conical segments

$$T(k) = < x_0 + \frac{k-1}{2k} ru , x_0 + \frac{k}{2k+2} ru> , \quad k = 1, 2, \ldots .$$

Below, in each $T(k)$ we shall select elements ξ_k, η_k such that

$$\xi_k, \eta_k \in T_k \cap \mathcal{D}(A); \ \xi_k \prec \eta_k; \ \|A\xi_k - A\eta_k\| \geq \gamma , \tag{24.4}$$

where $\gamma > 0$. Since

$$\eta_k \prec x_0 + \frac{k}{2k+2} ru \prec \xi_{k+1} ,$$

the sequence

$$\xi_1 \prec \eta_1 \prec \xi_2 \prec \eta_2 \prec \ldots \prec \xi_k \prec \eta_k \prec \ldots \tag{24.5}$$

is monotone. We shall prove that it is upper-bounded by some element $h \in \mathcal{D}(A)$. Then the monotone sequence

$$A\xi_1 \prec A\eta_1 \prec A\xi_2 \prec A\eta_2 \prec \ldots \prec A\xi_k \prec A\eta_k \prec \ldots$$

would also be upper-bounded and, by the regularity of the cone K_2 , convergent in the norm. This would imply the convergence $\|A\xi_k - A\eta_k\| \to 0$, contradicting the last of the inequalities (24.4).

Now, let us construct the points ξ_k, η_k . Divide the segment connecting the points

$$x_o + \frac{k-1}{2k} ru \quad \text{and} \quad x_o + \frac{k}{2k+2} ru$$

into four equal parts. Let y_1, y_2, y_3 be the dividing points (see Fig. 24.1). Take ξ_k as any arbitrary point of the intersection

$$\mathcal{D}(A) \cap \langle x_o + \frac{k-1}{2k} ru, y_1 \rangle ,$$

and η_k as any arbitrary point of the intersection

$$\mathcal{D}(A) \cap \langle y_3, x_o + \frac{k}{2k+2} ru \rangle ;$$

both these points do exist.

Since y_2 is an α-discontinuity point of the operator A , there exist points $v_k, w_k \in \mathcal{D}(A)$ in its neighbourhood $\langle y_1, y_3 \rangle$, such that $\|Av_k - Aw_k\| \geq \alpha$. By construction, $\xi_k \prec v_k \prec \eta_k$ and $\xi_k \prec w_k \prec \eta_k$. Thus, by the monotonicity of the operator A , we can

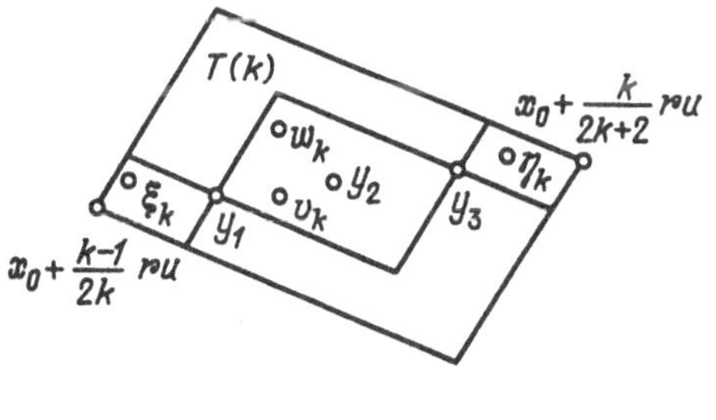

Fig. 24.1

conclude that $A\xi_k < Av_k < A\eta_k$ and $A\xi_k < Aw_k < A\eta_k$. Consequently,

$$||A\eta_k - Av_k|| \le \beta(K_2)\ ||A\eta_k - A\xi_k|| \quad ,$$

$$||A\eta_k - Aw_k|| \le \beta(K_2)\ ||A\eta_k - A\xi_k|| \quad ,$$

where $\beta(K_2)$ is the normality constant of the cone K_2 (let us remind that every regular cone is also normal) and, moreover,

$$\alpha \le ||Av_k - Aw_k|| \le ||Av_k - A\eta_k|| + ||A\eta_k - Aw_k|| \le 2\beta(K_2)\,||A\eta_k - A\xi_k|| \ .$$

All the conditions (24.4) are fulfilled.

To construct the majorant $h \in \mathcal{D}(A)$ of the sequence (24.5), note that the conical segment $< x_o + \frac{1}{2} ru\ ,\ x_o + ru >$ contains points of the ball B and, consequently, points of $\mathcal{D}(A)$. Any such point h is then the upper limit not only for the sequence (24.5) but also for all sets $T(k)$.

∎

Theorem 24.1 applies also if the space C is replaced by L_∞ , since the cone $K(T_1)$ remains solid in L_∞ .

Were $\mathfrak{f}$ considered as an operator from C into L_∞ , also in the case of a monotone characteristic the set of discontinuity points would be not necessarily meagre. An an example (at $N = M = 1$) we can take an element with the characteristic $f(t,u) = \operatorname{sign} u$.

24.5. Proof of Theorem 24.2

Suppose the assertion of Theorem 24.2 is false, then a certain ball in L_p would contain only α-discontinuity points of the operator $\mathfrak{f}$, for any $\alpha > 0$. Take a segment Γ in the above ball that connects the points x_o and $x_o + w$. Let w be a constant vector--function , with value chosen as an interior element of the cone $T_1 \subset R^N$ with non-zero components. We are going to derive the estimate

$$||\mathfrak{f}(x_o + \lambda w) - \mathfrak{f}(x_o + \mu w)|| \ge \alpha_1 \quad (0 \le \lambda < \mu \le 1) \quad , \tag{24.6}$$

where $\alpha_1 > 0$. Though, (24.6) contradicts the regularity of the cone $K(T_2)$ in L_q , as it implies the lack of convergence of the following monotone bounded sequence

$$\mathfrak{f}\left(x_o + \frac{n-1}{2n} w\right) , \quad w = 1, 2, \ldots .$$

Let us proceed to the proof of (24.6). Take $z = x_o + \frac{\lambda + \mu}{2} w$. Then $z \in \Gamma$ and there exist sequences ξ_k, η_k convergent to z in L_p , such that $\|\mathfrak{f}\xi_k - \mathfrak{f}\eta_k\| \geq \alpha$. Since the cone $K(T_1)$ is non-flattened in L_p , one can construct sequences u_k, v_k convergent to z , such that $\xi_k, \eta_k, z \in \langle u_k, v_k \rangle$, and by virtue of the monotonicity of the operator (24.3), $\mathfrak{f}\xi_k, \mathfrak{f}\eta_k, \mathfrak{f}z \in \langle \mathfrak{f}u_k, \mathfrak{f}v_k \rangle$. Therefore,

$$\|\mathfrak{f}\eta_k - \mathfrak{f}\xi_k\| \leq (1 + 2\beta) \|\mathfrak{f}u_k - \mathfrak{f}v_k\| ,$$

where β is the normality constant of the cone $K(T_2)$ in L_q , and furthermore

$$\|\mathfrak{f}v_k - \mathfrak{f}z\| + \|\mathfrak{f}u_k - \mathfrak{f}z\| \geq \frac{\alpha}{1 + 2\beta} .$$

By the last estimate, the monotonicity of operator $\mathfrak{f}$ and Theorem 23.5 imply the bound

$$\|\mathfrak{f}(z + \varepsilon w) - \mathfrak{f}z\| \geq \frac{\alpha}{2 + 4\beta} ,$$

which in turn yields the estimate (24.6). The proof of Theorem 24.2 has been completed.

∎

24.6. Remarks

a. Assume that the spaces E_1 and E_2 are partially ordered by cones K_1 and K_2 , respectively. Let us consider a monotone operator A acting from E_1 into E_2 , with the domain $\mathcal{D}(A)$ dense in E_1 . Not all operators of such type admit extension onto the whole E_1 with monotonicity preserved. Even upon extending an operator by continuity onto its continuity points, the operator may loose the monotonicity property! Nevertheless, *if* K_1 *is solid and* K_2 *is strongly minihedral, then a monotone extension can be defined, in particular, by one of the equalities*

$$A_1 x = \sup\{y : y = Au, u \prec x, u \in \mathcal{D}(A)\} , \tag{24.7}$$

$$A_2 x = \inf\{y : y = Av, v \prec x, v \in \mathcal{D}(A)\} . \tag{24.8}$$

b. A linearly ordered continuum F in the space E, partially ordered by a solid cone K will be called *regular*, if by $x, y \in F$, $x \prec y$ and $x \neq y$ it follows that $y - x$ is an interior element of the cone K.

Theorem 24.3. Suppose that the spaces E_1 and E_2 are partially ordered by a solid cone K_1 and a semi-regular cone K_2, respectively. Assume that the intersection of the domain $\mathcal{D}(A) \subset E_1$ of a monotone operator A acting from E_1 into E_2 with every ball contains a regular, linearly ordered continuum (for example, $\mathcal{D}(A)$ is linear and dense in E_1). Then the set of discontinuity points of the operator A is meagre.

Proof. Having assumed converse, we could indicate a regular, linearly ordered continuum F whose elements are the α-discontinuity points of the operator A, where $\alpha > 0$. We are going to establish the estimate

$$\|Au - Av\| \geq \gamma > 0 \quad (u, v \in F;\ u \neq v) . \tag{24.9}$$

Then the set AF, linearly ordered in E_2, would be a continuum and the uniform estimate (24.9) would contradict the semiregularity of the cone K_2.

In order to show (24.9), we can assume that the whole convex hull of the set F consists of the α-discontinuity points of the operator A. Therefore, for all $u, v \in F$, $u \prec v$, $u \neq v$, we can find sequences $\xi_k, \eta_k \in \mathcal{D}(A)$ such that $\|A\xi_k - A\eta_k\| \geq \alpha$ and both ξ_k and η_k are convergent to the element $\frac{1}{2}(u+v)$. Since the element $v - u$ belongs to the interior of the cone K_1, relations $u \prec \xi_k$ $\eta_k \prec v$ hold for sufficiently large k. In this case, $Au \prec A\xi_k \prec Av$ and $Au \prec A\eta_k \prec Av$, hence

$$\|Av - A\xi_k\| \leq \beta \|Av - Au\| \quad \text{and} \quad \|Av - A\eta_k\| \leq \beta \|Av - Au\| ,$$

where β is the normality constant of the cone K_2 . Consequently, the inequality

$$\alpha \leq \|A\xi_k - A\eta_k\| \leq 2\beta \|Av - Au\|$$

is satisfied, hence also the estimate (24.9) holds.

∎

We should remark that Theorem 24.3 is of minor importance at the study of static elements.

25. Elements with multi-valued characteristics

25.1. Selection problem

In the sequel we shall interpret t as time. Suppose that for every t and for any input u , the corresponding output x of a transducer W may assume more than just one value from the set $\{f(t,u)\}$. As a suitable example we can take a hysteron, despite of the fact that for hysterons, at any prescribed initial state (i.e., for any fixed output at an initial time instant) a function $x(t)$ which defines the output corresponding to a given variable input $u(t)$ is uniquely determined. For hysteron, some other questions are of interest. In particular, one looks for conditions which would guarantee that the values of output $x(t)$ at some t only "weakly" or even "hardly" affect its values at other time instants. We would say then that W is a *static element with multi-valued characteristic* $f(t,u)$. For a while we shall admit that every single-valued output $x(t) \in f[t,u(t)]$ is feasible at a given input $u(t)$ (such outputs $x(t)$ are called *selectors of the multi-valued function* $f[t,u(t)]$.

One can impose some additional requirements on the output signal, taking into consideration only those *selectors* which are *continuous*, *measurable*, *monotone*, etc. Admission of elements with multi-valued characteristics may be due to the lack of accurate quantitative data for the considered transducer; sometimes such admission results from a conceptual idealization, as a stage of mathematical construction, etc.

The choice of a concrete multi-valued characteristic should be done so that to ensure desired properties of its selectors. Before making such a choice, the number of all selectors of interest should be estimated, in particular. These problems are relatively complicated and require a more detailed exposition. We shall discuss them in the sequel.

In several cases, instead of working with selectors, some functions which are "almost selectors" can be used.

25.2. General theorems on selectors

Let us consider a multi-valued operator (function, mapping - all these terms are treated equivalently) F with the domain $\mathcal{D}(F)$ in a certain Banach space E_1, and with values in the form of sets $F(\xi)$ $(\xi \in \mathcal{D}(F))$ in a space E_2.

We shall use the notations: $F(U) = \bigcup_{\xi \in U} F(\xi)$ (for $U \subset \mathcal{D}(F)$) ; $F^{-1}(\eta) = \{\xi : \xi \in \mathcal{D}(F), \eta \in F(\xi)\}$ (for every $\eta \in E_2$) ; $\Gamma(F)$ - *graph of the operator* F, i.e., the set of pairs $\{\xi,\eta\} \in E_1 \times E_2$ $(\xi \in \mathcal{D}(F), \eta \in F(\xi))$. The operator F is called *closed*, if its graph $\Gamma(F)$ is a closed set in $E_1 \times E_2$. In a natural way, we also define operators which are *bounded-valued*, *convex-valued*, etc., as well as *bounded* and *locally bounded*, *compact* and *locally compact*, etc.

The operator F is *upper semicontinuous* in the point $\xi \in \mathcal{D}(F)$, if

$$\lim_{\|\zeta-\xi\|\to 0,\, \zeta\in\mathcal{D}(F)} \theta_\chi[F(\zeta), F(\xi)] = 0 \; ; \tag{25.1}$$

it is *lower semicontinuous* in the point $\xi \in \mathcal{D}(F)$, if

$$\lim_{\|\zeta-\xi\|\to 0,\, \zeta\in\mathcal{D}(F)} \theta_\chi[F(\xi), F(\zeta)] = 0 \; ; \tag{25.2}$$

and, finally, *continuous* in the point $\xi \in \mathcal{D}(F)$, if

$$\lim_{\|\zeta-\xi\|\to 0,\, \zeta\in\mathcal{D}(F)} \rho_\chi[F(\zeta), F(\xi)] = 0 \; . \tag{25.3}$$

The notions of semicontinuity and continuity on a set are introduced in a natural way. In equalities (25.1) - (25.3), $\theta_\chi(M,N)$ denotes the *Hausdorff deviation* of the set N from M, $\rho_\chi(M,N)$ is the *Hausdorff distance* between the sets M and N. In Figure 25.1, function F_1 is upper semicontinuous; function F_2 $(F_2(\beta) = \sin(1/\xi)$ at $\xi \neq 0,$

$F_2(0) = 0$) is not lower semicontinuous in the point $\xi = 0$; function F_3 is continuous. If F is closed-valued and locally compact, then it is upper semicontinuous if and only if the graph $\Gamma(F)$ is closed in $\mathcal{D}(F) \times E_2$.

F is said to satisfy *Michael's conditions*, if it is convex- and closed-valued and for each open $V \subset E_2$ the set $F^{-1}(V)$ is open in $\mathcal{D}(F)$. Obviously, a convex- and closed-valued function satisfies Michael's conditions, if it is lower semicontinuous.

Theorem 25.1. If the multi-valued function F satisfies Michael's conditions, then it has a continuous selector. ∎

As a closeness measure of two single-valued functions with common domain one usually assumes various norms of their difference. There are also other closeness measures which turn out convenient at studying multi-valued functions. Let us define χ*-distance* $\rho_\chi(F,G)$ between two operators F and G which act from E_1 into E_2 as the Hausdorff distance in $E_1 \times E_2$ between the graphs $\Gamma(F)$ and $\Gamma(G)$. In a similar way we can define a χ*-deviation* $\theta_\chi(F,G)$ of a function F from a function G as the Hausdorff deviation of the graph $\Gamma(F)$ from the graph $\Gamma(G)$.

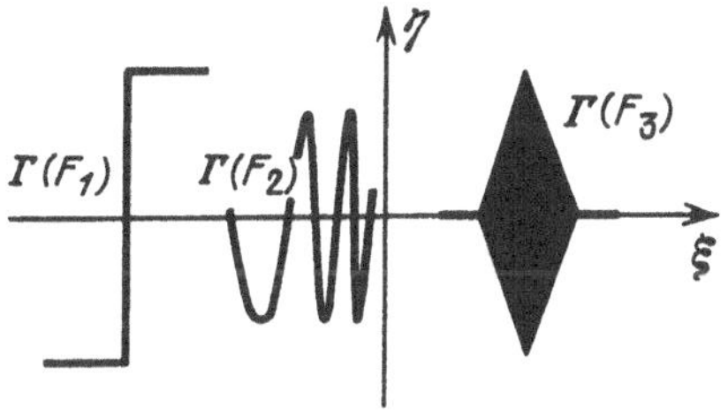

Fig. 25.1

A single-valued function $\eta = \varphi(\xi)$ will be called an ε*-selector* of the multi-valued function F , if $\theta_\chi(\varphi,F) < \varepsilon$ and $\mathcal{D}(\varphi) = \mathcal{D}(F)$. The notion of ε-selector admits a clear "physical" interpretation: for any point $\xi_0 \in \mathcal{D}(F)$ and its small neighbourhood there exists a point ξ such that the distance from $\varphi(\xi_0)$ to the set $F(\xi)$ is small. We shall remind a known criterion for the existence of an ε-selector.

Let $\operatorname{co} M$ be the convex hull of the set M and $\overline{\operatorname{co}} M$ its closure.

Theorem 25.2. Let the space E_1 be separable, and the operator F be convex-valued and upper semicontinuous. Then for any $\varepsilon > 0$, F admits a continuous ε-selector φ such that

$$\varphi(\xi) \in \mathrm{co}\{\eta\colon \eta \in F(\zeta),\ \zeta \in \mathcal{D}(F),\ \|\zeta-\xi\| \le \varepsilon\} \quad (\xi \in \mathcal{D}(F))\ . \tag{25.4}$$

Proof. Assume $\varepsilon > 0$ is given. By the semicontinuity of F, for each $\xi \in \mathcal{D}(F)$ there exists a positive $\delta(\xi;\varepsilon) < \frac{\varepsilon}{2}$ such that

$$\theta_X[F(\zeta),F(\xi)] < \frac{\varepsilon}{2}\ ,\quad \zeta \in \mathcal{D}(F)\ ,\quad \|\zeta-\xi\| < 2\delta(\xi,\varepsilon)\ .$$

The balls $T(\xi) = \{\zeta\colon \|\zeta-\xi\| < \delta(\xi,\varepsilon)\}$ cover the set $\mathcal{D}(F)$. In view of the separability of E_1, this covering contains a locally finite, countable subcovering with open sets U_j $(j = 1,2,\dots)$ which have non-empty intersections $U_j \cap \mathcal{D}(F)$. Being locally finite means here that each point $\xi \in \mathcal{D}(F)$ has a neighbourhood intersecting only a finite number of the sets U_j. Thus there exist non-negative and continuous functions $\alpha_j(\xi)$ defined on E_1, vanishing outside the corresponding U_j, and

$$\alpha_1(\xi) + \alpha_2(\xi) + \dots + \alpha_j(\xi) + \dots = 1\ ,\quad \xi \in \mathcal{D}(F)$$

(in this sum, for each ξ, only a finite number of terms is different from zero).

For each U_j, let us choose a point $\xi_j \in \mathcal{D}(F) \cap U_j$ and a point $\eta_j \in F(\xi_j)$. Then, the single-valued function

$$\varphi(\xi) = \sum_{j=1}^{\infty} \alpha_j(\xi)\eta_j \quad (\xi \in \mathcal{D}(F)) \tag{25.5}$$

will be an ε-selector of the multi-valued operator F. The continuity of function (25.5) follows from the continuity of functions $\alpha_j(\xi)$. Relation (25.4) is obvious.

∎

A multi- or single-valued function $\varphi(\xi)$ $(\xi \in \mathcal{D}(F))$ will be called an *ε-approximation of the function* F, if $\rho_X(\varphi,F) \le \varepsilon$.

Theorem 25.3. Assume that the hypotheses of Theorem 25.2 are fulfilled and, moreover, the operator F is locally compact, with the domain $\mathcal{D}(F)$ containing no isolated points. Then, for every $\varepsilon > 0$, F admits a continuous, single-valued ε-approximation having the property

(25.4).

Proof. We only need to modify slightly the scheme exploited at proof of Theorem 25.2 (that scheme corresponded to a principal part of the original proof given by Michael to his remarkable Theorem 25.1). ∎

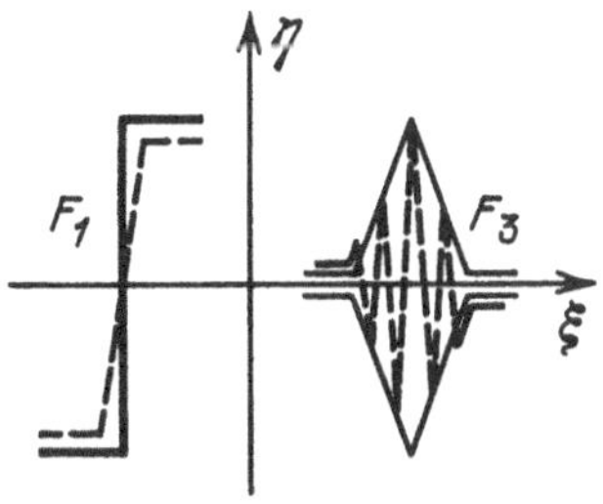

Fig. 25.2

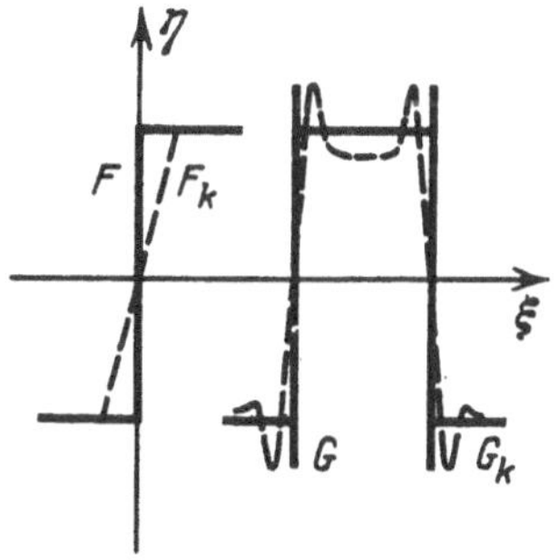

Fig. 25.3

The assumption of the local compactness of F is essential in Theorem 25.3. In Fig. 25.2 the dashed lines depict continuous ε-approximations of the functions F_1 and F_3 from Fig. 25.1.

The Hausdorff distance is not a metric, since $\rho_\chi(F,G) = 0$ does not imply that the operators F and G coincide. For example, always $\rho_\chi(F,\bar{F}) = 0$, where $\bar{F}$ denotes the *closure of operator* F (i.e., $\bar{F}$ is an operator whose graph coincides with the closure of the graph $\Gamma(F)$ in $E_1 \times E_2$). A sequence F_k will be called *Hausdorff-convergent* to F , if $\rho_\chi(F_k,F) \to 0$. The *Hausdorff limit* of F is non-unique, though all such limits admit a unique closure; we shall call it the *closed Hausdorff limit*.

The Hausdorff limit does not coincide with the usual pointwise limit of a sequence of functions. For example, the sequence

$$F_k(\xi) = \begin{cases} \operatorname{sign} \xi \ , & \text{for } |k\xi| \geq 1 \\ k\xi \ , & \text{for } |k\xi| \leq 1 \end{cases} \tag{25.6}$$

is pointwise convergent to the function $F(\xi) = \operatorname{sign} \xi$, while the Hausdorff limit of $F_k(\xi)$ is multi-valued (cf., Fig. 25.3). Another example: partial sums $G_k(\xi)$ of the Fourier series expansion for the function sign sin ξ converge pointwise to this function, while the suitable closed

Hausdorff limit G is (in view of the well-known *Gibbs effect*) a multi-valued function whose graph is shown in Fig. 25.3.

If $E_2 = R^1$, then the closed Hausdorff limit of a sequence of single-valued continuous functions is convex-valued. This does not necessarily hold if $E_2 = R^2$. For example, take the sequence $F_k(\xi)$ (cf., Figure 25.4) of vector functions that are equal to zero for $\xi \leq 0$ and for $k\xi \geq 1$, and whose graphs at $0 \leq k\xi \leq 1$ have the form of turns of the helices on the cylinder $\eta_1^2 - 2\eta_1 + \eta_2^2 = 0$. Its closed Hausdorff limit is a function F with $F(0)$ being a circle.

By Theorem 25.3, if E_1 is separable, then any convex-valued, locally compact and upper semicontinuous operator F is the Hausdorff limit of a sequence of continuous, single-valued functions defined on $\mathcal{D}(F)$. Let us note that the interest in determining the Hausdorff limits contributed to developing the study of multi-valued operators.

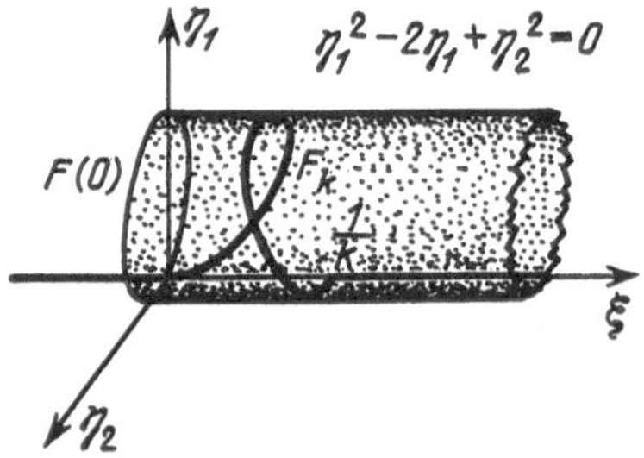

Fig. 25.4

25.3. Monotone selectors

Let E_1 and E_2 be partially ordered by cones K_1 and K_2. An operator F is *monotone* if $\xi_1 < \xi_2$ $(\xi_1, \xi_2 \in \mathcal{D}(F))$ and $u \in F(\xi_1)$, $v \in F(\xi_2)$ imply that the sets $\{\eta: \eta \in F(\xi_2), \eta > u\}$ and $\{\eta: \eta \in F(\xi_1), \eta < v\}$ are non-empty. The operator F is *completely monotone* if $\xi_1 < \xi_2$ $(\xi_1, \xi_2 \in \mathcal{D}(F), \xi_1 \neq \xi_2)$ implies that $u < v$ for all $u \in F(\xi_1)$, $v \in F(\xi_2)$.

The complete monotonicity of an operator F is equivalent to the monotonicity of all its selectors. However, monotone selectors do not exist for all monotone operators.

If a monotone operator F acts from E_1 into R^1 and is compact-valued, then the function $\varphi(\xi) = \max\{\eta: \eta \in F(\xi)\}$ is its monotone selector. If F acts from E_1 into R^2, is compact- and convex-valued, then the function

$$\varphi(\xi) = \frac{1}{2}\min\{\eta: \eta \in F(\xi)\} + \frac{1}{2}\max\{\eta: \eta \in F(\xi)\}$$

is its monotone selector. If F acts from E_1 into R^3 and is monotone, then (as shown by examples) from the compact- and convex-valuedness of F the existence of a monotone selector does not follow.

The situation becomes simpler in the case of monotone functions of one scalar variable.

Theorem 25.4. Let the compact-valued, monotone operator F act from R^1 into a Banach space E_2, partially ordered by a cone K_2. Then each set $F(\xi)$ $(\xi \in R^1)$ contains the values of monotone selectors $\varphi(\xi)$ of the operator F.

Proof. Let us choose points $\xi_* \in R^1$, $\eta_* \in F(\xi_*)$ and construct a monotone selector $\varphi(\xi)$ such that $\varphi(\xi_*) = \eta_*$.

Fix a certain number ξ_o and an element $\eta_o \in F(\xi_o)$. Assume $\xi > \xi_o$ and $\xi_o < \zeta_1 < \zeta_2 < \ldots < \zeta_k = \xi$, where k is any fixed number. By the monotonicity of F there exist $v_1 \in F(\zeta_1)$, $v_2 \in F(\zeta_2)$, $\ldots$, $v_k \in F(\zeta)$, such that $\eta_o \prec v_1 \prec v_2 \prec \ldots \prec v_k$. Denote the set of the last elements v_k by $G(\xi_o, \eta_o; \zeta_1, \zeta_2, \ldots, \zeta_k)$; this set is closed. The intersection of any finite number of the constructed sets (they are dependent on the number k and the selected numbers $\zeta_1, \zeta_2, \ldots, \zeta_{k-1}$) is non-empty, i.e., the constructed sets form a centered system and, by compactness of $F(\xi)$, have a non-empty intersection $G_+(\xi_o, \eta_o, \xi)$. The function $G_+(\xi_o, \eta_o, \xi)$ $(\xi > \xi_o)$ is compact-valued and monotone, and any element $\eta \in G_+(\xi_o, \eta_o, \xi)$ at $\xi > \xi_o$ satisfies the relation $\eta \prec \eta_o$.

Analogously, we define the compact-valued and monotone function $G_-(\xi_o, \eta_o, \xi)$ $(\xi < \xi_o)$ for which from $\eta \in G_-(\xi_o, \eta_o, \xi)$ at $\xi < \xi_o$ it follows that $\eta \prec \eta_o$. Let $G(\xi, \xi_o, \eta_o; F)$ be the multi-valued function coinciding with $G_-(\xi_o, \eta_o, \xi)$ at $\xi < \xi_o$, coinciding with $G_+(\xi_o, \eta_o, \xi)$ at $\xi > \xi_o$ and having the set $\{\eta_o\}$ as the value corresponding to $\xi = \xi_o$.

Suppose now that ξ_j is a dense sequence in R^1, with $\xi_1 = \xi_*$. Choose $F_1(\xi) = G(\xi, \xi_*, \eta_*; F)$ and construct by induction the sequence of functions $F_j(\xi)$ according to the rule $F_j(\xi) = F(\xi, \xi_j, \eta_j; F_{j-1})$, where η_j is any arbitrary point of the set $F_{j-1}(\xi_j)$. Each of the functions $F_j(\xi)$ constructed in such way is monotone and compact-valued, with $F_j(\xi) \subset F_{j-1}(\xi)$ for all $\xi \in R^1$. Therefore, for every $\xi \in R^1$ the intersection $H(\xi) = F_1(\xi) \cap F_2(\xi) \cap \ldots \cap F_j(\xi) \cap \ldots$ is non-empty. The value of the function $H(\xi)$ in any point ξ_j will be the one-

-element set $\{\eta_j\}$; hence, the function $H(\xi)$ is completely monotone. Every selector $\varphi(\xi)$ of this function will be monotone and satisfy the condition $\varphi(\xi_*) = \eta_*$.

∎

25.4. Measurable selectors

We shall now consider multi-valued functions F with the domain $\mathcal{D}(F)$ and values in finite-dimensional Euclidean spaces E_1 and E_2 . We shall assume that $\mathcal{D}(F)$ is Borel measurable and has positive measure. We shall be concerned with Lebesgue measurable selectors of the functions F .

A multi-valued function F is called a *Borel function* if its graph $\Gamma(F)$ is Borel measurable in $E_1 \times E_2$. The function F is called *measurable* if $F(\xi) = G(\xi)$ for almost all $\xi \in \mathcal{D}(F)$, where G is a Borel function.

For further convenience we shall formulate, as a separate theorem, a certain special case of the well-known Lusin-Yankov principle.

Theorem 25.5 (*measurable selection principle*). Every multi-valued measurable function has measurable selectors.

The measurable selectors provide a sufficiently complete representation of a multi-valued function. For example, every measurable function F admits a sequence of measurable selectors $\varphi_j(\xi)$ $(j = 1,2,\ldots)$, such that at almost every $\xi \in \mathcal{D}(F)$ the sequence of points $\varphi_j(\xi)$ is dense in $F(\xi)$. The hypothesis on the existence of the above sequence of selectors may be taken as a basis for defining a multi-valued measurable function; the measurability interpreted in this sense does not imply, however, the measurability as understood throughout this book. Nevertheless, for closed-valued functions both measurability notions are equivalent.

A natural question concerns the existence of Borel measurable selectors for Borelian multi-valued functions. It has been shown in P.S. Novikov's celebrated papers that such selectors simply may not exist!

25.5. Input-output relations

Let us return to studying a static element W with multi-valued characteristic $f(t,u)$ $(t \in \Omega,\ u \in R^N)$ and with values in R^M. As usual, Ω is here a bounded closed set in a finite-dimensional space.

Extending the definitions of Chapter 23, we shall call the characteristic $f(t,u)$ *superpositionally measurable*, if for every measurable (in the Lebesgue sense) single-valued function $u(t)$ with values in R^N, the multi-valued function $f[t,u(t)]$ is measurable. Clearly, all multi-valued characteristics $f(t,u)$ with graphs closed in $R^1 \times R^N \times R^M$ are superpositionally measurable.

Recall that a function $f(t,u)$ is *Borelian (mod 0)* if its graph, possibly only for almost all time instants t, is a Borel set.

Let us now assume that $f(t,u)$ is a superpositionally measurable characteristic. Then, for any measurable $u(t)$ we can define the set $\mathfrak{f}u(t)$ of all measurable selectors to the multi-valued function $f[t,u(t)]$. By virtue of Theorem 25.5, the set $\mathfrak{f}u(t)$ is non-empty. The multi-valued superposition operator $\mathfrak{f}$ determines input-output relations for the static element W.

In order to avoid complications of the same kind as already faced at studying properties of static elements with single-valued characteristics in Chapter 23, we shall confine our further considerations to elements with multi-valued characteristics $f(t,u)$ that are Borelian (mod 0) functions.

Let us recall some of the properties of the operator $\mathfrak{f}$. It acts from L_p into L_q $(1 \le p,q < \infty)$, provided that

$$\sup_{x \in f(t,u)} |x| \le a(t) + b|u|^{p/q}, \tag{25.7}$$

where $a(t) \in L_q$. If the estimate (25.7) holds and $\mathfrak{f}$ is treated as a multi-valued operator from L_p into L_q, then it suffices for its continuity that its characteristic is closed-valued, measurable in t for fixed u, and continuous in u at almost all t. If $f(t,u)$ is at least upper semicontinuous with respect to u (as a multi-valued function), then also $\mathfrak{f}$ is upper semicontinuous; if $f(t,u)$ is monotone with respect to u at almost every t, with the spaces R^N and R^M partially ordered by cones T_1 and T_2, respectively, then $\mathfrak{f}$

is monotone in the spaces L_p and L_q partially ordered by cones $K(T_1)$ and $K(T_2)$, accordingly, etc.

As one could naturally expect, close characteristics generate close superposition operators. Consider this problem in a more detail, assuming the Hausdorff distances between the involved multi-valued functions (i.e., the Hausdorff distances between their graphs) as the closeness measures.

For any specified $t \in \Omega$, by $\chi(t,f,g)$ denote the Hausdorff distance between multi-valued functions $f(t,u)$ and $g(t,u)$. If both operators $\mathfrak{f}$ and $\mathfrak{g}$ of the input-output relationships generated by the functions $f(t,u)$ and $g(t,u)$ act from E_1 into E_2, then the Hausdorff distance in $E_1 \times E_2$ between the graphs $\Gamma(\mathfrak{f})$ and $\Gamma(\mathfrak{g})$ will be denoted by $\rho_\chi(\mathfrak{f},\mathfrak{g}; E_1 \to E_2)$. The most interesting is the situation when

$$\operatorname*{vrai\,max}_{t\in\Omega} \chi(t;f,g) = a < \infty \quad . \tag{25.8}$$

If, in such a case, operators $\mathfrak{f}$ and $\mathfrak{g}$ act from L_∞ into L_∞, then

$$\rho_\chi(\mathfrak{f}, \mathfrak{g}; L_\infty \to L_\infty) = a \quad . \tag{25.9}$$

If, instead, $\mathfrak{f}$ and $\mathfrak{g}$ operate from L_p into L_q where $1 \leq p,q < \infty$, then

$$\rho_\chi(\mathfrak{f}, \mathfrak{g}; L_p \to L_q) \leq a \max\{(\operatorname{mes}\Omega)^{1/p}, (\operatorname{mes}\Omega)^{1/q}\} \quad . \tag{25.10}$$

Estimate (25.10) admits various sharpened versions. In particular, we can set

$$\theta_\chi(A, B; \alpha, \beta) = \sup_{\{u,x\}\in A} \inf_{\{v,y\}\in B} \max\{\alpha|u-v|, \beta|x-y|\} \tag{25.11}$$

and

$$\rho_\chi(A, B; \alpha, \beta) = \max\{\theta_\chi(A, B; \alpha, \beta), \theta_\chi(B, A; \alpha, \beta)\} \quad , \tag{25.12}$$

where $\alpha, \beta > 0$. For any fixed t, let $A(t)$ and $B(t)$ be the graphs of multi-valued functions $f(t,u)$ and $g(t,u)$ of variable u . Choose positive functions $\alpha = \alpha(t)$, $\beta = \beta(t)$ and take

$$\chi(t;\ f,\ g;\ \alpha,\ \beta) = \rho_\chi[A(t),\ B(t);\ \alpha(t),\ \beta(t)] \quad (t \in \Omega) \quad . \tag{25.13}$$

If the operators $\mathfrak{f}$ and $\mathfrak{g}$ act from L_p into L_q $(1 \le p,q < \infty)$, and furthermore

$$\chi(t;\ f,\ g;\ \alpha,\ \beta) \le 1 \quad (t \in \Omega) \quad , \tag{25.14}$$

then the estimate

$$\rho_\chi(\mathfrak{f},\ \mathfrak{g};\ L_p \to L_q) \le \max\left\{\|\alpha(t)\|_{L_p},\ \|\beta(t)\|_{L_q}\right\} \tag{25.15}$$

is true. Upon an appropriate choice of the functions $\alpha(t)$ and $\beta(t)$ estimate (25.15) may become optimal.

26. Closures of static element

26.1. Closure of transducer

We continue the study of operators F (both single- and multi--valued), acting from the space E_1 into E_2 . Let us remind that an operator $\bar{F}$ is called a *closure of operator* F , if the graph $\Gamma(\bar{F})$ coincides with the closure $\overline{\Gamma(F)}$ of the graph $\Gamma(F)$ in $E_1 \times E_2$.

A transfer from the operator F to its closure $\bar{F}$ may be connected with an extension of the appropriate domain. The same may happen for the whole range of operator or for some its parts: the closure of a single-valued operator may be multi-valued, etc.

An operator F is called *closed* , if it coincides with its closure. Each value $F(\xi)$ $(\xi \in \mathcal{D}(F))$ of a closed operator is itself a closed set, whereas the domain of a closed operator may be not closed.

As a simplest example of the closed operator we can take a continuous operator with closed domain.

The notion of closure admits a simple interpretation in physical terms. Suppose that F describes the input-output relations of an element W , possible subject to some noises at the input which produce disturbances at the output. Let the perturbations be equal ε and δ (measured in the norms of E_1 and E_2), respectively.

The simplest situation corresponds to the case $\mathcal{D}(F) = E_1$. Then, at any fixed input $\xi_0 \in E_1$, the attainable outputs create the set

$$A(\xi_0;\ \varepsilon,\ \delta) = \left\{x:\ x \in F(\xi) + z;\quad \|\xi - \xi_0\|_{E_1} < \varepsilon,\ \|z\|_{E_2} < \delta\right\}. \tag{26.1}$$

By definition of the closure $\bar{F}$, the equality

$$\bar{F}(\xi_0) = \bigcap_{\varepsilon,\delta>0} A(\xi_0;\ \varepsilon,\ \delta) \tag{26.2}$$

is true. Thus, *the closure* $\bar{F}$ *comprises all those outputs which are attainable subject to noises with any arbitrary low value.*

Analogous arguments can be used in the case of $\mathcal{D}(F)$ which is a proper subset of E_1 . In this case, the set of outputs that are attainable subject to the noises with values ε and δ , respectively, can be defined as

$$A(\xi_0;\ \varepsilon,\ \delta) = \left\{x:\ x \in F(\xi) + z;\ \xi \in \mathcal{D}(F),\ \|\xi - \xi_0\|_{E_1} < \varepsilon,\ \|z\|_{E_2} < \delta\right\}$$

$$(\xi_0 \in \mathcal{D}(\bar{F}))\ , \tag{26.3}$$

similar to (26.1) . Equality (26.2) remains then valid for all $\xi_0 \in \mathcal{D}(\bar{F})$. A transducer $\bar{W}$ that assigns to every input $\xi_0 \in \mathcal{D}(\bar{F})$ outputs from the set $\bar{F}(\xi_0)$ will be called a *closure of the element* W . Hence, $\bar{W}$ represents an element with input-output relations described by the operator $\bar{F}$.

26.2. Characteristic of the closure

Let us now consider a static element W . As usual, we shall assume that a characteristic $f(t,u)$ $(t \in \Omega,\ u \in R^N)$ has been given (values of the function $f(t,u)$ belong to R^M , the set Ω is bounded and closed in an Euclidean space). Input-output relations of the element W are defined by the superposition operator

$$\mathfrak{f}\,u(t) = f[t,u(t)]\ . \tag{26.4}$$

The closure $\bar{W}$ of a static element W is characterized by the pair of spaces E_1 and E_2, the former including inputs, the latter with the corresponding outputs. There is a question whether the closure $\bar{W}$ is also a static element. Were $\bar{W}$ a static element, whether one could construct its characteristic at a given characteristic $f(t,u)$ of element W? Is the characteristic of such an element $\bar{W}$ dependent (and, if yes, then how ?) on a choice of the spaces E_1 and E_2? Quite often, we are able to give answers to the above questions.

Suppose that a characteristic $f(t,u)$ is Borelian (mod 0). Then for each fixed $t = t_*$, we can construct the closure $\bar{f}_u(t_*,u)$ of the function $f(t_*,u)$ with respect to u. The closure $\bar{f}_u(t,u)$ is then also a Borelian (mod 0) function that generates the superposition operator

$$\bar{\mathfrak{f}}_u\, u(t) = \bar{f}_u[t,u(t)] \quad . \tag{26.5}$$

The main result of this section is given in the form of the following simple theorem.

Theorem 26.1. Assume that a characteristic $f(t,u)$ is Borelian (mod 0) and the operator $\mathfrak{f}$ acts from L_p into L_q, where $1 \le p,q \le \infty$. Then also the operator $\bar{\mathfrak{f}}_u$ acts from L_p into L_q and $\bar{\mathfrak{f}}_u = \bar{\mathfrak{f}}$.

Proof. It is enough to show the inclusions

$$\bar{\mathfrak{f}}\, u_o(t) \subset \bar{\mathfrak{f}}_u\, u_o(t) \quad , \quad \bar{\mathfrak{f}}_u\, u_o(t) \subset \bar{\mathfrak{f}}\, u_o(t) \tag{26.6}$$

for any fixed function $u_o(t) \in L_p$.

Let $x_o(t) \in \bar{\mathfrak{f}}\, u_o(t)$. By definition of the closure of an operator, we can find sequences $u_k(t)$ and $x_k(t) \in \mathfrak{f}\, u_k(t)$, respectively convergent to $u_o(t)$ and $x_o(t)$ (in the norms of the spaces E_1 and E_2, correspondingly).

These sequences can be considered as convergent at almost every $t \in \Omega$. Therefore for almost every t the relation $x_o(t) \in \bar{f}_u[t,u_o(t)]$ holds, i.e., $x_o(t) \in \bar{\mathfrak{f}}_u\, u_o(t)$.

The first of the inclusions (26.6) has been proved.

Now assume $x_o(t) \in \bar{\mathfrak{f}}_u\, u_o(t)$. For each $k = 1, 2, \ldots$ define the function

$$h_k(t) = \left\{\{u,x\} : |u - u_o(t)| \le \frac{1}{k} , x \in f(t,u), |x - x_o(t)| \le \frac{1}{k}\right\}.$$

This function is Borelian (mod 0) and defined for almost all $t \in \Omega$. By Theorem 25.5, each function $h_k(t)$ has a measurable selector $\{u_k(t), x_k(t)\}$.

By the definition of the function $h_k(t)$, the components of these selectors satisfy the inequalities

$$|u_k(t) - u_o(t)| \le \frac{1}{k} , \quad |x_k(t) - x_o(t)| \le \frac{1}{k} \quad (k = 1, 2, \ldots) .$$

Hence ,

$$\lim_{k\to\infty} \|u_k(t) - u_o(t)\|_{L_p} = \lim_{k\to\infty} \|x_k(t) - x_o(t)\|_{L_q} = 0 ,$$

and, because $x_k(t) \in \mathfrak{f}\, u_k(t)$, $x_o(t) \in \bar{\mathfrak{f}}\, u_o(t)$. The latter of inclusions (26.6) has been proved.

∎

If $\mathfrak{f}$ is considered as an operator from C into L_p , where $1 \le p < \infty$, then the equality $\bar{\mathfrak{f}}_u = \bar{\mathfrak{f}}$ is still true. However, if $\mathfrak{f}$ is treated as an operator from C into L_∞ , then only the former of inclusions (26.6) remains true, the latter may fail !

As an example consider a static element with the characteristic $f(t,u) = \operatorname{sign} u$ $(0 \le t \le 1, u \in R^1)$. The function $\bar{f}_u(t,u)$ is single-valued at $u \ne 0$ and coincides with sign u ; the set $\bar{f}_u(t,0)$ contains three elements: -1, 0, 1. Thus, the set $\bar{\mathfrak{f}}_u u_o(t)$, where $u_o(t) \equiv 0$, comprises all functions measurable on [0,1] which assume three values -1, 0, 1. At the same time, the set of zeros of each function from $\bar{\mathfrak{f}}\, u_o(t)$, if $\mathfrak{f}$ is treated as an operator from C into L_∞ , is closed (with the accuracy up to a set of measure zero).

Theorem 26.1 and the further considerations bring answers to the questions stated at the begin of this section, concerning static elements with Borelian characteristics.

In the case of a non-Borelian (but still superpositionally measurable) characteristic $f(t,u)$, the closure $\bar{\mathfrak{f}}$ of an operator $\mathfrak{f}$ acting from L_p into L_q $(1 \le p,q < \infty)$ will also be a static element. A direct construction of the characteristic of element $\bar{\mathfrak{f}}$ as generated by a function $f(t,u)$ (without passing to operators in function spaces) is apparently difficult.

26.3. Closure modulo a negligence class

In many constructions it turns out useful to neglect values of some functions in single points, on countable sets, on sets of measure zero, etc. Sometimes, the operations of taking closure to functions and operators are also treated in a similar way.

Consider a function F with the domain $\mathcal{D}(F) \subset E_1$ and with values in E_2. Take a system N of sets $D \subset \mathcal{D}(F)$, having the following properties: every subset $D \in N$ has no interior points; $D \in N$ implies that every subset of D also belongs to N; the union of every countable family of sets from N belongs to N, as well. Denote by $\Gamma(F;N)$ the set of those pairs $\{\xi,x\} \in E_1 \times E_2$ for which the intersection of every neighbourhood $S(\xi) \subset E_1$ of the point ξ with the whole pre-image $F^{-1}[S(x)]$ of any neighbourhood $S(x) \subset E_2$ of the point x does not belong to N. The set $\Gamma(F;N)$ will be the graph of a certain single- or multi-valued function $\bar{F}$ (mod N), called a *closure of the function* F *modulo negligence class* N.

Every closure $\bar{F}$ (mod N) is a closed operator. It is upper semi-continuous in the Hausdorff metric (as was the usual closure $\bar{F}$). By the local compactness of F, $\bar{F}$ (mod N) is locally compact, etc. The closure $\bar{F}$ coincides with closure $\bar{F}$ (mod N), if N consists only of the empty set.

Denote by H the set of those Borelian (mod 0) functions H(t) $(t \in \Omega)$ with values in R^N, whose graphs $\Gamma(H)$ are of measure zero in $\Omega \times R^N$. By $D_p(H)$ $(H \in H)$ denote the set of those single-valued functions $u(t) \in L_p$ which coincide with a certain selector of the function H(t) on some set of positive measure. The system N_p^o of sets $D_p(H)$ $(H \in H)$ fulfils all the requirements assumed for the negligence classes.

The hypothesis that the function H(t) is Borelian turns out essential. For example, a monster (see Section 23.2) gives rise to a construction of a non-Borelian, superpositionally measurable function H(t) for which $D_p(H)$ would coincide with the whole L_p and the sets from the negligence class would not have interior points.

Let us return to the study of the operator $\mathfrak{f}$ defined according to (26.4) on the basis of the Borelian (mod 0) function f(t,u). Also the function

$$g(t,u) = \bigcap_{\varepsilon>0} \bigcap_{S \subset R^N,\ \mathrm{mes}\, S = 0} \{x: x = f(t,v);\ v \notin S,\ |v-u| < \varepsilon\} \quad (26.7)$$

will then be Borelian (mod 0) . The function (26.7) and the corresponding superposition operator g may be multi-valued in this case.

Theorem 26.2. Suppose that $f(t,u)$ is Borelian (mod 0) and $\mathfrak{f}$ operates from L_p into L_q , where $1 \le p,q \le \infty$. Then the closure $\bar{\mathfrak{f}}$ (mod N_p^o) of operator $\mathfrak{f}$ modulo negligence class N_p^o coincides with the operator g constructed on the basis of the function (26.7).

Proof. The assertion of Theorem 26.2 follows almost directly from Theorem 25.5 on measurable selection.

∎

Theorem 26.2 says that the closure of a static element with Borelian characteristic modulo the negligence class N_p^o is also a static element (with characteristic (26.7)).

26.4. Comments

a. If the characteristic $f(t,u)$ is treated as a function on the set $\Omega_1 = \Omega \times R^N$, then its closure would be some function $\tilde{f}(t,u)$. If $f(t,u)$ does not depend on t , then $\tilde{f}(t,u)$ coincides with $\bar{f}_u(t,u)$ (see Section 26.1); in the general situation these functions are different. The inclusion $\bar{f}_u(t,u) \subset \tilde{f}(t,u)$ is evident.

b. The measurability of the function $f(t,u)$ does not imply (see, Chapter 23) its superpositional measurability. In this connection, if only the measurability of the function $f(t,u)$ is known, then one can guarantee the measurability of functions $f[t,u(t)]$ only for some measurable $u(t)$. The set of these measurable $u(t)$ will be assumed as the domain of operator $\mathfrak{f}$. If $\mathfrak{f}$ acts from L_p into L_q (i.e., by $u(t) \in L_p$ and the measurability of $f[t,u(t)]$ it follows that $f[t,u(t)] \in L_q$), then we might construct its closures in a various sense. In particular, the closure modulo negligence class would coincide with the operator g generated by the Borelian (mod 0) function (26.7).

c. Consider a single-valued operator F, acting from a Banach space E_1 into a Banach space E_2. Denote by F_C the restriction of F to the set of its continuity points. The graph $\Gamma(\overline{F_C})$ of the closure $\overline{F_C}$ of operator F_C is contained in the graph of the closure $\bar{F}$ (mod N) of operator F modulo any arbitrary negligence class N.

If the corresponding set of discontinuity points is meagre and N^* denotes the class of meagre sets, then $\overline{F_C} = \bar{F}$ (mod N^*).

Assume that the characteristic $f(t,u)$ is monotone and the operator $\mathfrak{f}$ acts from the space C or L_p, where $1 \leq p \leq \infty$, into the space L_q where $1 \leq q < \infty$. By virtue of Theorems 24.1 and 24.2, the set of the discontinuity points of $\mathfrak{f}$ is then meagre. Thus, $\overline{\mathfrak{f}_C} = \bar{\mathfrak{f}}(\text{mod } N^*)$. Furthermore, $\overline{\mathfrak{f}_C}$ coincides with $\bar{\mathfrak{f}}$ (mod N_p^0) and, by Theorem 26.2, with the operator g generated by the function (26.7). If $\mathfrak{f}$ is defined as an operator from C into L_q, then $\overline{\mathfrak{f}_C}$ coincides with the closure of the operator $\mathfrak{f}$ modulo the class of sets which have Wiener measure zero (the last statement refers to the case $\Omega = [a,b]$).

27. Weak closures and convexification procedure

27.1. Weak closures

Important properties of nonlinear transducers, in general, and static elements, in particular, are reflected by their closures in various weak topologies. Here we shall need only simplest concepts and statements related to the weak topology and the weak convergence (see also [42] for some more details, in particular).

Let E be a Banach space and $S \subset E$. By definition, a point $x_* \in E$ belongs to the *closure of the set* S *in the weak topology*, if for every finite set $\ell_1, \ldots, \ell_r$ of linear functionals from E^* and every $\varepsilon > 0$, the intersection of the set S with the *weak neighbourhood*

$$U(x_*; \ell_1, \ldots, \ell_r; \varepsilon) = \{x: |\ell_j(x - x_*)| < \varepsilon;\ j = 1, \ldots, r\}$$

of the point x_* is non-empty. A point x_* is called the *weak limit of the sequence* $x_k \in E$, and we write $x_k \rightharpoonup x_*$, if $\ell(x_k - x_*) \to 0$ for every $\ell \in E^*$. The closure of a set S in the sense of weak limit is, in general, different from its closure in weak topology (in the case of a finite-dimensional space E, both procedures are identical with taking the standard closure in norm).

Suppose that, within a certain study (experiment, measurement, etc.), an element $x \in E$ is described by a fixed number of scalar characteristics $\ell_j(x)$. Then, provided $x_* \in \bar{S}$, it follows that there exist elements of S whose characteristics are arbitrarily close to the relevant characteristics of the element x_*. The weak convergence $x_k \to x_*$ yields the convergence of all chosen characteristics of the elements x_k to the corresponding characteristics of the element x_*. If E denotes a function space, then one can take averaged (with various weights) values of the functions, values of the functions in separate points, etc., as the characteristics $\ell_j(x)$ (linear functionals).

The above "physical" considerations by no means exhaust the role of weak closures.

Assume now that two Banach spaces E_1 and E_2 are given, together with a set $\Gamma \subset E_1 \times E_2$. By $\vec{\Gamma}$ we shall denote the set of pairs $\{u_*, x_*\} \in E_1 \times E_2$ for which there exist $\{u_k, x_k\} \in \Gamma$ such that $\|u_k - u_*\| \to 0$ and $x_k \to x_*$ (the sequence x_k converges weakly to x_* in the space E_2). By $\overset{\twoheadrightarrow}{\Gamma} \subset E_1 \times E_2$ we shall denote the set of pairs $\{u_*, x_*\}$ such that for every finite set $\ell_1, \ldots, \ell_r$ of the linear functionals from E_2^* and for every $\varepsilon > 0$, Γ has a non-empty intersection with the neighbourhood

$$U(u_*, x_*; \ell_1, \ldots, \ell_r; \varepsilon) =$$

$$= \{\{u,x\}: \|u - u_*\|_{E_1} < \varepsilon;\ |\ell_j(x - x_*)| < \varepsilon;\ j = 1, \ldots, r\}. \qquad (27.1)$$

Evidently, the following relations hold:

$$\Gamma \subset \bar{\Gamma} \subset \vec{\Gamma} \subset \overset{\twoheadrightarrow}{\Gamma} \subset \overline{\mathrm{co}}\,\Gamma. \qquad (27.2)$$

Let F be an operator with the domain $\mathcal{D}(F) \subset E_1$ and with values in E_2. Then the closures $\vec{F}$ and $\overset{\twoheadrightarrow}{F}$ of the graph $\Gamma(F)$ represent the graphs of some operators $\vec{F}$ and $\overset{\twoheadrightarrow}{F}$. We shall call them a *sequential weak closure* and a *weak closure* of operator F, respectively. By (27.2), $\vec{F}$ is an *extension* of operator $\bar{F}$ (i.e., $\mathcal{D}(\bar{F}) \subset \mathcal{D}(\vec{F})$ and $\bar{F}u \subset \vec{F}u$ for all $u \in \mathcal{D}(\bar{F})$), whereas $\overset{\twoheadrightarrow}{F}$ is an extension of operator $\vec{F}$.

If the operator F is locally bounded and from every sequence bounded in E_2 one can select a weakly convergent subsequence (for example, $E_2 = L_q$ with $1 < q < \infty$), then

$$\mathcal{D}(\vec{F}) = \mathcal{D}(\overset{\twoheadrightarrow}{F}) = \overline{\mathcal{D}(F)} \quad . \tag{27.3}$$

In all these cases, the operator $\overset{\twoheadrightarrow}{F}$ is closed, i.e., its graph $\Gamma(\overset{\twoheadrightarrow}{F})$ is closed in $E_1 \times E_2$ in the natural norm). If F is locally bounded and the space E_2^* is separable (e.g., $E_2 = L_q$ with $1 < q < \infty$), then also $\vec{F}$ is a closed operator.

Let F define the input-output relations of a transducer W. Then the operators $\vec{F}$ and $\overset{\twoheadrightarrow}{F}$ define the input-output relations for transducers $\vec{W}$ and $\overset{\twoheadrightarrow}{W}$ which represent a *sequential weak closure* and a *weak closure of the transducer* W, respectively. In the sequel, we shall study properties of the sequential weak closures and the weak closures of static elements. Are those closures static elements themselves ? If yes, then how can one construct their characteristics ? Are those characteristics dependent upon a choice of E_1 and E_2 ? ...

27.2. Convexification

Let an operator F (single- or multi-valued) act from E_1 into E_2. By its *convexification* we shall mean

$$F^{\square}u = \bigcap_{\varepsilon>0} \overline{\mathrm{co}}\,\{y\colon y \in Fv\,,\ v \in \mathcal{D}(F)\ ,\ \|v-u\|_{E_1} < \varepsilon\}\ , \tag{27.4}$$

defined for all those $u \in E_1$ for which the set (27.4) is non-empty.

Directly by the above definition, numerous properties of the convexification can be concluded. Every operator $F^{\square}$ is closed; for any F, $\mathcal{D}(\bar{F}) \subset \mathcal{D}(F^{\square}) \subset \overline{\mathcal{D}(F)}$; the operator $F^{\square}$ is an extension of the weak closure $\overset{\twoheadrightarrow}{F}$ to the operator F. The equalities $F^{\square} = (F^{\square})^{\square} = (\bar{F})^{\square}$ hold. If F is locally compact, then $\mathcal{D}(F^{\square}) = \overline{\mathcal{D}(F)}$ and $F^{\square}$ is upper semicontinuous, etc.

Sometimes a different from (27.4), convex-valued (but not necessarily closed) extension

$$F_1^{\square}\,u = \overline{\mathrm{co}}\,\{y\colon y \in \bar{F}u\} \tag{27.5}$$

of the operator F is considered, too. Assume that $F_2^\square$ is the minimal closed, convex-valued extension of the operator (27.5). If F is locally compact, then $F_2^\square = F_1^\square = F^\square$.

If $F(u) = \operatorname{sign} u \quad (u \in R^1)$, then $F^\square(u) = F(u)$ at $u \neq 0$, and $F^\square(0)$ is the whole segment $[-1,1]$.

If $F(u)$ is the characteristic function of the set of all irrational numbers $u \in R^1$, then $\bar{F}(u)$ for every u comprises two points 0 and 1, and $F^\square(u)$ is the segment $[0,1]$.

Let F describe the input-output relations of a transducer W. Then $F^\square$ describes the input-output relations of the transducer $W^\square$ which is the *convexification of transducer* W.

In the sequel we shall analyze properties of the convexification of static elements.

27.3. Weak closures and convexification of static element

Consider a characteristic $f(t,u) \quad (t \in \Omega;\ u \in R^N)$ with values in R^M, Borelian (mod 0) and locally bounded with respect to u for almost all t. By $f_u^\square(t,u) \quad (t \in \Omega;\ u \in R^N)$ we shall denote its convexification with respect to u (for any fixed value of t). The characteristic $f_u^\square(t,u)$ will also be Borelian (mod 0); for almost all $t \in \Omega$ it will be upper semicontinuous in u. The characteristics $f(t,u)$ and $f_u^\square(t,u)$ generate the superposition operators $\mathfrak{f}$ and $\mathfrak{f}_u^\square$.

Throughout this section we assume that $\mathfrak{f}$ acts from L_p into L_q, where for the time being p and q are arbitrary. Then, $\mathfrak{f}_u^\square$ acts also from L_p into L_q; it is closed and convex-valued. If $q < \infty$, then $\mathfrak{f}_u^\square$ is upper semicontinuous.

The most important properties of the sequential weak closures, weak closures and convexifications that correspond to static elements are given by the following theorems.

Theorem 27.1. Assume that a characteristic $f(t,u)$ is Borelian (mod 0) and the operator $\mathfrak{f}$ acts from L_p into L_q, where $1 \leq p \leq \infty$ and $1 \leq q \leq \infty$. Then

$$\mathfrak{f}^\square = \mathfrak{f}_u^\square \quad . \tag{27.6}$$

The equality (27.6) is true also in the case when $\mathfrak{f}$ is considered as an operator acting from the space C of continuous functions into the space L_q where $1 \leq q \leq \infty$. If, however, $\mathfrak{f}$ is treated as an operator from C into L_∞, then $\mathfrak{f}^{\square}$ may be different from $\mathfrak{f}_u^{\square}$.

Theorem 27.2. Assume that a characteristic $f(t,u)$ is Borelian (mod 0) and the operator $\mathfrak{f}$ acts from L_p into L_q, where $1 \leq p \leq \infty$ and $1 \leq q < \infty$. Then

$$\vec{\mathfrak{f}} = \overset{\rightharpoonup}{\mathfrak{f}} = \mathfrak{f}^{\square} = \mathfrak{f}_u^{\square} . \tag{27.7}$$

Theorems 27.1 and 27.2 permit, in particular, to omit indices p and q in the notions of the sequential weak closure, weak closure and convexification to the operator $\mathfrak{f}$.

27.4. Proof of Theorem 27.1

First we establish the inclusion

$$\mathfrak{f}^{\square} u_*(t) \subset \mathfrak{f}_u^{\square} u_*(t) \quad (u_*(t) \in L_p) , \tag{27.8}$$

where $\mathfrak{f}^{\square}$ is the convexification of $\mathfrak{f}$ treated as an operator from L_p into L_q. The right-hand side of inclusion (27.8) is independent of q and its left-hand side may only expand while q decreases. Therefore it suffices to consider solely the case $q = 1$.

But for $q = 1$ the operator $\mathfrak{z}$ is upper semicontinuous and, due to Theorem 26.1, the equality $\mathfrak{z} = \mathfrak{z}_u$ holds. Consequently,

$$\overline{co}\, \mathfrak{z}\, u_*(t) = \overline{co}\, \mathfrak{z}_u\, u_*(t) \subset \mathfrak{f}_u^{\square} u_*(t) \quad (u_*(t) \in L_p) , \tag{27.9}$$

and (27.8) follows by the upper semicontinuity of the operator $\mathfrak{z}$.

To complete the proof, it remains to show that also the converse is true, i.e.,

$$\mathfrak{f}_u^{\square} u_*(t) \subset \mathfrak{f}^{\square} u_*(t) \quad (u_*(t) \in L_p) . \tag{27.10}$$

Let us assume that $q < \infty$. Both sides of inclusion (27.10) are convex, bounded and closed sets. Hence (and by virtue of the Hahn-Banach theorem), inclusion (27.10) is equivalent to the inequality

$$\sup_{x(t)\in \mathfrak{f}_u^{\square} u_*(t)} \int_\Omega (x(s), w(s))dx \leq$$

$$\leq \sup_{y(t) \in \mathfrak{f}^{\square} u_*(t)} \int_\Omega (y(s), w(s))dx \quad (u_*(t) \in L_p) ,$$

for any continuous function $w(t)$ $(t \in \Omega)$. To prove this inequality, for any given continuous function $w(t)$, we construct a function $y_*(t) \in \bar{\mathfrak{f}} u_*(t)$ such that

$$\int_\Omega (x(s), w(s))ds \leq \int_\Omega (y_*(s), w(s))ds \quad (x(t) \in \mathfrak{f}_u^{\square} u_*(t)) .$$

This inequality would be fulfilled, if any arbitrary selector of the Borelian (mod 0) function

$$\Pi(t) = \bar{f}_u[t, u_*(t)] \cap \{z: z \in R^M: \ (z,w(t)) \geq \alpha(t)\} \ (t \in \Omega) ,$$

with

$$\alpha(t) = \max \{\alpha: \alpha = (x, w(t)), \ x \in f_u^{\square}[t, u_*(t)]\} \tag{27.11}$$

will be chosen as equal $y_*(t)$.

For almost all t, the set $\Pi(t)$ is non-empty. It contains the point x of the convex compact set $f_u^{\square}[t,u_*(t)] \subset R^M$ which realizes the maximum in (27.11).

It remains to prove inclusion (27.10) for $q = \infty$. In this case we have to show that every $x_*(t) \in \bar{\mathfrak{f}}_u^{\square} u_*(t) \subset L_\infty$ is an element of the set $\mathfrak{f}^{\square} u_*(t)$, at the same time.

At almost all $t \in \Omega$, the set $f_u^{\square}[t, u_*(t)]$ is compact in R^M. Let us fix the function $x_*(t)$ and denote by $S(t)$ the set of all vectors

$$\xi = \{x_1, x_2, \dots, x_{M+1}; \alpha_1, \alpha_2, \dots, \alpha_{M+1}\}$$

such that each of their first $M+1$ components is an extremal point of the convex compact set $f_u^{\square}[t, u_*(t)]$, the remaining components α_j are non-negative scalars, and

$$\alpha_1 + \alpha_2 + \dots + \alpha_{M+1} = 1 \quad \text{and} \quad \alpha_1 x_1 + \alpha_2 x_2 + \dots + \alpha_{M+1} x_{M+1} = x_*(t) .$$

By the classical Minkowski theorem, every point of a convex compact set in R^M is a convex combination of at most $M+1$ extremal points of that compact set. Thus, the set $S(t)$ is non-empty. The function $S(t)$, defined for almost all $t \in \Omega$, is Borelian (mod 0). Moreover, by Theorem 25.5, this function has a measurable selector

$$\xi(t) = \{x_1(t), \dots, x_{M+1}(t); \alpha_1(t), \dots, \alpha_{M+1}(t)\} \quad \{t \in \Omega\} .$$

Therefore, the function $x_*(t)$ admits the representation

$$x_*(t) = \alpha_1(t)\, x_1(t) + \dots + \alpha_{M+1}(t)\, x_{M+1}(t) \quad (t \in \Omega) , \qquad (27.12)$$

with scalar functions $\alpha_j(t)$, measurable and non-negative, for almost all $t \in \Omega$ satisfying the equality

$$\alpha_1(t) + \alpha_2(t) + \dots + \alpha_{M+1}(t) = 1 ,$$

and with vector functions $x_j(t)$ belonging to the set $\mathfrak{f}_u^{\square} u_*(t)$, bounded in L_∞. Moreover, since values of the vector functions $x_j(t)$ are extremal points of the compact sets $f_u^{\square}[t, u_*(t)]$, they belong to the set $\mathfrak{f}_u[t, u_*(t)]$, thus $x_j(t) \in \mathfrak{f}_u u_*(t) \subset L_\infty$.

Now let us fix the number $\varepsilon > 0$ and construct finite-valued, measurable scalar functions $\beta_j(t)$ $(t \in \Omega;\ j = 1, 2, \dots, M+1)$ with rational values such that

$$0 \le \beta_j(t) \le 1 , \quad \beta_1(t) + \beta_2(t) + \dots + \beta_{M+1}(t) = 1 ,$$

and for almost all $t \in \Omega$, the following estimates hold:

$$|\alpha_j(t)\, x_j(t) - \beta_j(t)\, x_j(t)| < \frac{\varepsilon}{M+1} \quad (j = 1, 2, \dots, M+1) .$$

Then by (27.12) it follows that

$$\|x_*(t) - \beta_1(t)\,x_1(t) - \dots - \beta_{M+1}(t)\,x_{M+1}(t)\|_{L_\infty} < \varepsilon \ . \qquad (27.13)$$

Let us divide the set Ω into a finite number of measurable subsets $\Omega_1, \dots, \Omega_r$ such that all functions $\beta_j(t)$ are constant on each of these subsets,

$$\beta_j(t) = \frac{\gamma_{jk}}{\ell} \quad (t \in \Omega_k;\ j = 1, \dots, M+1;\ k = 1, \dots, r) \ ;$$

ℓ is there some positive integer and γ_{jk} are non-negative integers such that, for any fixed k,

$$\gamma_{1k} + \gamma_{2k} + \dots + \gamma_{\overline{M+1}\,k} = \ell \qquad (k = 1, \dots, r) \ .$$

On each set Ω_k we define a family which comprises ℓ functions $z_{1k}(t), z_{2k}(t), \dots, z_{\ell k}(t)$ such that the first γ_{1k} of them coincide with $x_1(t)$, γ_{2k} equal $x_2(t)$, etc. Eventually, the last $\gamma_{\overline{M+1}\,k}$ equal $x_{M+1}(t)$. Next, define the functions $y_1(t), \dots, y_\ell(t)$ on the whole Ω by

$$y_i(t) = z_{ik}(t) \quad (t \in \Omega_k;\ k = 1, \dots, r;\ i = 1, \dots, \ell).$$

By construction, these functions belong to the set $\mathfrak{z}_u\, u_*(t)$ and satisfy the equality

$$\beta_1(t)\ x_1(t) + \dots + \beta_{M+1}(t)\ x_{M+1}(t) = \frac{1}{\ell}\,[y_1(t) + \dots + y_\ell(t)] \quad (t \in \Omega) \ ,$$

i.e.,

$$\beta_1(t)\,x_1(t) + \dots + \beta_{M+1}(t)\,x_{M+1}(t) \in (\mathfrak{z}_u)^\square u_*(t) = (\mathfrak{z})^\square u_*(t) = \mathfrak{z}^\square u_*(t) \ .$$

Therefore, due to (27.13) and the arbitrary choice of ε, the function $x_*(t)$ belongs to the set $\mathfrak{z}^\square\, u_*(t)$.

Theorem 27.1 has been completely proved.

∎

In the proof of Theorem 27.1, inclusion (27.9) was used. It is also possible to give a simple proof of the following stronger equality

$$\overline{\mathrm{co}}\ \bar{f}\ u_*(t) = \mathfrak{f}_u^{\square}\ u_*(t) \quad (u_*(t) \in L_p) \quad . \tag{27.14}$$

In Theorem 27.1 we have assumed that the characteristic $f(t,u)$ is Borelian. This hypothesis turns out crucial. A relevant example may be easily constructed for a monster (cf., Section 23.2).

For a non-Borelian, but superpositionally measurable characteristic $f(t,u)$, we can construct a Borelian characteristic $g(t,u)$ such that

$$\mathfrak{f}^{\square} = \mathfrak{g} = \bar{\mathfrak{g}} = \mathfrak{g}^{\square} = \mathfrak{g}_u^{\square} \quad .$$

If $\mathfrak{f}$ is treated there as an operator from L_p into L_q, where $1 \le p \le \infty$ and $1 \le q < \infty$, then $\mathfrak{f}^{\square}$ is upper semicontinuous.

27.5. Proof of Theorem 27.2

At first let us consider any arbitrary operator F acting from a Banach space E_1 into a Banach space E_2.

Lemma 27.1. Assume that the operator F is locally bounded, the space E_2 (e.g., $E_2 = L_q$, $1 < q < \infty$) is separable and from every bounded sequence of elements of E_2 one can choose a weakly convergent subsequence. Let the set $\vec{F}\,u_*$ be convex for any fixed element u_* of the space E_1. Then, $\vec{F}u_* = F^{\square}u_*$.

Proof. Let us assume the converse. Then, because $F^{\square}$ is an extension of the operator $\vec{F}$, it is possible to find an element $x_* \in F^{\square}u_*$ that does not belong to the closed convex set $\vec{F}\,u_*$. By the Hahn-Banach theorem, there exists a linear functional $\ell \in E_2^*$ such that

$$\ell(x) - \ell(x_*) \ge \alpha > 0 \quad (x \in \vec{F}\,u_*) \quad . \tag{27.15}$$

Simultaneously, by definition of the operator $F^{\square}$, there must be points in the neighbourhood $\{x\colon \ell(x) < \ell(x_*) + \alpha/2\}$ that belong to all sets $\overline{\mathrm{co}}\,\{x\colon x \in F\,u\ ,\ \|u - u_*\| < \varepsilon\}$, thus also to all sets $\{x\colon x \in F\,u\ ,\ \|u - u_*\| < \varepsilon\}$. Therefore, we can construct a sequence $\{u_k, x_k\} \in \Gamma(F)$ so that $\|u_k - u_*\| \to 0$ and $\ell(x_k) < \ell(x_*) + \alpha/2$. By the local boundedness of the operator F, the sequence x_k is bounded and it can be regarded as weakly convergent to a certain point

$y_* \in E_2$.

∎

Now we are prepared to give a proof of Theorem 27.2. We shall confine ourselves to the case of single-valued characteristics $f(t,u)$ and $1 < q < \infty$.

Since $\vec{F}$ is an extension of the operator $\vec{F}$ and $F^{\square}$ is an extension of the operator $\vec{F}$, by Lemma 27.1 it is enough to establish the convexity of every set $\vec{f}\, u_*(t)$ for $u_*(t) \in L_p$.

The set $\vec{f}\, u_*(t)$ is closed, thus it is sufficient to take arbitrary functions $x(t)$ and $y(t)$ such that $x(t),\, y(t) \in \vec{f}\, u_*(t) \subset L_q$, and to show that

$$\frac{1}{2}\, x(t) + \frac{1}{2}\, y(t) \in \vec{f}\, u_*(t) \quad .$$

For this purpose, let us suppose that there exist two sequences $u_k(t),\, v_k(t) \in L_p$ $(k = 1, 2, \ldots)$, convergent to $u_*(t)$ in the norm of L_p and such that $f\, u_k(t) \rightharpoonup x(t)$, $f\, v_k(t) \rightharpoonup y(t)$. It is sufficient to construct a new sequence $w_k(t)$ of functions assuming (at a fixed k and for every $t \in \Omega$) either the value $u_k(t)$ or the value $v_k(t)$, and convergent. Then,

$$f\, w_k(t) \rightharpoonup \frac{1}{2}\, x(t) + \frac{1}{2}\, y(t) \quad . \tag{27.16}$$

$L_{q'}$ is the dual space to L_q , provided that

$$\frac{1}{q} + \frac{1}{q'} = 1 \quad .$$

Let the sequence of functions $h_1(t), h_2(t), \ldots$ be dense in $L_{q'}$. Then it suffices for ensuring the convergence (27.16) that the following estimates hold:

$$\left| \int_{\Omega} \left(f\, w_k(t) - \frac{1}{2}\, x(t) - \frac{1}{2}\, y(t),\; h_j(t) \right) dt \right| < \frac{1}{k} \quad (j = 1, \ldots, k) \quad . \tag{27.17}$$

Let $\Omega \subset R^{N_0}$ and a certain system of coordinates be chosen in R^{N_0}.

Denote by $\nu(t)$ the first coordinate of the point $t \in \Omega$. For all $k, r = 1,2, \ldots$, define the function

$$\beta_{k,r}(t) = \begin{cases} u_k(t) & , \text{ if } j - 1/2 \leq r\nu(t) < j , \\ v_k(t) & , \text{ if } j \leq r\nu(t) < j + 1/2 \end{cases}$$

where j assumes integer values. Estimates (27.17) will be true, if a function $r = r(k)$ with sufficiently fast asymptotic growth has been taken and if the functions

$$w_k(t) = \beta_{k,r(k)}(t) \quad (t \in \Omega;\ k = 1, 2, \ldots)$$

have been defined.

Theorem 27.2 has been completely proved.

∎

In the case of a non-Borelian characteristic $f(t,u)$, the operators $\vec{f}$ and $\vec{\mathfrak{f}}$ may be different from $f_u^{\square}$. We can only claim that for every operator f acting from L_p into L_q (where $q < \infty$), the operators $\vec{f}$ and $\vec{\mathfrak{f}}$ coincide with some operator g having Borelian (mod 0) characteristic, and such that $g = g^{\square} = \bar{g}$.

27.6. Convexification of static element modulo negligence class

This section is closely related to Section 26.3.

Let F be an operator from E_1 into E_2. Let us introduce a certain class N of negligible sets (which have all properties specified in Section 26.3). Then the operator

$$F^{\square}(\mathrm{mod}\ N)\, u = \bigcap_{\varepsilon>0} \ \bigcap_{D\in N} \overline{\mathrm{co}}\, \{y : y \in Fv,\ v \in \mathcal{D}(F) \setminus D,\ \|v - u\| < \varepsilon\} \tag{27.18}$$

represents a *convexification of the operator* F *modulo the negligence class* N.

Let us continue the study of a static element with characteristic $f(t,u)$. Set

$$h(t,u) = \bigcap_{\varepsilon>0} \ \bigcap_{S\subset R^N,\ \mathrm{mes}\ S=0} H(t,u;S,\varepsilon) \quad (t \in \Omega,\ u \in R^N), \qquad (27.19)$$

where

$$H(t,u;S,\varepsilon) = \overline{\mathrm{co}}\ \{x:\ x = f(t,v) \in R^M,\ v \notin S,\ |v-u| < \varepsilon\} \ . \qquad (27.20)$$

For almost all $t \in \Omega$ and arbitrary $u \in R^N$, the set (27.19) which corresponds to the pair $\{t,u\}$ is non-empty. If $\mathfrak{f}$ acts from L_p into L_q $(1 \leq p,q < \infty)$, then also the operator h generated by the function (27.19) acts from L_p into L_q. For this operator,

$$h = \bar{h} = \bar{h}_u = h_u^{\square} = \bar{h}^{\square} \ (\mathrm{mod}\ N_p^o) \ . \qquad (27.21)$$

In the formula (27.21), the negligence class N_p^o coincides with the class used in Theorem 26.2.

It is also possible to extend the notions of the sequential weak closure and weak closure of an operator by defining them modulo some negligence classes.

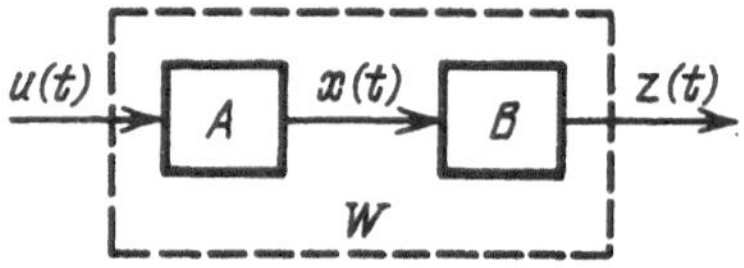

Fig. 27.1

27.7. Examples of open nonlinear systems composed of static elements

Let W be a complex transducer which is composed of two elements A and B, *connected in a cascade* as shown in Figure 27.1. We shall use the same notations A and B for the defining operators of static elements A and B. Natural questions arise whether it is possible to construct closures and convexifications of the operator

$$W = BA \qquad (27.22)$$

at given closures and convexifications of operators A and B, and, if it is possible, then how it could be performed. These questions are non-trivial and often bring quite unexpected answers. We shall give here two simple examples.

a. Let $A = \mathfrak{f}$ be a transducer with Borelian (mod 0) characteristic $f(t,u)$ ($t \in \Omega$, $u \in R^N$) such that $\mathfrak{f}$ act from L_p into L_q ($1 \leq p \leq \infty$, $1 \leq q < \infty$). Let B be a linear compact operator which acts from L_q into a certain function space E. Recall that classical examples of linear compact operators are the integral operators and the inverse operators in the boundary value problems of mathematical physics, vibration theory, etc. The linear compact operators transform weakly convergent sequences into sequences which are convergent in an appropriate norm. Thus, according to Theorem 27.2,

$$\bar{W} = \overline{B\mathfrak{f}} = (B\,\mathfrak{f})^{\square} = B\vec{\mathfrak{f}} = B\vec{\mathfrak{f}} = B\,\mathfrak{f}^{\square} = B\,\mathfrak{f}_u \quad . \tag{27.23}$$

We should necessarily point out here that in case of discontinuous characteristics, the operators $\overline{B\mathfrak{f}}$ and $B\,\bar{\mathfrak{f}}$ do not coincide, in general.

b. For the treatment of more complex systems, with block-diagrams comprehending several elements with discontinuous characteristics, one has to introduce new closure and convexification operations. This is connected with the necessity of taking into account small noises not only at the input and output of the system but also in some internal channels of information processing.

As an important example, let us consider a system W composed of two static elements $\mathfrak{f}$, $\mathfrak{g}$ with Borelian (mod 0) characteristics $f(t,u)$, $g(t,u)$ ($t \in \Omega$, $u \in R^N$, $x \in R^M$) and one linear compact element B, connected in a cascade. For definiteness, let us assume that the inputs and outputs of all the elements of system W are treated as elements of the appropriate spaces L_2, and smallness of the noises is to be understood as measured by the norms of the perturbation at the input. An operator $\bar{\bar{W}}$ that describes the whole system W subject to small noises, should for each fixed input $u_o(t)$ specify the set of limits in L_2-norm of all the convergent sequences

$$z_n(t) = B\,g\{t, f[t, u_o(t) + \xi_n(t)] + \eta_n(t)\} \quad ,$$

where

$$\lim_{n\to\infty} \|\xi_n(t)\| = \lim_{n\to\infty} \|\eta_n(t)\| = 0 .$$

One can easily find that

$$\bar{\bar{W}} = B\, h^{\square} , \tag{27.24}$$

where h is the superposition operator with the multi-valued characteristic

$$h(t,u) = \bar{g}_u[t, \bar{f}_u(t,u)] \quad (t \in \Omega,\ u \in R^N) .$$

In a natural way, the transfer from $B\,g\,f$ to the operator (27.24) can be treated as taking the closure of the system composed of the elements f, g and B connected in a cascade.

In general, (27.24) coincides neither with the operator $B\,g^{\square} f^{\square}$ nor with $B f_1^{\square}$, where $f_1(t,u) = g[t, f(t,u)]$.

28. Relay

28.1. Ideal relay

We shall use the same notion of an *ideal relay with threshold value* α for two different static elements: an element $r_-(\alpha)$ with the characteristic

$$r_-(u,\alpha) = \begin{cases} 0 , & \text{if } u \le \alpha , \\ 1 , & \text{if } u > \alpha \end{cases} \tag{28.1}$$

and an element $r_+(\alpha)$ with the characteristic

$$r_+(u,\alpha) = \begin{cases} 0 , & \text{if } u < \alpha , \\ 1 , & \text{if } u \ge \alpha . \end{cases} \tag{28.2}$$

The characteristics (28.1) and (28.2) define superposition operators $r_-(\alpha)$ and $r_+(\alpha)$. Sometimes, they are referred to as *ideal relays*. Each of these operators acts from any L_p $(1 \le p \le \infty)$ into the space L_∞ .

Characteristics (28.1) and (28.2) admit the same closure

$$\bar{r}(u,\alpha) = \bar{r}_u(u,\alpha) = \begin{cases} 0 \ , & \text{if} \quad u < \alpha \ , \\ \text{two-element set} \quad \{0,1\} \ , & \text{if} \quad u = \alpha \ , \\ 1 \ , & \text{if} \quad u > \alpha \end{cases} \qquad (28.3)$$

and convexification

$$r^{\sqcap}(u,\alpha) = r_u^{\square}(u,\alpha) = \begin{cases} 0 \ , & \text{if} \quad u < \alpha \ , \\ \text{interval} \quad [0,1] \ , & \text{if} \quad u = \alpha \ , \\ 1 \ , & \text{if} \quad u > \alpha \ . \end{cases} \qquad (28.4)$$

An analysis of ideal relays may take advantage of all theorems formulated in this chapter, since both (28.1) and (28.2) are Borel functions.

Ideal relays with characteristics (28.1) and (28.2) are completely monotone. Their closures and convexifications (i.e., static elements with characteristics (28.3) and (28.4)) are monotone but not completely monotone.

28.2. Non-ideal relay

From now on up to the end of this chapter, we shall consider only transducers with scalar inputs $u(t)$ and scalar outputs $x(t)$.

The non-ideal relays we consider are no longer static elements. However, they can still be taken as representative examples of transducers with input-output relations described by discontinuous operators.

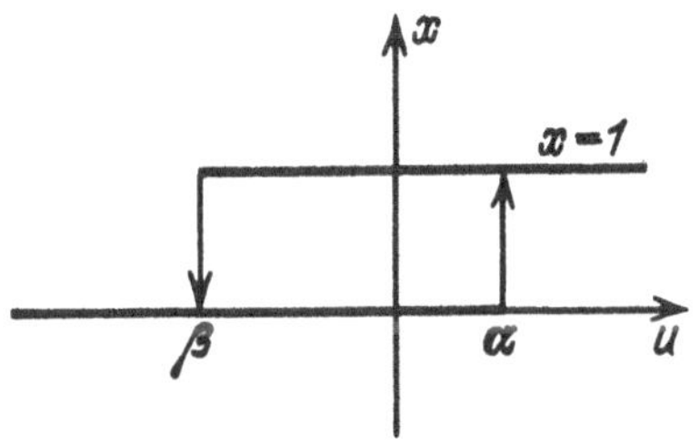

Fig. 28.1

The set of points $\{u,x\}$ of the plane, contained between two half-lines: $x = 0$ for $u < \alpha$ and $x = 1$ for $u > \beta$ defines the *domain* $\Omega(\alpha,\beta) = \Omega(R(\alpha,\beta))$ *of feasible states of the non-ideal relay* $R(\alpha,\beta)$ *with the threshold values* α *and* β $(\beta < \alpha)$. In Figure 28.1, the set $\Omega(\alpha,\beta)$ is indicated by the thickened line.

Sometimes the state of a non-ideal relay is defined by the value of its output. The set of feasible states comprises then two elements: 0 and 1 . For our exposition, such situation seems to be of less interest.

For any initial state $\{u_0,x_0\} \in \Omega(\alpha,\beta)$, continuous functions $u(t)$ $(t \geq t_0)$ that satisfy the condition

$$u(t_0) = u_0 \ , \tag{28.5}$$

can be taken as admissible inputs at $t = t_0$.

For any fixed admissible input $u(t)$, the corresponding output

$$x(t) = R[t_0,\ x_0;\ \alpha,\ \beta]\ u(t) \qquad (t \geq t_0) \tag{28.6}$$

can be given various descriptions. The simplest possibility is offered by the following explicit (although quite awkward) formula:

$$x(t) = \begin{cases} 0 \ , & \text{if } u(t) \leq \beta \ , \\ 1 \ , & \text{if } u(t) \geq \alpha \ , \\ x_0 \ , & \text{if } u(\tau) \in (\beta,\alpha) \text{ for all } \tau \in [t_0,t] \ , \\ 0 \ , & \text{if } u(t) \in (\beta,\alpha) \text{ and there exists } t_1 \in [t_0,t) \text{ such} \\ & \quad \text{that } u(t_1) = \beta \text{ and } u(\tau) \in (\beta,\alpha) \text{ for all } \tau \in [t_1,t], \\ 1 \ , & \text{if } u(t) \in (\beta,\alpha) \text{ and there exists } t_1 \in [t_0,t) \text{ such} \\ & \quad \text{that } u(t_1) = \alpha \text{ and } u(\tau) \in (\beta,\alpha) \text{ for all } \tau \in [t_1,t]. \end{cases} \tag{28.7}$$

Hence, the non-ideal relay is an example of a deterministic transducer defined for all continuous inputs. Clearly, the non-ideal relay is static and controllable. The suitable semigroup identity assumes the standard form

$$R[r_0,x_0;\alpha,\beta]u(t) = R[t_1,R[t_0,x_0;\alpha,\beta]u(t_1);\alpha,\beta]u(t) \qquad (t_0 \leq t_1 \leq t) \ . \tag{28.8}$$

The values of the output (28.7), corresponding to the continuous input $u(t)$ $(t \geq t_0)$, can be completely characterized on a closed interval $[t_1,t_2]$ by the following rule: if either $x(t_1) = 0$ and $u(t) < \alpha$ for $t \in [t_1,t_2]$, or $x(t_1) = 1$ and $u(t) > \beta$ for $t \in [t_1,t_2]$, then $x(t)$ remains constant for $t \in [t_1,t_2]$. We say that the above rule *excludes superfluous switchings*.

An important property of the non-ideal relay is its *monotonicity with respect to inputs:* if $\{u(t_0),x_0\}$, $\{v(t_0),y_0\} \in \Omega(\alpha,\beta)$, $x_0 \leq y_0$ and

$$u(t) \leq v(t) \quad (t \geq t_0) , \tag{28.9}$$

then

$$R[t_0,x_0;\alpha,\beta]u(t) \leq R[t_0,y_0;\alpha,\beta]v(t) \quad (t \geq t_0) . \tag{28.10}$$

We can also use the monotonicity with respect to inputs in order to define a non-ideal relay. To this purpose, first we must define the outputs that correspond to monotone inputs. As a next step, by using the semigroup identity we define the outputs for continuous inputs which are only piecewise monotone. In turn, due to the postulated monotonicity with respect to inputs, it is possible to construct a unique monotone extension of the input-output relations, so far defined on piecewise monotone inputs, to the whole space of continuous functions. The extension constructed in such a way represents the operator (28.6) defined over all continuous inputs.

As an additional kind of monotonicity of a relay, let us observe its natural monotonicity with respect to the threshold values.

The range of operator (28.6) has the form of the two-element set $\{0,1\}$. Thus, (28.6) can be treated as an operator from the space $C = C(t_0,t_1)$ into any space $L_q = L_q(0,1)$ where $1 \leq q \leq \infty$. Directly by definition of relay, we can conclude the following compactness property.

Theorem 28.1. Every $R[t_0,x_0;\alpha,\beta]$ $(\beta < \alpha)$ is locally compact as an operator from $C(t_0,t_1)$ into $L_q(t_0,t_1)$, where $1 \leq q < \infty$.

Proof. It is enough to take a fixed function $u_*(t) \in C(t_0,t_1)$ which fulfils the condition $\{u_*(t_0),x_0\} \in \Omega(\alpha,\beta)$ and observe that the number of discontinuity points of the outputs $R[t_0,x_0;\alpha,\beta]u(t)$ $(t_0 \leq t \leq t_1)$ is uniformly bounded, if all the outputs $u(t)$ under consideration belong to a small neighbourhood of the function $u_*(t)$ in $C(t_0,t_1)$. ∎

Suppose that the operator $R[t_0,0;\alpha,\beta]$ with values in L_∞ is defined for functions $u(t) \in C$ such that $u(t_0) < \beta$. Then it will be discontinuous for all those $u(t)$ which assume value β at some $\tau \in (t_0,t_1]$. Analogously, $R[t_0,1;\alpha,\beta]$ as an operator from C into L_∞ is discontinuous in the point $u(t) \in C$ $(u(t_0) > \alpha)$, if $u(\tau) = \alpha$ at least for one $\tau \in (t_0,t_1]$. Therefore, the set of discontinuity points of the operator $R[t_0,x_0;\alpha,\beta]$ acting from C into L_∞ contains open domains.

The situation becomes different, if $R[t_0,x_0;\alpha,\beta]$ is an operator from C into L_q , where $1 \leq q < \infty$. In this case, the relevant set of the discontinuity points is meagre.

Let Figure 28.1 represent the graph of a multi-valued characteristic of some static element. Then such an element will be different from the relay $R(\alpha,\beta)$.

To conclude this section, let us remark that the threshold values α and β are often referred to as switch-on and switch-off currents, respectively. If the output assumes value 1 , one says sometimes that the relay is switched on; correspondingly, at the output equal 0 the relay is switched off.

28.3. Periodic inputs

First we are going to extend the operator (28.6) to all continuous inputs $u(t)$ $(t \geq t_0)$, without maintaining the requirement that $\{u(t_0),x_0\} \in \Omega(\alpha,\beta)$. To do this, in addition to (28.7) we shall assume that

$$R[t_0,1;\alpha,\beta]u(t) \equiv R[t_0,0;\alpha,\beta]u(t) , \quad \text{if } u(t_0) \leq \beta , \tag{28.11}$$

and

$$R[t_0,0;\alpha,\beta]u(t) \equiv R[t_0,1;\alpha,\beta]u(t) , \quad \text{if } u(t_0) \geq \alpha . \tag{28.12}$$

The operators (28.11) and (28.12), extended onto all continuous inputs, can be characterized in the case of monotone inputs by the following intuitively clear formulas

$$R[t_0,1;\alpha,\beta]u(t) = \begin{cases} r_-[u(t),\beta] , & \text{if } u(t_0) > \beta , \\ r_+[u(t),\alpha] , & \text{if } u(t_0) \leq \beta , \end{cases} \tag{28.13}$$

and

$$R[t_0,0;\alpha,\beta]u(t) = \begin{cases} r_+[u(t),\alpha] , & \text{if } u(t_0) < \alpha , \\ r_-[u(t),\beta] , & \text{if } u(t_0) \geq \alpha , \end{cases} \tag{28.14}$$

incorporating the functions (28.1) and (28.2). Similarly simple and easily interpretable is the formula which determines the outputs for piecewise monotone inputs, provided that the system (28.13), (28.14) is complemented by the semigroup identity.

For any continuous input $u(t)$ $(-\infty < t < \infty)$, let us define, also for all t , two continuous functions

$$R_0\, u(t) = \lim_{\tau \to -\infty} R[\tau,0;\alpha,\beta]u(t) \qquad (-\infty < t < \infty) \tag{28.15}$$

and

$$R_1\, u(t) = \lim_{\tau \to -\infty} R[\tau,1;\alpha,\beta]u(t) \qquad (-\infty < t < \infty) . \tag{28.16}$$

The existence of both limits is evident. Furthermore, for each fixed t_* there exists τ_* such that for $\tau \leq \tau_*$ all the functions $R[\tau,0;\alpha,\beta]u(t)$ (analogously, all the functions $R[\tau,1;\alpha,\beta]u(t)$) assume identical values at $t \geq t_*$. The functions (28.15) and (28.16) do not coincide if and only if $u(t) \in (\beta,\alpha)$ for all $t \leq t_{**}$, where t_{**} is some finite constant. We shall write

$$y(t) = R(-\infty;\alpha,\beta)u(t) \qquad (-\infty < t < \infty) , \tag{28.17}$$

treating the right-hand side as an operator which ascribes two functions (28.15) and (28.16) to the input $u(t)$ $(-\infty < t < \infty)$. As already said, the same function may correspond to different inputs as a value of the operator (28.17).

In a more detail we shall consider the case of a periodic input $u(t)$ with some period T. Then also the functions (28.15) and (28.16) are periodic, with the same period. They are different only in the case of $u(t) \in (\alpha,\beta)$; in that case $R_0\, u(t) \equiv 0$ and $R_1\, u(t) \equiv 1$.

For the T-periodic input $u(t)$, the corresponding output $x(t) = R[t_0,x_0;\alpha,\beta]\, u(t)$ is periodic at $t \geq T + t_0$, i.e., $x(t + T) = x(t)$ at $t \geq t_0 + T$. Upon extending this output by periodicity from the half-axis $t \geq t_0 + T$ onto the whole real axis, the resulting function will coincide with $R_{x_0}\, u(t)$.

28.4. Closure of relay

The closure $\bar{\Omega}(\alpha,\beta)$ of the domain of feasible states $\Omega(\alpha,\beta)$ of the relay $R(\alpha,\beta)$ can be obtained by completing $\Omega(\alpha,\beta)$ with two points: $\{\alpha,0\}$ and $\{\beta,1\}$ (see Fig. 28.1).

For any continuous input $u(t)$ $(t \geq t_0)$, we define two corresponding sets $\bar{R}[t_0,x_0;\alpha,\beta]u(t)$ $(x_0 = 0,1)$ of the functions $x(t)$ with two values $0, 1$, such that the following hypotheses are satisfied:

a. If $u(t) > \beta$ at $t \in [t_1,t_2] \subset [t_0,\infty)$, then $x(t)$ is non-decreasing on $[t_1,t_2]$.

b. If $u(t) < \alpha$ at $t \in [t_1,t_2] \subset [t_0,\infty)$, then $x(t)$ is non-decreasing on $[t_1,t_2]$.

c. $\{u(t),x(t)\} \in \Omega(\alpha,\beta)$ at $t \geq t_0$ and $x(t_0) = R[t_0,x_0;\alpha,\beta]u(t_0)$.

Theorem 28.2. Assume that $R[t_0,x_0;\alpha,\beta]$ are treated as operators from $C(t_0,t_1)$ into $L_q(t_0,t_1)$ where $1 \leq q < \infty$. Then the operators $\bar{R}[t_0,x_0;\alpha,\beta]$ will represent their closures.

Proof. The assertion follows as a direct consequence of the appropriate definitions.

∎

The closures $\bar{R}[t_0,x_0;\alpha,\beta]$ differ in their properties from the operators $R[t_0,x_0;\alpha,\beta]$. For example, the semigroup property admits there the following interpretation: if a function $x(t)$ $(t \geq t_o)$ belongs to $\bar{R}[t_o,x_o;\alpha,\beta]u(t)$, then its restriction to the interval $t \geq t_1$ $(t_1 > t_o)$ can only be claimed to belong to some set $R[t_1,x_1;\alpha,\beta]u(t)$ where $x_1 \in \bar{R}[t_o,x_o;\alpha,\beta]u(t_1)$.

Consider the relay $R(-1,1)$ and input $u(t) = \sin t$ $(t \geq 0)$. The set $\bar{R}[0,0;1,-1]$ is then countable and the restriction of every function from this set to $[0,\pi]$ will coincide with one of the following three functions: $x_1(t) \equiv 0$; $x_2(t) = 0$ for $0 \leq t \leq \pi/2$ and $x_2(t) = 1$ for $\pi/2 < t \leq \pi$; $x_3(t) = 0$ for $0 \leq t < \pi/2$ and $x_3(t) = 1$ for $\pi/2 \leq t \leq \pi$. Note that the last two functions are indistinguishable as elements of any L_q.

28.5. Convexification of relay

Consider $R[t_o,x_o;\alpha,\beta]$ as an operator acting from the space $C(t_o,t_1)$ into some space $L_q = L_q(t_o,t_1)$. A *convexification of the relay* is defined by the standard equality

$$R^{\square}[t_o,x_o;\alpha,\beta]u(t) = \bigcap_{\varepsilon>0} \overline{\text{co}}\, \{x(t) : x(t) = R[t_o,x_o;\alpha,\beta]v(t), \ \|v(t)-u(t)\|_{t_o,t_1} < \varepsilon\} . \tag{28.18}$$

The following theorem is a direct consequence of this definition.

Theorem 28.3. Every set $R^{\square}[t_o,x_o;\alpha,\beta]u(t)$ consists of functions $x(t)$ satisfying the hypotheses :

(a) If $u(t) > \beta$ for $t \in [\tau_1,\tau_2] \subset [t_o,\infty)$, then $x(t)$ is non-decreasing on $[\tau_1,\tau_2]$.

(b) If $u(t) < \alpha$ for $t \in [\tau_1,\tau_2] \subset [t_o,\infty)$, then $x(t)$ is non-increasing on $[\tau_1,\tau_2]$.

(c) $x(t_o) = R[t_o,x_o;\alpha,\beta]u(t_o)$ and $\{u(t),x(t)\} \in \Omega^{\square}(\alpha,\beta)$ for all $t \geq t_o$, where the domain $\Omega^{\square}(\alpha,\beta)$ of the feasible states of the convexified relay comprises the original domain $\Omega(\alpha,\beta)$ and the rectangle $\beta \leq u \leq \alpha$, $0 \leq x \leq 1$. ∎

If $u(t) \in (\beta,\alpha)$ for $\tau_1 \leq t \leq \tau_2$, then by virtue of Theorem 28.3, every function $x(t)$ from $R^{\square}[t_o,x_o;\alpha,\beta]u(t)$ assumes on $[\tau_1,\tau_2]$

some constant value $x(\tau_1) \in [0,1]$.

The closures and convexifications of the operators $R[t_0,x_0;\alpha,\beta]$ are multi-valued. It seems natural to develop a study of non-ideal relays along the following lines: at the very beginning introduce a multi-valued operator $R[t_0;\alpha,\beta]$ whose values comprise two elements, $R[t_0,0;\alpha,\beta]u(t)$ and $R[t_0,1;\alpha,\beta]u(t)$, then construct the closure $\bar{R}[t_0;\alpha,\beta]$ and the convexification $R^{\square}[t_0;\alpha,\beta]$. If $R[t_0;\alpha,\beta]$ is treated as an operator from C into any L_q , then for every input the set $\bar{R}[t_0;\alpha,\beta]u(t)$ is intersection of the sets $\bar{R}[t_0,0;\alpha,\beta]u(t)$ and $\bar{R}[t_0,1;\alpha,\beta]u(t)$. If $R[t_0;\alpha,\beta]$ is considered as an operator from C into L_q , where $1 \le q < \infty$, then every set $R^{\square}[t_0;\alpha,\beta]u(t)$ consists of functions $x(t)$ which satisfy hypotheses (a), (b) of Theorem 28.3, and hypothesis (c) of that theorem, relaxed by removing the condition $x(t_0) = R[t_0,x_0;\alpha,\beta]u(t_0)$. Analogues of the semigroup identity can be formulated for the operator $R^{\square}[t_0;\alpha,\beta]$ more naturally and simpler than it was for the operators $R^{\square}[t_0,x_0;\alpha,\beta]$.

Now consider the sequence $L(\Gamma_\ell^k,\Gamma_r^k)$ $(k = 1, 2, \dots)$ of the generalized plays (cf., Section 2.2). By $L(t_0;\Gamma_\ell^k,\Gamma_r^k)$ denote the operator that assigns to any continuous input $u(t)$ $(t \ge t_0)$ the set of functions $L[t_0,x_0;\Gamma_\ell^k,\Gamma_r^k]u(t)$ where $\Gamma_r^k[u(t_0)] \le x_0 \le \Gamma_\ell^k[u(t_0)]$. Let $\theta_\ell(k) + \theta_r(k) \to 0$, where $\theta_\ell(k)$ is the Hausdorff distance between the graphs of functions $\Gamma_\ell^k(u)$ and $r^{\square}(u,\beta)$, and $\theta_r(k)$ is the same distance between the graphs of $\Gamma_r^k(u)$ and $r^{\square}(u,\alpha)$. Then it turns out that *the Hausdorff distance in* $C(t_0,t_1) \times L_q(t_0,t_1)$ (where $1 \le q < \infty$) *between the graphs of the operators* $L[t_0,\Gamma^k,\Gamma_r^k]$ *and* $R^{\square}[t_0;\alpha,\beta]$ *converges to zero.* This explains why the multi-valued operator $R^{\square}[t_0;\alpha,\beta]$ can be taken as a reasonable approximation (idealization) for the generalized plays under the imposed conditions.

The standard operators $R[t_0,x_0;\alpha,\beta]$ also may serve as natural idealizations of the plays $L[t_0,x_0;\Gamma_\ell,\Gamma_r]$, provided that the graphs of functions $\Gamma_\ell(u)$ and $\Gamma_r(u)$ are close to the graphs of functions $r^{\square}(u,\beta)$ and $r^{\square}(u,\alpha)$, respectively.

28.6. Relay and "slow" controls

Assume that the equation

$$f(x,u) = 0 \tag{28.19}$$

describes the curve Γ shown in Fig. 28.2. This curve has one point common with the straight line $u = u_0$, if $u_0 < \beta$ or $u_0 > \alpha$, and three common points if $\beta < u_0 < \alpha$. The curve Γ divides the plane into two parts. Suppose that the function $f(x,u)$ is negative in the upper part and positive in the lower one. Consider the equation

$$\frac{dx}{dt} = f(x,u) \tag{28.20}$$

with control $u = u(t)$ which can vary only rather slowly ("slow" control).

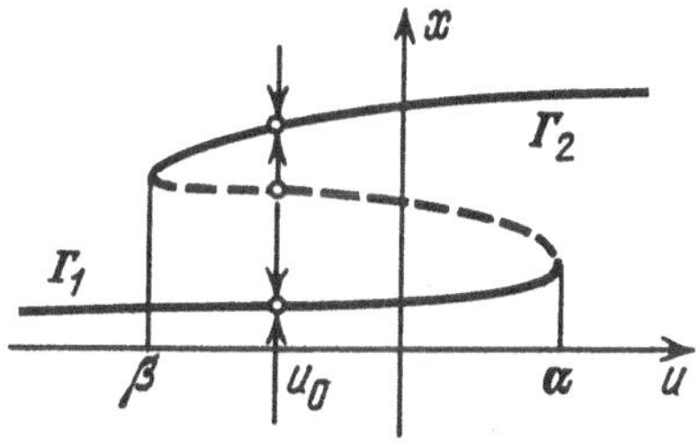

Fig. 28.2

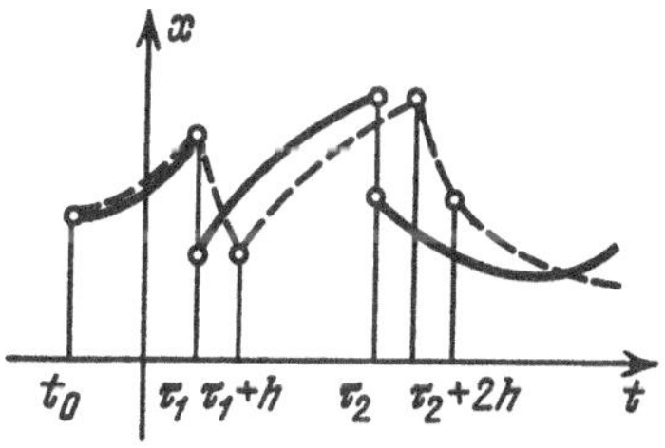

Fig. 28.3

If $u(t) \equiv u_0$, then equation (28.20) describes the motion of a point along the vertical straight line $u = u_0$. In Figure 28.2, the direction of motion along one of such lines is indicated by the arrows. The intersection points of the straight-line $u = u_0$ and curve Γ represent equilibrium states. The dashed line points the part of the curve Γ which contains the unstable equilibrium points, whereas the solid line shows the asymptotically stable equilibrium points. The solid line consists of the graphs Γ_1 and Γ_2 of some functions $\gamma_1(u)$ and $\gamma_2(u)$.

The above information turns out sufficient for describing a qualitative behaviour of the solutions $x(t)$ to equation (28.20) in the case of slowly varying controls $u(t)$ $(t \geq t_0)$. Except for a short time interval (further negligible), the point $\{u(t),x(t)\}$ will fall, in general, into such a small neighbourhood of either of the points $\{u(t),\gamma_1[u(t)]\}$, $\{u(t),\gamma_2[u(t)]\}$ that it can be regarded as coinciding either with $\{u(t),\gamma_1[u(t)]\}$ or $\{u(t),\gamma_2[u(t)]\}$. For definiteness, assume that $u(t_1) < \alpha$ and $x(t_1) \approx \gamma_1[u(t_1)]$. Then for the values $t \in (t_1,t_2)$ such that $u(t) < \alpha$, the point $\{u(t),x(t)\}$ will not leave a small neighbourhood of the curve Γ_1 and we can consider the equality $x(t) = \gamma_1[u(t)]$ as satisfied. If $u(t_2) = \alpha$ and $u(t)$ is increasing at $t = t_2$, then within a short time interval (on which the variation of $u(t)$ is also small) the point $\{u(t),x(t)\}$ will fall into a small neighbourhood of the curve Γ_2 and the equality $x(t) = \gamma_2[u(t)]$ can be considered as satisfied. Carrying on this reasoning, we would give a description of the solution $x(t)$. Upon excluding all superfluous switchings according to the introduced rule, this solution will coincide with the output of a non-ideal relay with threshold values α and β , provided that Γ_1 is the half-line $x = 0$ $(u < \alpha)$ and Γ_2 the half-line $x = 1$ $(u > \beta)$.

Nonlinearities of relay type are also quite typical for the problems related to singular perturbations for differential equations, to the catastrophe theory, etc.

28.7. Discontinuous inputs

In complex systems which contain some elements treated as non-ideal relays, it may be necessary to consider discontinuous input signals as admissible. For this, one has to define the operators $R[t_0,x_0;\alpha,\beta]$ on some classes of discontinuous inputs. Below we propose one of the constructions applicable in this connection.

A function $u(t)$ will be called *piecewise continuous*, if it has only a finite number of discontinuity points on any finite interval $[t_o,t_1]$, all of them of the first kind. In other words, if $\tau_1,\ldots,\tau_{n-1}$ are the discontinuity points of a piecewise continuous function $u(t)$ $(t_o \le t \le t_1)$, then on each interval (τ_{j-1},τ_j) $u(t)$ coincides with some function $v_j(t)$ continuous on $[\tau_{j-1},\tau_j]$. We are not interested in values of the function in its discontinuity point. Nevertheless, it will be convenient to assume that the function is right continuous at the initial time moment, i.e., $u(t_o) = v_1(t_0)$.

Let us take any $h > 0$ and, for the piecewise continuous function $u(t)$ introduced above, define a function $w_h(t)$ which is continuous on the interval $[t_o,t_1+(n-1)h]$, moreover, is linear on each of the intervals $[\tau_j+(j-1)h, \tau_j+jh]$ and $w_h(t) = v_j(t-(j-1)h)$ on each of the intervals $[\tau_{j-1}+(j-1)h, \tau_j+(j-1)h]$. In Figure 28.3, the solid line represents the graph of function $u(t)$ and the dashed line refers to the graph of $w_h(t)$. Clearly, piecewise continuous function $u(t)$ can be taken as a "good" approximation of the function $w_h(t)$ if h is small enough.

By using the functions $w_h(t)$, we can extend operators $R[t_o,x_o;\alpha,\beta]$ to the class of piecewise continuous inputs $u(t)$. This follows from the representation

$$R[t_o,x_o;\alpha,\beta]u(t) = \xi_h[t+(j-1)h] \quad (\tau_{j-1} \le t \le \tau_j;\ j=1,\ldots,n) \quad (28.21)$$

where

$$\xi_h(t) = R[t_o,x_o;\alpha,\beta]w_h(t) \quad (t_o \le t \le t_1+(n-1)h) . \quad (28.22)$$

Because all relays are static elements, the output value (28.21) is independent of the value of parameter h used in the construction.

Almost without any change, the above construction can be applied to any discontinuous input $u(t)$ with finite variation on any bounded interval $[t_o,t_1]$.

Part 6. Self-magnetization phenomenon

Of course, this problem has two aspects.
Let us consider the latter one.
O'Henry

29. Madelung's hysterons

29.1. Non-correct prehysteron

In this chapter, we shall consider systems with hysteresis which are not vibro-correct. Such systems arise in many real-life problems. For instance, in magnetism the self-magnetization phenomena are not vibro--correct.

Let $\Omega(W)$ be the domain of the feasible states $\{u,x\}$ for a certain prehysteron W, and the intersection of $\Omega(W)$ with any straight line $u = u_0$ be an interval. Then, each point $\{u_0,x_0\} \in \Omega(W)$ belongs to the graphs of two continuous functions

$$x = T_\ell(u;u_0,x_0) \quad (u \leq u_0) , \qquad x = T_r(u;u_0,x_0) \quad (u \geq u_0) , \tag{29.1}$$

contained in $\Omega(W)$. Furthermore, the identities

$$T_\ell(u_0;u_0,x_0) \equiv x_0 , \qquad T_r(u_0;u_0,x_0) \equiv x_0$$

hold. By virtue of

$$u_1 \leq u_0 , \quad u_1 \leq v_0 , \qquad T_\ell(u_1;u_0,x_0) = T_\ell(u_1;v_0,y_0)$$

we conclude the equality

$$T_\ell(u;u_o,x_o) = T_\ell(u;v_o,y_o) \quad (u \leq u_1) \quad ,$$

and in view of

$$u_o \leq u_2, \quad v_o \leq u_2, \quad T_r(u_2;u_o,x_o) = T_r(u_2;v_o,y_o) \, ,$$

the equality

$$T_r(u;u_o,x_o) = T_r(u;v_o,y_o) \quad (u \geq u_2)$$

is true. For a given initial state $\{u_o,x_o\}$ and a monotone, continuous input $u(t)$ $(t \geq t_o)$ satisfying

$$u(t_o) = u_o \quad , \tag{29.2}$$

the corresponding output

$$x(t) = W[t_o,x_o]u(t) \quad (t \geq t_o) \tag{29.3}$$

is defined by the equality

$$x(t) = \begin{cases} T_\ell[u(t);u_o,x_o] \, , & \text{if } u(t) \text{ is non-increasing,} \\ T_r[u(t);u_o,x_o] \, , & \text{if } u(t) \text{ is non-decreasing.} \end{cases} \tag{29.4}$$

To admit all piecewise monotone continuous inputs $u(t)$, we use the semigroup identity

$$W[t_o,x_o]u(t) = W[t_1,W[t_o,x_o]u(t_1)]u(t) \quad (t_o \leq t_1 \leq t) \quad , \tag{29.5}$$

as usual.

By Theorem 8.1, the constructed prehysteron W does not have the vibro-correctness property (if at least one interior point $\{u_o,x_o\}$ of some vertical segment contained in $\Omega(W)$ is simultaneously an interior point of the graphs of functions $T_\ell(u;u_1,x_1)$, $T_r(u;u_2,x_2)$ which do not coincide on any neighbourhood $(u_o - \varepsilon, u_o + \varepsilon)$ of the point $u = u_o$). The prehysteron W is only weakly vibro-correct: for every sequence of monotone inputs $u_n(t)$ $(t \geq t_o)$ convergent to the input $u_*(t)$ $(t \geq t_o)$ uniformly on each finite time interval, and for every sequence of the initial states $\{u_n(t_o), x_n\} \in \Omega(W)$ convergent to $\{u_*(t_o), x_*\} \in \Omega(W)$, the equality

$$\lim_{n\to\infty} W[t_o, x_n]u_n(t) = W[t_o, x_*]u_*(t) \quad (t \geq t_o) \tag{29.6}$$

holds and the convergence in (29.6) is also uniform on any finite time interval.

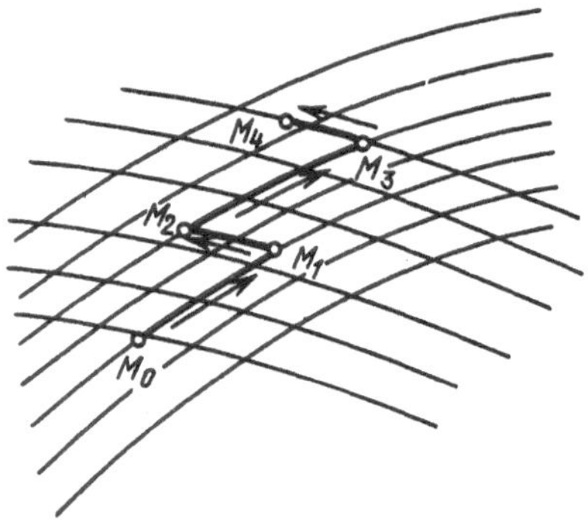

Fig. 29.1

In Fig. 29.1, a part of the domain $\Omega(W)$ of the feasible states of a certain prehysteron W is shown. The thin lines represent the graphs of the functions T_ℓ and T_r. The thickened line refers to one of the trajectories $\{u(t), W[t_o, x_o]u(t)\}$; each section $M_{i-1} M_i$ of this trajectory corresponds to a monotonicity interval of the input $u(t)$.

As an important example of the prehysteron one can take a transducer W defined by two standard hysterons W_1 and W_2 , with the same domain of feasible states. Let us set $\Omega(W) = \Omega(W_1) = \Omega(W_2)$ and, for a non-decreasing input $u(t)$, define the corresponding output $x(t)$ by the equality

$$W[t_o, x_o]u(t) = W_1[t_o, x_o]u(t) \quad (t \geq t_o) \quad ; \tag{29.7}$$

for the non-increasing input , define it by the equality

$$W[t_o, x_o]u(t) = W_2[t_o, x_o]u(t) \quad (t \geq t_o) \quad , \tag{29.8}$$

and for piecewise monotone inputs by the semigroup identity (29.5).

29.2. Periodic inputs

In the case of a general prehysteron, a periodic input generates not always a periodic output. Nevertheless, for a large class of prehysterons the output becomes periodic in a finite time. If a periodic input $u(t)$ $(t \geq t_0)$ produces the output $W[t_0,x_0]u(t)$, periodic at $t \geq t_1 \geq t_0$, then for $t \geq t_1$ the variable states $\{u(t), W[t_0,x_0]u(t)\}$ describe a closed curve - *hysteresis loop*.

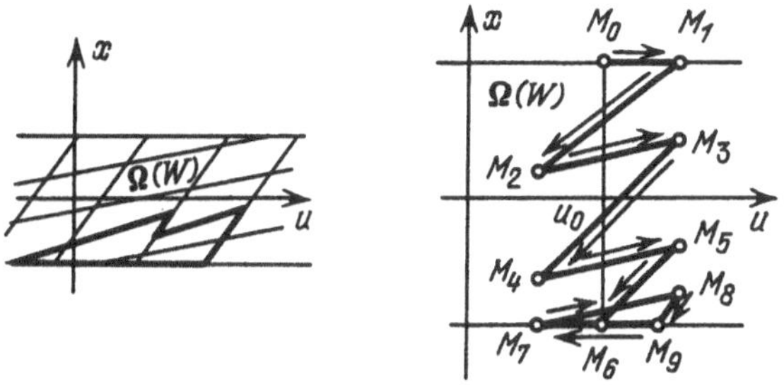

(a) (b)

Fig. 29.2

As an example, let us consider the prehysteron W , defined (cf., equalities (29.7) and (29.8)) by two plays W_1 and W_2 , having the same domain of the feasible states but different elasticity moduli. In Fig. 29.2,a , one of the hysteresis loops (the thickened line) is shown. In Fig. 29.2,b , the broken line $M_0\, M_1 \ldots M_9$ represents an example of the periodic output, corresponding to the initial state M_0 and some periodic input; the triangle $M_7\, M_8\, M_9$ forms there a hysteresis loop. The same Fig. 29.2,b shows an example of the system which is non-vibro-correct: if $M_0 = \{u_0, x_0\}$ and $u_0(t) = u_0$, then for the variable input $u_\varepsilon(t) = u_0 + \varepsilon \sin t$ and any small value $\varepsilon > 0$, the output $W[t_0,x_0]u_\varepsilon(t)$ is no longer close to the output $W[t_0,x_0]u_0(t) \equiv x_0$.

In the general case, a periodic input $u(t)$ always induces either asymptotically periodic or unbounded outputs.

29.3. Madelung's prehysteron

In the sequel, the functions (29.1) are assumed to be solutions of the differential equations

$$\frac{dx}{du} = f_\ell(x,u) \quad , \quad \frac{dx}{du} = f_r(x,u) \quad , \tag{29.9}$$

satisfying the initial condition

$$x(u_0) = x_0 \quad . \tag{29.10}$$

The functions $f_\ell(x,u)$ and $f_r(x,u)$ are often smooth in the interior points of the domain $\Omega(W)$; on the boundary of $\Omega(W)$, as a rule, they are discontinuous. The functions $f_\ell(x,u)$ and $f_r(x,u)$ are assumed to be Borelian, locally bounded and satisfying the unilateral Lipschitz conditions with respect to the first variable:

$$(x - y)\ [f_\ell(x,u) - f_\ell(y,u)] \geq -\lambda(u)\ (x - y)^2 \quad , \tag{29.11}$$

$$(x - y)\ [f_r(x,u) - f_r(y,u)] \leq \lambda(u)\ (x - y)^2 \quad , \tag{29.12}$$

where $\lambda(u)$ is a certain continuous, non-negative function. The solutions of equations (29.9) are defined as absolutely continuous functions, at almost all u satisfying the inclusions

$$\frac{dx(u)}{du} \in f_\ell^{\square}[x(u),u] \quad , \quad \frac{dx(u)}{du} \in f_r^{\square}[x(u),u] \quad , \tag{29.13}$$

respectively. In (29.13), the convexifications (cf., Section 27.3) are constructed with respect to the first variable. Under the assumed hypotheses, the solutions of (29.1) exist, are unique and can be globally extended along the trajectory of u .

The reader who prefers to avoid the treatment of (29.13) as differential inclusions, may assume the right-hand sides of equations (29.9) regular enough so as to ensure existence of classical solutions to (29.13). This is exactly the case in all basic applications (cf., Section 30.1).

If the functions (29.1) defining the prehysteron W are constructed as solutions of equations (29.9) , then the piecewise monotone, continuous and piecewise smooth input $u(t)$ $(t \geq t_0)$ determines the output (29.3) as a solution of the differential inclusion

$$\frac{dx}{dt} \in F^{\square}[x,u(t),\ u'(t)] \quad , \tag{29.14}$$

where

$$F(x,u,v) = \begin{cases} f_\ell(x,u)v\ , & \text{at} \quad v \leq 0\ , \\ f_r(x,u)v\ , & \text{at} \quad v \geq 0\ . \end{cases} \tag{29.15}$$

This solution, certainly, has to satisfy the initial condition (29.2).

In the case of piecewise monotone inputs, passing from the system (29.9) of two equations to single equation (29.14) reflects only assuming an alternative notation form. However, equation (29.14) can be used for determining the response of a system to any piecewise regular, continuous input (not necessarily piecewise monotone). The same equation specifies the response to any absolutely continuous input. Indeed, for each absolutely continuous input $u(t)$ $(t \geq t_0)$, the solution of (29.14) subject to the initial condition (29.10) exists, is unique and is defined globally in time.

The prehysterons described in this section are referred to as *Madelung's prehysterons*.

29.4. Properties of Madelung's prehysteron

As already mentioned, the operators $W[t_0,x_0]$ that correspond to Madelung's prehysteron W usually are not vibro-correct, i.e., they are not extendable by continuity to operators acting from $C(t_0,t_1)$ into $C(t_0,t_1)$.

A natural problem arises which concerns methods of studying properties of the operators $W[t_0,x_0]$ in function spaces different from $C(t_0,t_1)$. Further, the space $S = S(t_0,t_1)$ of the absolutely continuous functions, with finite norm

$$\|u(t)\|_S = |u(t_0)| + \int_{t_0}^{t_1} |u'(t)|\, dt \quad , \tag{29.16}$$

will play a fundamental role for that study.

As remarked in the preceding section, operators $W[t_0,x_0]$ are defined for all functions $u(t) \in S$ satisfying the relation $\{u(t_0),x_0\} \in \Omega(W)$.

Theorem 29.1. Every operator $W[t_0,x_0]$ describing the input-output relations of Madelung's prehysteron, acts in the space S , and is continuous and bounded on any ball in this space.

A proof will be given in the next chapter.

If the right-hand sides of the differential equations (29.9) are assumed more regular with respect to u (Lipschitz or Hölder continuous, etc.), then these properties hold also for the operators $W[t_0,x_0]$ treated as acting from S into S. Here we formulate one of the relevant results.

Theorem 29.2. Suppose that the functions $f_\ell(x,u)$ and $f_r(x,u)$ satisfy the local Hölder condition

$$|f(x,u) - f(x,v)| \leq c(\rho)\ |u - v|^\alpha \quad (|x|,\ |u|,\ |v| \leq \rho) \tag{29.17}$$

with respect to the second variable. Then the operators $W[t_0,x_0]$, acting from S into S , on each ball in S satisfy Hölder condition with the same Hölder index α :

$$\|W[t_0,x_0]u(t) - W[t_0,x_0]v(t)\|_S \leq c_1(\rho)\ \|u(t) - v(t)\|_S^\alpha$$
$$(\|u(t)\|_S,\ \|v(t)\|_S \leq \rho) \quad . \tag{29.18}$$

A proof of this theorem will be given in the next chapter.

By Theorem 23.1, $W[t_0,x_0]$ are continuous and locally bounded as operators from S into C . Under the assumptions of Theorem 29.2, $W[t_0,x_0]$ treated as operators from S into C fulfil the local Hölder condition with index α . Both these assertions are obvious by virtue of the continuous embedding of S in C :

$$\|x(t)\|_{t_0,t_1} \leq \|x(t)\|_S \qquad (x(t) \in S) \quad .$$

One could expect that $W[t_0,x_0]$, treated as operators from S into C, exhibit various "good" properties also under less restrictive hypotheses than those which are necessary for providing analogous properties of these operators treated as assuming values in the space S . Such an expectation turns out correct. Let us restrict ourselves to the case of the Lipschitz condition.

We shall say that $f(x,u)$ satisfies the *upper* or respectively *lower* *(*)-Lipschitz condition*, if the equality $|x - y| \geq \lambda_1 |u - v|$ implies that

$$(x - y)[f(x,u)-f(y,v)] \leq \lambda_2 (x - y)^2 \tag{29.19}$$

or, respectively,

$$(x - y)[f(x,u)-f(y,v)] \geq -\lambda_2 (x - y)^2 \ . \tag{29.20}$$

The values $\lambda_1 > 0$ and $\lambda_2 \geq 0$ are assumed to be given. In a natural way, we could also define local (*)-Lipschitz conditions.

Theorem 29.3. Suppose that the function $f_r(x,u)$ satisfies an upper (*)-Lipschitz condition and the function $f_\ell(x,u)$ a lower (*)--Lipschitz condition in every bounded domain. Then $W[t_0,x_0]$, treated as an operator from S into C, satisfies the Lipschitz condition on every ball in the space S.

Proof. This theorem is a direct consequence of standard results on differential inequalities.

∎

If $\lambda_2 = 0$ in the estimates (29.19), (29.20), then under the assumptions of Theorem 29.3, $W[t_0,x_0]$, treated as an operator from S into C, satisfies the global Lipschitz condition.

29.5. Madelung's hysteron

The notion *Madelung's hysteron* will be used to distinguish an important special class of Madelung's prehysterons. A use of the name "hysteron" is justified there due to a similarity of characterization of such an element to the description of hysterons by vibro-correct equations with constraints. However, in many important cases Madelung's hysteron is quite different from the hysterons we have considered in Parts 1-4. In particular, Madelung's hysteron is not vibro-correct.

In order to construct Madelung's hysteron, we introduce two

absolutely continuous functions

$$\gamma_-(u), \quad \gamma_+(u) \qquad (-\infty < u < \infty) \quad , \tag{29.21}$$

satisfying the inequality

$$\gamma_-(u) \leqq \gamma_+(u) \qquad (-\infty < u < \infty) \quad , \tag{29.22}$$

and two functions

$$\varphi_\ell(x,u), \quad \varphi_r(x,u) \qquad (\gamma_-(u) \leqq x \leqq \gamma_+(u);\ -\infty < u < \infty) \quad , \tag{29.23}$$

jointly continuous in both arguments and differentiable with respect to x .

Functions (29.21) and (29.23) will be called *characteristics of* the corresponding *Madelung's hysteron.* The domain $\Omega(W)$ of the feasible states for Madelung's hysteron with characteristics (29.21) and (29.23) is given by

$$\Omega(W) = \{\{u,x\} : \gamma_-(u) \leqq x \leqq \gamma_+(u);\ -\infty < u < \infty\} \quad . \tag{29.24}$$

The output

$$x(t) = W[t_0,x_0]u(t) \qquad (t \geqq t_0) \quad , \tag{29.25}$$

corresponding to an absolutely continuous input $u(t)$ $(t \geqq t_0)$, is the solution of the differential equation

$$\frac{dx}{dt} = \Phi[x,u(t),u'(t)] \quad ,$$

subject to the constraints (29.21), where

$$\Phi(x,u,v) = \begin{cases} \varphi_\ell(x,u)v \ , & \text{if} \quad v \leqq 0 \ , \\ \varphi_r(x,u)v \ , & \text{if} \quad v \geqq 0 \ . \end{cases}$$

For many years, Madelung's hysterons were already studied in physics and mechanics. Mathematically, they still require a more detailed treatment.

Clearly, Madelung's hysteron can be treated as Madelung's pre-hysteron. The functions $f_\ell(x,u)$ and $f_r(x,u)$ can be defined by the equalities

$$f_\ell(x,u) = \begin{cases} \varphi_\ell(x,u) , & \text{if } \gamma_-(u) < x < \gamma_+(u) , \\ \max\{\gamma_+'(u), \varphi_\ell(x,u)\} , & \text{if } x = \gamma_+(u) > \gamma_-(u) , \\ \min\{\gamma_-'(u), \varphi_\ell(x,u)\} , & \text{if } x = \gamma_-(u) < \gamma_+(u) , \\ \gamma_-'(u) , & \text{if } x = \gamma_-(u) = \gamma_+(u) \end{cases} \tag{29.26}$$

and

$$f_r(x,u) = \begin{cases} \varphi_r(x,u) , & \text{if } \gamma_-(u) < x < \gamma_+(u) , \\ \max\{\gamma_-'(u), \varphi_r(x,u)\} , & \text{if } x = \gamma_-(u) < \gamma_+(u) , \\ \min\{\gamma_+'(u), \varphi_r(x,u)\} , & \text{if } x = \gamma_+(u) > \gamma_-(u) , \\ \gamma_+(u) , & \text{if } x = \gamma_-(u) = \gamma_+(u) . \end{cases} \tag{29.27}$$

Hence, Theorems 29.1 - 29.3 are also true for Madelung's hysterons.

Theorem 29.4. Assume that the derivatives $\gamma_-'(u)$ and $\gamma_+'(u)$ of the functions (29.21) and (29.23) are locally Lipschitz continuous with respect to u. Let the inequality (29.22) be fulfilled in the sharpened form

$$\gamma_-(u) < \gamma_+(u) \qquad (-\infty < u < \infty) . \tag{29.28}$$

Then $W[t_0,x_0]$, treated as operators from S into S, satisfy the Lipschitz condition in each ball in that space.

Proof. Set $\gamma_-^*(u) \equiv 0$, $\gamma_+^*(u) \equiv 1$ and

$$\varphi_\ell^*(x,u) = \frac{\varphi_\ell(x,u) - \gamma_-(u)}{\gamma_+(u) - \gamma_-(u)} - [x - \gamma_-(u)] \, \frac{\gamma_+'(u) - \gamma_-'(u)}{\gamma_+(u) - \gamma_-(u)} ,$$

$$\varphi_r^*(x,u) = \frac{\varphi_r(x,u) - \gamma_-(u)}{\gamma_+(u) - \gamma_-(u)} - [x - \gamma_-(u)] \frac{\gamma_+'(u) - \gamma_-'(u)}{\gamma_+(u) - \gamma_-(u)} .$$

These functions define Madelung's hysteron W^* . By Theorem 29.2, every operator $W^*[t_o,x_o]$ is locally Lipschitz continuous. It remains to note that

$$W[t_o,x_o]u(t) = \{\gamma_+[u(t)] - \gamma_-[u(t)]\} x^*(t) + \gamma_-[u(t)] ,$$

where

$$x^*(t) = W^*[t_o,x_o^*]u(t)$$

and

$$x_o^* = \frac{x_o - \gamma_-[u(t_o)]}{\gamma_+[u(t_o)] - \gamma_-[u(t_o)]} .$$

In a relaxed version of Theorem 29.4, with the sharp inequality (29.28) removed, we only could claim (due to Theorem 29.3) that $W[t_o,x_o]$, treated as operators from S into C, are Lipschitz continuous on each ball. Rough estimates on the coefficients in the Lipschitz and Hölder conditions for the operators $W[t_o,x_o]$ (acting from S into S , or from S into C) may be easily derived. We would be interested in some more accurate estimates - they play an important role in the analysis of closed-loop systems including Madelung's prehysteron and hysteron.

29.6. Discontinuous inputs with bounded variation

Sometimes we have to accept discontinuity of the inputs. Let us return to the prehysteron W , considered in Section 29.1. We could assume that the formulas (29.4) determine the output (29.3) for every (not necessarily continuous) monotone input $u(t)$, then make use of the semigroup identity (29.5) for determining the outputs which correspond to piecewise monotone inputs. The question arises concerning a possibility of passing to more general classes of discontinuous inputs, for example, to inputs with bounded variation.

As it turns out, every operator $W[t_o,x_o]$ admits an extension by continuity to an operator acting from any arbitrary space $V = V(t_0,t_1)$ of the inputs with bounded variation into the same space V, provided that W is Madelung's prehysteron. This property reminds the passage to the inputs with bounded variation, considered in Section 6.6 in the case of hysterons; also the proof may be performed in an analogous way. The operators $W[t_0,x_0]$, extended onto the space V, preserve the properties of these operators, acting from S into S, considered in the preceding sections.

30. Proofs of Theorems 29.1 and 29.2

30.1. Passage to classical solutions

Let us consider the differential inclusion (29.14). As already mentioned, due to the unilateral Lipschitz conditions (29.11) and (29.12), for any absolutely continuous input $u(t)$ this inclusion admits a unique solution

$$x(t) = W[t_0,x_0]u(t) \quad (t \geq t_0) \tag{30.1}$$

which satisfies the initial condition

$$x(t_0) = x_0 \ . \tag{30.2}$$

Certainly, the initial condition $\{u(t_0),x_0\}$ should belong to $\Omega(W)$. In this section we are going to show that the functions (30.1) may be determined as the classical solutions of some ordinary differential equation.

Because function $f_\ell(x,u)$ satisfies the unilateral Lipschitz condition (29.11), the function

$$g_\ell(x,u) = f_\ell(x,u) + x + \lambda(u)\, x \tag{30.3}$$

is strictly increasing in x. Hence, all discontinuity points of the function $f_\ell(x,u)$ with respect to x belong to the graphs of the functions

$$x_n(t) = \sup\{y\colon g_\ell(y,u) < r_n\} \quad (u \in R^1) \ , \tag{30.4}$$

where r_n is any fixed sequence of reals, dense in R^1. Every function (30.4) is Borelian. Denote by E_n the set of numbers u_* for which there exist $x_n^o(u_*)$ such that at every $\varepsilon > 0$

$$\lim_{\delta \to 0} \frac{1}{\delta} \operatorname{mes}\{u: |u-u_*| \le \delta, \left| \frac{x_n(u) - x_n(u_*)}{u - u_*} - x_n^o(u_*)\right| > \varepsilon\} = 0 \quad . \tag{30.5}$$

The set E_n includes, in particular, all differentiability points of the function $x_n(u)$. The sets E_n with different indices may have common points. If $u_* \in E_n \cap E_m$ and $x_n(u_*) = x_m(u_*)$, then the equality $x_n^o(u_*) = x_m^o(u_*)$ holds (since $x_n(u) \geqq x_m(u)$ at $r_n > r_m$).

The sets E_n are measurable. Denote by E_n^o a Borelian part of the set E_n, such that $\operatorname{mes}(E_n \setminus E_n^o) = 0$. Define

$$h_\ell(x,u) = \begin{cases} x_n^o(u) \ , & \text{if } u \in E_n^o, \ x = x_n(u), \\ & f_\ell(x-0,u) \le x_n^o(u) \le f_\ell(x+0,u) \ , \\ f_\ell(x,u), & \text{in other points } \{x,u\} \ . \end{cases} \tag{30.6}$$

Here

$$f_\ell(x-0,u) = \lim_{\substack{y \to x \\ y < x}} f_\ell(y,u), \quad f_\ell(x+0,u) = \lim_{\substack{y \to x \\ y > x}} f_\ell(y,u) \quad .$$

Analogously, we can also define the function $h_r(x,u)$. By construction, functions $h_\ell(x,u)$ and $h_r(x,u)$ are Borelian.

Let us now consider the differential equation

$$\frac{dx}{dt} = H[x,u(t),u'(t)] \quad , \tag{30.7}$$

where

$$H(x,u,v) = \begin{cases} h_\ell(x,u)v \ , & \text{if } v \le 0 \ , \\ h_r(x,u)v \ , & \text{if } v \geqq 0 \ . \end{cases} \tag{30.8}$$

Lemma 30.1. The function (30.1) coincides with the unique classical solution of equation (30.7) that satisfies the initial condition (30.2).

Proof. By (30.6), the function $h_\ell(x,u)$ satisfies the unilateral Lipschitz condition (29.11) with the same coefficient $\lambda(u)$. Analogously, function $h_r(x,u)$ satisfies the unilateral Lipschitz condition (29.12). Thus, the solutions of the Cauchy problem for equation (30.7) are uniquely defined forwards in time. To complete the proof, it suffices to show that $x(t) = W[t_0,x_0]u(t)$ is a solution of equation (30.7). We shall confine ourselves to the case of piecewise monotone inputs, or equivalently, to monotone inputs.

For definiteness, let the input $u(t)$ $(t \geq t_0)$ be non-increasing. We need to show that the function (30.1) is a classical solution of the differential equation

$$\frac{dx}{dt} = h_\ell[x,u(t)]u'(t) \quad . \tag{30.9}$$

By $z(u)$ $(u \leq u(t_0))$ we shall denote the solution of the Cauchy problem

$$\frac{dx}{du} \in f_\ell^\square(x,u), \quad x[u(t_0)] = x_0 \quad ;$$

the uniqueness of this solution follows in view of (29.11). The function (30.6) has been constructed so that, for almost all u, $z(u)$ satisfies the equation

$$\frac{dx}{du} = h_\ell(x,u) \quad , \tag{30.10}$$

i.e., $z(u)$ is the unique classical solution of equation (30.10) for which $z[u(t_0)] = x_0$. By F we shall denote the set of those values of u for which either $z(u)$ is non-differentiable or $z'(u) \neq h_\ell[z(u),u]$. Since mes $F \neq 0$, the equality $u'(t) = 0$ takes place for almost all t such that $u(t) \in F$. Therefore, the function $x(t) = z[u(t)]$ almost everywhere satisfies equation (30.9). It remains to note that $z[u(t)]$ coincides with the function (30.1).

∎

The function $h_\ell(x,u)$ defined by equality (30.6) and the analogous function $h_r(x,u)$ preserve all the properties of functions $f_\ell(x,u)$ and $f_r(x,u)$, important from our view-point. As it has been already mentioned, they are Borelian and satisfy the unilateral Lipschitz conditions (29.11) and (29.12); if the functions $f_\ell(x,u)$ and $f_r(x,u)$ are Lipschitz or Hölder continuous with respect to u, then the analogous property holds also for the functions $h_\ell(x,u)$ and $h_r(x,u)$, etc.

In this connection, we can assume in the proofs of Theorems 29.1 and 29.2 (as well as in other analogous situations) that in the case of absolutely continuous $u(t)$, there exists a classical solution of equation (29.14).

30.2. Lemma on differential inequalities

In the sequel we shall use some special estimates of the solutions of the differential inequalities

$$z' \leq a|w_0(t)|z + b_1|w_1(t)| + b_2|w_2(t)| \quad , \tag{30.11}$$

where $w_0(t)$, $w_1(t)$, $w_2(t) \in L_1 = L_1(a,b)$ and a, b_1, b_2 are non-negative reals.

Standard theorems on differential inequalities give uniform estimates of the solutions to the relevant equations. Our main interest is focused on the estimates in the norm of the space S.

Lemma 30.2. Suppose that a non-negative, absolutely continuous function $z(t)$ $(t_0 \leq t \leq t_1)$ almost everywhere satisfies the inequality (30.11). Then

$$\|z(t)\|_S \leq 2e^{a\|w_0(t)\|_{L_1}} [|z(t_0)| + b_1\|w_1(t)\|_{L_1} + b_2\|w_2(t)\|_{L_1}] \; . \tag{30.12}$$

Proof. Set

$$z_+(t) = z(t_0) + \int_{[t_0,t]\cap E_+} z'(s)ds \quad ,$$

where

$$E_+ = \{t: z'(t) \geq 0,\ t_o \leq t \leq t_1\} \ .$$

In view of the non-negativeness of $z(t_1)$, the inequality

$$\int_{[t_o,t_1]\setminus E_+} z'(t)dt \leq z(t_o) + \int_{E_+} z'(t)dt$$

holds, hence

$$\mathrm{Var}_{t_o}^{t_1} z(t) \leq 2z_+(t_1) - z(t_o) \ . \tag{30.13}$$

But $z_+(t_o) = z(t_o)$ and almost everywhere

$$z_+'(t) \leq a|w_o(t)|z_+(t) + b_1|w_1(t)| + b_2|w_2(t)| \quad ,$$

whence (by standard theorems for differential inequalities)

$$z_+(t_1) \leq e^{a\|w_o(t)\|_{L_1}} [z(t_o) + b_1\|w_1(t)\| + b_2\|w_2(t)\|_{L_1}] \ .$$

Consequently, (30.12) follows from (30.11).

∎

30.3. Proof of Theorem 29.1

The local boundedness of the operators $W[t_o,x_o]$ follows directly by Lemma 30.2. We need to prove their continuity.

Consider a certain partitioning Λ of the interval $[t_o,t_1]$ by the points

$$t_o = \tau_o < \tau_1 < \ldots < \tau_m = t_1 \ . \tag{30.14}$$

For any continuous function $u(t)$, let us define the function

$$x(t) = B(\Lambda)u(t) \quad , \tag{30.15}$$

whose values in the points (30.14) coincide with the values of the function $u(t)$ and which is defined on each interval (τ_{i-1},τ_i) by the equality (see Fig. 30.1)

$$x(t) = \begin{cases} \max\{u(\tau_i) \; , \; \min\limits_{\tau_{i-1} \leq s \leq t} u(s)\} \; , & \text{if } u(\tau_i) \leq u(\tau_{i-1}) \; , \\ \min\{u(\tau_i) \; , \; \max\limits_{\tau_{i-1} \leq s \leq t} u(s)\} \; , & \text{if } u(\tau_i) \geq u(\tau_{i-1}) \; . \end{cases}$$

The function $x(t)$ is continuous on $[t_0,t_1]$; on each interval $[\tau_{i-1},\tau_i]$ it is monotone. An elementary (although rather awkward) reasoning shows that the operator $B(\Lambda)$ is non-expanding in the spaces C and S :

$$\|B(\Lambda)u(t) - B(\Lambda)v(t)\|_{t_0,t_1} \leq \|u(t) - v(t)\|_{t_0,t_1} \tag{30.16}$$

and

$$\|B(\Lambda)u(t) - B(\Lambda)v(t)\|_S \leq \|u(t) - v(t)\|_S \quad . \tag{30.17}$$

For the proof of Theorem 29.1, we need the equality

$$\lim_{n\to\infty} \|x_n(t) - x_0(t)\|_S = 0 \quad , \tag{30.18}$$

where

$$x_n(t) = W[t_0,x_0]u_n(t), \qquad x_0(t) = W[t_0,x_0]u_0(t) \tag{30.19}$$

and the functions $u_n(t)$ form a sequence convergent in S to a certain given function $u_0(t)$.

Let us introduce the notations

$$u_n(t;\Lambda) = B(\Lambda)u_n(t) \quad (n = 0,\ 1,\ 2,\ \ldots)$$

and

$$x_n(t;\Lambda) = W[t_o,x_o]u_n(t;\Lambda) \quad (n = 0,\ 1,\ 2,\ \ldots) \quad .$$

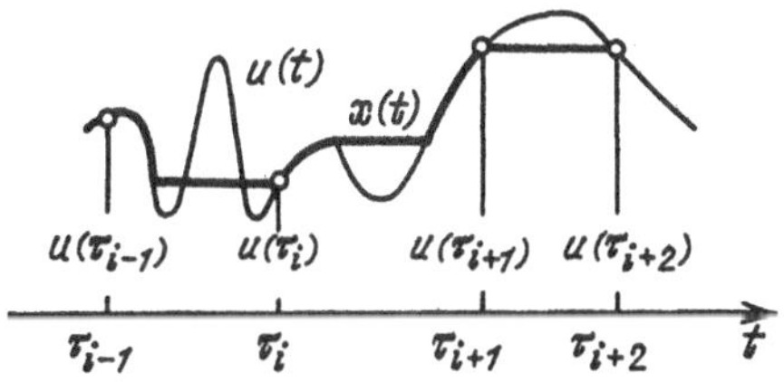

Fig. 30.1

Because

$$\|x_n(t) - x_o(t)\|_S \le \|x_n(t) - x_n(t;\Lambda)\|_S +$$

$$+ \|x_n(t;\Lambda) - x_o(t;\Lambda)\|_S + \|x_o(t;\Lambda) - x_o(t)\|_S \quad ,$$

for the proof of (30.18) first we need to establish the following convergence

$$\lim_{n\to\infty} \|x_n(t;\Lambda) - x_o(t;\Lambda)\|_S = 0 \tag{30.20}$$

at any partitioning Λ, and secondly to show for each $\varepsilon > 0$ the existence of a partitioning $\Lambda = \Lambda(\varepsilon)$ such that

$$\|x_n[t;\Lambda(\varepsilon)] - x_n(t)\|_S < \varepsilon \quad (n = 0,\ 1,\ 2,\ \ldots) \quad . \tag{30.21}$$

Consider a sequence of monotone functions $v_n(t) \in S$ and some reals y_n such that

$$\lim_{n\to\infty} \|v_n(t) - v_o(t)\|_S = 0 \quad , \quad \lim_{n\to\infty} y_n = y_o \quad . \tag{30.22}$$

As a consequence of Lemma 30.2, we get the equality

$$\lim_{n\to\infty} \|W[t_o,y_n]v_n(t) - W[t_o,y_o]v_n(t)\|_S = 0 \quad . \tag{30.23}$$

Suppose that the function $v_o(t)$ is non-increasing and non-constant (other cases can be considered in an analogous way). Then, for sufficiently large n, also the functions $v_n(t)$ are non-increasing. Therefore, the equalities

$$W[t_o,y_o]v_n(t) = T_\ell[v_n(t)] \quad (n = 0, 1, 2, \ldots)$$

take place, with a certain function $T_\ell(u)$ satisfying Lipschitz condition on $[t_o,t_1]$. The function $T_\ell(u)$ generates the superposition operator

$$\mathcal{T}_\ell\, v(t) = T_\ell[v(t)] \quad .$$

This operator is continuous in S. Hence (and due to (30.23)), by (30.22) it follows that

$$\lim_{n\to\infty} \|W[t_o,y_o]v_n(t) - W[t_o,y_o]v_o(t)\|_S = 0 \quad .$$

This, in turn, yields (30.20), since the functions $u_n(t;\Lambda)$ are monotone on each interval $[\tau_{j-1},\tau_j]$ and, due to (30.17),

$$\lim_{n\to\infty} \|u_n(t;\Lambda) - u_o(t;\Lambda)\|_S = 0 \quad .$$

Since the functions $f_\ell(x,u)$ and $f_r(x,u)$ are locally bounded, the estimates

$$|f_\ell[x_n(t),u_n(t)]| \;, \quad |f_r[x_n(t),u_n(t)]| \le b \;,$$

$$|f_\ell[x_n(t;\Lambda),u_n(t;\Lambda)]| \;, \quad |f_r[x_n(t;\Lambda),u_n(t;\Lambda)]| \le b$$

follow with a certain constant b independent of the partitioning Λ. Let us take

$$z_n(t;\Lambda) = |x_n(t;\Lambda) - x_n(t)| \quad (t_o \le t \le t_1) \quad .$$

For almost all $t \in [t_o,t_1]$ such that $z_n(t;\Lambda) = 0$,

$$\frac{d}{dt} z_n(t;\Lambda) = 0 \quad .$$

For almost all $t \in [t_o,t_1]$ such that $u_n(t;\Lambda) = u_n(t)$, the estimate

$$\frac{d}{dt} z_n(t;\Lambda) \leq a|u_n'(t)|z_n(t;\Lambda)$$

holds with some constant a. Eventually, for almost all $t \in [t_o,t_1]$ such that $u_n(t;\Lambda) \neq u_n(t)$, the estimate

$$\frac{d}{dt} z_n(t;\Lambda) \leq b|u_n'(t)|$$

takes place, and because of the equality $u_n'(t;\Lambda) = 0$, for almost all $t \in [t_o,t_1]$ we have

$$\frac{d}{dt} z_n(t;\Lambda) \leq b|u_n'(t) - u_n'(t;\Lambda)| \quad .$$

This implies for almost all $t \in [t_o,t_1]$ the following differential inequality

$$\frac{d}{dt} z_n(t;\Lambda) \leq a|u_n'(t;\Lambda)|z_n(t;\Lambda) + b|u_n'(t) - u_n'(t;\Lambda)| \quad .$$

As a consequence of the last inequality (since $z_n(0;\Lambda) = 0$) it follows that

$$\|z_n(t;\Lambda)\|_S \leq 2\, b\, e^{a\|u_n(t)\|_S} \|u_n(t) - u_n(t;\Lambda)\|_S \quad .$$

In this connection, the estimate (30.21) is satisfied, provided a sufficiently fine partitioning Λ has been taken.

Theorem 29.1 has been proved.

∎

30.4. Proof of Theorem 29.2

Assume

$$x(t) = W[t_o,x_o]u(t), \quad y(t) = W[t_o,x_o]v(t),$$

where $u(t), v(t) \in S$; $\|u(t)\|_S$, $\|v(t)\|_S \leq \rho$. The functions $f_\ell(x,u)$ and $f_r(x,u)$ are Hölder continuous with a certain index $\alpha \in (0,1]$. Thus, as a consequence of the relation

$$\frac{d}{dt}|x(t) - y(t)| \leq |F[x(t),u(t),u'(t)] - F[y(t),u(t),u'(t)]| +$$

$$+ |F[y(t),u(t),u'(t)] - F[y(t),v(t),u'(t)]| +$$

$$+ |F[y(t),v(t),u'(t)] - F[y(t),v(t),v'(t)]|,$$

we get the differential inequality

$$\frac{d}{dt}|x(t) - y(t)| \leq a|u'(t)|\cdot|x(t) - y(t)| +$$

$$+ b_1|u'(t)|\cdot\|u(t) - v(t)\|^{\alpha}_{t_o,t_1} + b_2|u'(t) - v'(t)| \quad ,$$

where a, b_1, b_2 depend only upon ρ . Hence, by Lemma 30.2 we can conclude the estimate

$$\|x(t) - y(t)\|_S \leq a_1[\|u(t) - v(t)\|_S + \|u(t) + v(t)\|_S] \quad , \qquad (30.24)$$

where a_1 is some positive constant. The assertion of Theorem 29.2 follows directly from (30.24).

∎

31. Response to small perturbations of the input

31.1. General scheme

In this section we continue the study of properties of a transducer W with scalar input u and scalar output x . We shall do this for a given class Σ of input signals $v(t)$ $(t_o \leq t \leq t_1)$. Let the operator $x(t) = Av(t)$ describe the input-output relations of transducer W at some given initial state.

Let us take a function $u(t)$ $(t_o \leq t \leq t_1)$ which not necessarily belongs to Σ . We shall treat the function $u(t)$ as an input of a certain transducer, if it admits an arbitrarily accurate approximation (in the uniform norm) by functions from Σ . Let a function $v(t) \in \Sigma$ approximate the input $u(t)$ with the error $\delta(t) = v(t) - u(t)$. Then, the smaller the error $\delta(t)$, the better $Av(t)$ approximates the output that corresponds to $u(t)$.

The above scheme can be quite easily implemented, provided that functions $Av(t)$ uniformly converge to a certain limit $x(t)$ as $\|\delta(t)\|_{t_0,t_1} \to 0$. In this case, $x(t) = Au(t)$ if $u(t) \in \Sigma$. If $u(t) \notin \Sigma$, it is convenient to introduce a suitable extension to the operator A (also then we shall write $x(t) = Au(t)$). This scheme can be used for constructing various vibro-correct hysterons.

In this chapter, we shall also consider transducers which are not vibro-correct. From many existing classes of such transducers we shall choose two representatives that can be used as models of the self-magnetization effect. The notion of an *intensity of function* $v(t)$ plays an important role at describing those classes. For this, we can use the mean variation on an arbitrarily long interval, the mean-square variation (energy), etc.; precise definitions will be given below.

Let us consider the situation when instead of a desired input $u(t)$ some close inputs $v(t) = u(t) + \delta(t)$ are taken and, moreover, it is known that either a certain intensity of the error $\delta(t)$ or a certain intensity of the input $v(t)$ approaches a prescribed value κ when the error $\delta(t)$ decreases. If the outputs $A[u(t) + \delta_n(t)]$ converge to some function $x(t)$ as $\|\delta_n(t)\|_{t_0,t_1} \to 0$ and appropriately the intensity either of the errors $\delta_n(t)$ or inputs $u_n(t) = u(t) + \delta_n(t)$ converges to κ , then $x(t)$ is a natural choice for the output that corresponds to the input $u(t)$. For a fixed κ , in order to determine the limit $x(t)$, only the input $u(t)$ has to be known. Hence the natural notation $x(t) = W(\kappa)\, u(t)$. The values of the operator $W(\kappa)$ over the inputs $u(t) \in \Sigma$ may be different from the suitable values of operator A . In the case we consider only those functions, $u(t)$ which are elements of the domain of $W(\kappa)$ can be taken as the admissible inputs. The input-output relations can be then described by the operator $W(\kappa)$ (rather than by A).

So far we have not specified a character of the convergence of the sequence $Av_n(t)$ to $x(t) = W(\kappa)\, u(t)$. We are going to find conditions which guarantee the uniform convergence either on the whole interval $[t_0,t_1]$ under consideration or on any subinterval $[t_0+\alpha,t_1]$ $(\alpha > 0)$. In the second case, a boundary layer problem arises (as in the theory of singular perturbations for ordinary differential equations), but this problem remains out of the scope of the present book. If an appropriate interpretation of the intensity follows clearly from the context, values of the operator $W(\kappa)$ will be referred to as *κ-outputs* .

31.2. Intensities

a. Consider the space $V = V(t_0,t_1)$ of functions $v(t)$ $(t_0 \leq t \leq t_1)$ with bounded variation. We shall say that *the sequence* $v_n(t) \in V$ *has an asymptotic* 1*-intensity* κ, if the equalities

$$\lim_{n\to\infty} \Phi_1[v_n(t);\tau_1,\tau_2] = (\tau_2 - \tau_1) \quad (t_0 \leq \tau_1 < \tau_2 \leq t_1) \tag{31.1}$$

are fulfilled with

$$\Phi_1[u(t);\tau_1,\tau_2] = \operatorname{Var}_{\tau_1}^{\tau_2} u(t) \quad . \tag{31.2}$$

In the sequel, we shall equally use the notions of finite and infinite asymptotic stability.

Input signal sequences $v_n(t)$ with a prescribed asymptotic 1-intensity arise, for example, in a study of systems with variable structure.

Suppose that the error $\delta(t)$ is characterized as the mean value

$$\delta(t) = \frac{1}{N}[\xi_1(t) + \ldots + \xi_N(t)] \quad (t_0 \leq t \leq t_1) \tag{31.3}$$

of the outputs $\xi_j(t)$ of some independent static elements. Let each output $\xi_j(t)$ be piecewise constant, with values -1 or 1 . For every j, the switching time moments for the output $\xi_j(t)$ will be assumed as random variables with the Poisson distribution and mean expectation time κ . Since the elements are independent, for every $\varepsilon > 0$ the functions (31.3) with high probability will satisfy the inequalities

$$\max|\delta(t)| < \varepsilon, \quad |\Phi_1[\delta(t);\tau_1,\tau_2] - \kappa(\tau_2-\tau_1)| < \varepsilon \quad , \tag{31.4}$$

provided $N = N(\varepsilon)$ is large enough. In this connection, it is reasonable to approximate the functions $u(t) + \delta(t)$ by means of the sequences $u(t) + \delta_n(t)$, where the errors $\delta_n(t)$ satisfy conditions $\|\delta_n(t)\|_{t_o,t_1} \to 0$ and $\Phi_1[\delta_n(t);\tau_1,\tau_2] \to \kappa(\tau_2 - \tau_1)$.

b. Let Σ be the class of lower semicontinuous, piecewise constant functions $v(t)$ $(t_o \le t \le t_1)$. Assume that $\sigma_1, \sigma_2, \ldots, \sigma_k$ represent all the discontinuity points of the function $v(t) \in \Sigma$, where

$$t_o \leqq \sigma_1 < \sigma_2 < \ldots < \sigma_k < t_1 . \tag{31.5}$$

Set

$$\Phi_2[v(t);\tau_1,\tau_2] = \sum_{\tau_1 \leqq \sigma_i < \tau_2} |v(\sigma_i + 0) - v(\sigma_i)|^2 . \tag{31.6}$$

By definition, the sequence $v_n(t) \in \Sigma$ has an *asymptotic* 2*-intensity* κ, if the equality

$$\lim_{n\to\infty} \Phi_2[v_n(t);\tau_1,\tau_2] = \kappa(\tau_2 - \tau_1) , \tag{31.7}$$

similar to (31.1), holds for $t_o \leqq \tau_1 < \tau_2 \leqq t_1$.

Taking piecewise constant approximations of the continuous functions $u(t) \in C = C(t_o,t_1)$ is one of the possible methods of constructing the sequences $v_n(t) \in \Sigma$. To construct any approximation $v(t)$, one must select some points (31.5) and assume $v(t) = u(\sigma_i)$ for $\sigma_i < t \leqq \sigma_{i+1}$ (in Figure 31.1, the graph of function $u(t)$ is indicated by the thickened line). If some refined partitions are used at the construction of functions $v_n(t)$, then for almost every (with respect to the Wiener measure) function $u(t)$, the asymptotic 2-intensity of the corresponding sequence $v_n(t)$ equals 1.

Assume that two functions $u(t), u^o(t) \in C(t_o,t_1)$ are given and denote by $v_n(t), v_n^o(t) \in \Sigma$ the relevant approximating sequences, both constructed with the same partition points (31.5). Then, by (31.7) and due to sufficiently high regularity of the difference $u^o(t) - u(t)$ (for example, by the boundedness of its variation), the sequence $v_n^o(t)$ has the asymptotic 2-intensity κ.

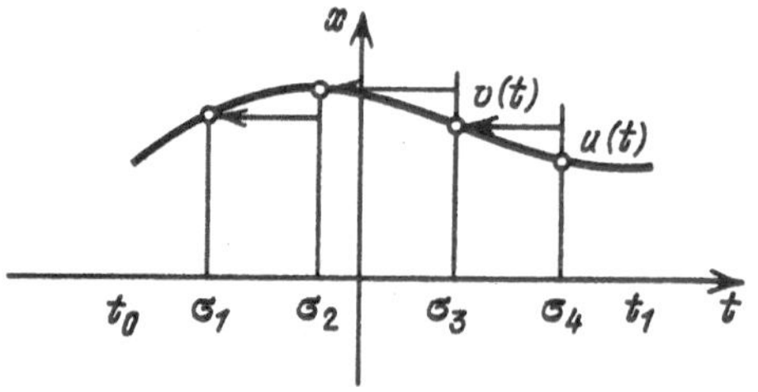

Fig. 31.1

Now let the interval $[t_o, t_1]$ be divided by the points

$$t_o = \sigma_o < \sigma_1 < \ldots < \sigma_N = t_1$$

into N equal parts. Consider a random motion described by functions $v(t)$ which are constant on the intervals $[\sigma_i, \sigma_{i+1}]$ and in each point σ_i have a jump, with probability 0.5 either equal to $-(\kappa N)^{-1/2}$ or to $(\kappa N)^{-1/2}$. For any $\varepsilon > 0$, the function $v(t)$ with some high probability satisfies (cf., (31.4)) the inequality

$$|\Phi_2[v(t);\tau_1,\tau_2] - \kappa(\tau_2 - \tau_1)| < \varepsilon , \tag{31.8}$$

provided $N = N(\varepsilon)$ is large enough. Such functions $u(t)$ in a natural way can be approximated by sequences $v_n(t) \in \Sigma$ which have the asymptotic 2-intensity κ and are uniformly convergent to some continuous function $u(t)$.

c. For piecewise constant functions $v(t)$, we can introduce the characteristic

$$\Phi_\alpha[v(t);\tau_1,\tau_2] = \sum_{\tau_1 \leq \sigma_i < \tau_2} |v(\sigma_i + 0) - v(\sigma_i - 0)|^\alpha, \tag{31.9}$$

with any $\alpha > 0$. For $\alpha = 1$, (31.9) represents the variation of function $v(t)$, for $\alpha = 2$ it assumes the form of the quadratic variation (31.6). If

$$\lim_{n\to\infty} \Phi_\alpha[v_n(t);\tau_1,\tau_2] = \kappa(\tau_2 - \tau_1) , \tag{31.10}$$

then the sequence $v_n(t)$ is said to have an asymptotic α-*intensity*.

In the definition of the intensity, also functionals of a more general form than (31.9) can be admitted (in particular, power functions $u^\alpha(t)$ can be replaced by some more general monotone functions). In many constructions, it turns out useful to introduce notions of variable intensities, parametrized intensities, etc.

A modification of the characteristics (31.6) is also necessary, if $v(t)$ that are different from piecewise constant functions are admissible. For instance, in the case of piecewise linear continuous functions $v(t)$ it is useful to take

$$\Phi_2^{o}[v(t);\tau_1,\tau_2] = \sum_{\tau_1 \leq \sigma_i \leq \tau_2} |v(\sigma_i) - v(\sigma_{i-1})|^2 , \tag{31.11}$$

where σ_i represent ordered discontinuity points of the derivative $v'(t)$. In turn, for any continuous function $v(t)$ one can define

$$\Phi_2^{oo}[v(t);\tau_1,\tau_2] = \sup \sum_{\tau_1 \leq \sigma_i \leq \tau_2} |v(\sigma_i) - v(\sigma_{i-1})|^2 , \tag{31.12}$$

where the upper bound is taken over all possible partitionings of the interval by points σ_i. There is an essential difference between the characteristics (31.11) and (31.12).

31.3. Construction of κ-outputs to Madelung's hysteron

Let us continue the study of Madelung's hysteron described (cf., Chapter 29) by the differential equation

$$\frac{dx}{dt} = \Phi[x,u(t),u'(t)] \tag{31.13}$$

subject to constraints

$$\gamma_-(u),\ \gamma_+(u) \quad (-\infty < u < \infty) , \tag{31.14}$$

where

$$\Phi(x,u,v) = \begin{cases} \varphi_\ell(x,u)v\ , & \text{at} \quad v \leq 0\ , \\ \varphi_r(x,u)v\ , & \text{at} \quad v \geq 0\ . \end{cases} \tag{31.15}$$

Any absolutely continuous input $u(t) \in S = S(t_o,t_1)$, together with the initial condition

$$x(t_o) = x_o \qquad (\gamma_-[u(t_o)] \leq x_o \leq \gamma_+[u(t_o)])\ , \tag{31.16}$$

determine the corresponding output

$$Au(t) = W[t_o,x_o]u(t) \qquad (t_o \leq t \leq t_1)\ . \tag{31.17}$$

The operator A is defined on all inputs with bounded variation. This operator is continuous in the space S, but does not admit any extension by continuity to a continuous operator defined in the space C (because Madelung's hysteron is not vibro-correct).

a. Denote by S^κ the set of functions $u(t)$, absolutely continuous on $[t_o,t_1]$, for which

$$\operatorname{Var}_{\tau_1}^{\tau_2} u(t) \leq \kappa(\tau_2 - \tau_1) \qquad (t_o \leq \tau_1 < \tau_2 \leq t_1)\ . \tag{31.18}$$

For each function $u(t) \in S^\kappa$, we can construct a sequence $v_n(t) \in S$ with the asymptotic 1-intensity κ, uniformly convergent to $u(t)$,

$$\lim_{n\to\infty} \|v_n(t) - u(t)\|_{t_o,t_1} = 0\ . \tag{31.19}$$

The suitable functions $Av_n(t)$ are also uniformly convergent; *the limit*

$$x(t) = W(\kappa)\, u(t) \tag{31.20}$$

is the κ-output that corresponds to the input $u(t)$.

Theorem 31.1. If $u(t) \in S^\kappa$, then the κ-output (31.20) is the unique solution of the differential equation

$$\frac{dx}{dt} = \Phi[x,u(t),u'(t)] + [\kappa - |u'(t)|]\{\varphi_r[x,u(t)] - \varphi_\ell[x,u(t)]\} \tag{31.21}$$

subject to the constraints (31.14) together with the initial condition (31.16).

Theorem 31.1 is a consequence of the following assertion which, although simple, requires a quite tedious proof.

Lemma 31.1. Let a sequence $v_n(t)$ of functions with bounded variation satisfy the condition (31.19), where $u(t)$ is absolutely continuous. Suppose that

$$\lim_{n\to\infty} \operatorname{Var}_{t_0}^{t} v_n(t) = \operatorname{Var}_{t_0}^{t} u(t) + \psi(t) \qquad (t_0 \leq t \leq t_1) \;, \tag{31.22}$$

where $\psi(t)$ is also absolutely continuous. Then the sequence $x_n(t) = Av_n(t)$ $(n = 1, 2, \dots)$ converges uniformly and its limit $x(t)$ is the unique solution of the differential equation

$$\frac{dx}{dt} = \Phi[x,u(t),u'(t)] + \psi(t)\{\varphi_r[x,u(t)] - \varphi_\ell[x,u(t)]\} \tag{31.23}$$

subject to the constraints (31.14), together with the initial condition (31.16).

b. Suppose that only inputs in the form $v(t) = u(t) + \delta(t)$, with piecewise constant functions $u(t) \in S$ and $\delta(t)$, are realizable. Consider a sequence of the inputs $v_n(t) = u(t) + \delta(t)$. Assume that $v_n(t)$ is uniformly convergent to $u(t)$ and that $\delta(t)$ has the asymptotic 1-intensity κ. Under these conditions, the sequence $x_n(t) = Av_n(t)$ is uniformly convergent and its limit

$$x(t) = W^{o}(\kappa)\, u(t) \tag{31.24}$$

is the κ-output corresponding to the input $u(t)$. Clearly, the κ-outputs (31.20) and (31.24) are different, in general.

By Lemma 31.1, we can conclude the following statement, similar to Theorem 31.1.

Theorem 31.2. If $u(t) \in S$, then the κ-output (31.25) is the unique solution of the differential equation

$$\frac{dx}{dt} = \Phi[x,u(t),u'(t)] + \kappa\{\varphi_r[x,u(t)] - \varphi_\ell[x,u(t)]\} \tag{31.25}$$

subject to the constraints (31.14), together with the initial condition (31.16).

Suppose that the sequence of inputs $v_n(t) = u(t) + \delta_n(t)$ uniformly converges to $u(t) \in S$ and the asymptotic 1-intensity of $\delta(t)$ is infinite:

$$\lim_{n\to\infty} \Phi_1[\delta_n(t);\tau_1,\tau_2] = \infty \qquad (t_o \leqq \tau_1 < \tau_2 \leqq t_1) \ . \tag{31.26}$$

In particular, equality (31.26) is true if the sequence $\delta_n(t)$ has a non-trivial asymptotic α-intensity (cf., (31.10)) where $\alpha > 1$.

If, in addition, the condition

$$\varphi_r(x,u) - \varphi_\ell(x,u) > 0 \qquad (\gamma_-(u) \leqq x \leqq \gamma_+(u)) \tag{31.27}$$

holds (see Fig. 31.2), *then under the above hypotheses the sequence* $x_n(t) = Av_n(t)$ *uniformly on each interval* $[t_*,t_1]$ $(t_* > t_o)$ *converges to a function* $\gamma_+[u(t)]$ *which may be treated as an* ∞*-output corresponding to the input* $u(t)$.

If the inequality

$$\varphi_r(x,u) - \varphi_\ell(x,u) < 0 \qquad (\gamma_-(u) \leqq x \leqq \gamma_+(u)) \ , \tag{31.28}$$

reverse to (31.27) holds, then the ∞-output will be equal to $\gamma_-[u(t)]$.

c. Let us now consider the case of admissible inputs in the form $v(t) = u(t) + \delta(t)$, where $u(t) \in S$ and the functions $\delta(t)$ are continuous and piecewise linear, with the derivatives $\delta'(t)$ assuming only three values $-a$, b and 0 $(-a < 0 < b)$. Consider, as usual, a sequence of the inputs $v_n(t) = u(t) + \delta_n(t)$. Suppose that $v_n(t)$ uniformly converge to $u(t) \in S$ and the sequence $\delta_n(t)$ has an asymptotic 1-intensity. Note that in the case under consideration

$$0 \leq \kappa \leq \frac{2ab}{a+b} \quad .$$

The sequence $x_n(t) = Av_n(t)$ converges now to a new κ-output

$$x(t) = W^{oo}[t_o,x_o;\kappa]u(t) \quad .$$

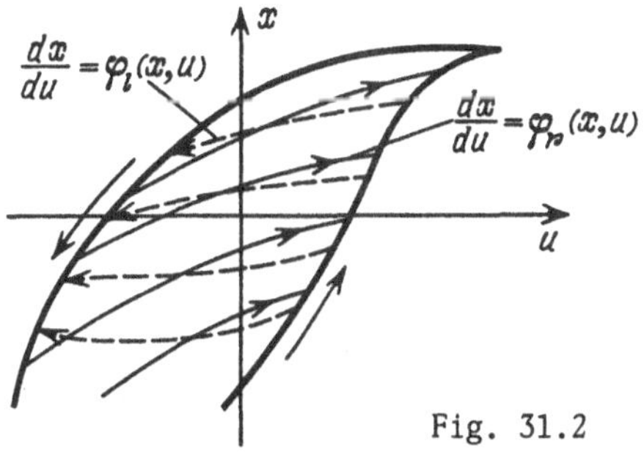

Fig. 31.2

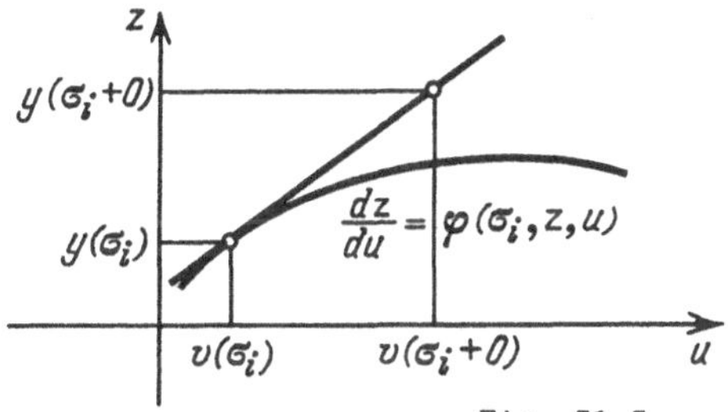

Fig. 31.3

By Lemma 31.1, for the κ-output (31.29) we can conclude an assertion similar to Theorems 31.1 and 31.2, but with (31.21) and (31.25) replaced by the equation

$$\frac{dx}{dt} = \Phi[x,u(t),u'(t)] + \kappa\,\beta[u'(t)]\,\{\varphi_r[x,u(t)] - \varphi_\ell[x,u(t)]\}, \quad (31.30)$$

with

$$\beta(v) = \begin{cases} 0\ , & \text{for} \quad v \notin [-a,b]\ , \\ 1+v/a\ , & \text{for} \quad v \in [-a,0]\ , \\ 1-v/a\ , & \text{for} \quad v \in [0,b]\ . \end{cases}$$

31.4. Construction of κ-vibrosolutions to differential equations

Let us consider the differential equation

$$\frac{dx}{dt} = \varphi[t,x,u(t)]\, u'(t) \quad . \tag{31.31}$$

Assume that $\varphi(t,x,u)$ is twice continuously differentiable, with $\varphi(t,x,u)$, $\varphi'_x(t,x,u)$, $\varphi'_u(t,x,u)$ bounded. Then (cf., Chapter 13) the equation (31.31) is vibro-correct.

In this section we shall construct some special solutions to the equations (31.31), different from the vibro-solutions.

Denote by Σ the class of continuous, piecewise left-constant inputs $v(t)$ $(t_0 \leqq t \leqq t_1)$. For each input $v(t) \in \Sigma$ with the discontinuity points $\sigma_0 = t_0 < \sigma_1 < \ldots < \sigma_k < \sigma_{k+1} = t_1$, there is an output $y(t)$ in the same class Σ, with the same discontinuity points and the values

$$y(t) = y(\sigma_i) + \varphi[\sigma_i, z(\sigma_i), v(\sigma_i)][v(\sigma_i + 0) - v(\sigma_i)]$$

$$(\sigma_i < t \leqq \sigma_{i+1};\ i = 0, 1, \ldots, k). \tag{31.32}$$

End-points of the interval $[t_0,t_1]$ are assumed to be the discontinuity points. With this, (31.32) will make sense for all $t \in [t_0,t_1]$; the discontinuities at t_0 and t_1 can be equal to zero. In the construction of function (31.32), we prescribe the initial value $y(t_0) = y_0$. Then, formula (31.32) defines the operator

$$y(t) = Av(t) \quad (v(t) \in \Sigma) \quad . \tag{31.33}$$

If the equation (31.31) describes the dynamics of some transducer W, then definition (31.32) admits a simple physical interpretation. It implies that as long as the input of a transducer is constant, its state $\{v(t), y(t)\}$ remains constant, too. Jumps of the output are proportional to those at the input, with proportionality coefficient dependent only upon the current state of the transducer. In Figure 31.3, a construction which yields a jump at the output is shown. The transducer we have described is not-vibro-correct on the class Σ. In particular, if all solutions of the equation

$$\frac{dx}{dt} = \varphi(t_*,x,u) \tag{31.34}$$

are concave, then the function $y(t)$ is decreasing also for $v(t)$ which admit small oscillations.

Now let us take a continuous function $u(t)$ and a sequence $v_n(t) \in \in \Sigma$, convergent to $u(t)$ uniformly on $[t_o,t_1]$. If the function $u(t)$ is smooth and the sequence $v_n(t)$ has an asymptotic 1-intensity, then the outputs $Av_n(t)$ uniformly converge to the standard solution $x(t)$ of the equation (31.31), satisfying the initial condition

$$x(t_o) = y(t_o) \quad . \tag{31.35}$$

If the asymptotic 1-intensity is infinite, then various situations may occur. The most important case refers to the sequence $v_n(t)$ which has finite 2-intensity κ. In that case, the sequence $Av_n(t)$ will converge uniformly on $[t_o,t_1]$ to a certain continuous function

$$x(t) = W[t_o,y_o;\kappa]u(t) \qquad (t_o \leq t \leq t_1) \quad . \tag{31.36}$$

The function (31.36) will be referred to as a κ-*vibrosolution of equation* (31.31), *satisfying the initial condition* (31.35).

Theorem 31.3. Each κ-vibrosolution (31.36) (corresponding to any arbitrary input $u(t)$) is the standard vibro-solution of the differential equation

$$\frac{dx}{dt} = \varphi[t,x,u(t)]u'(t) + \kappa\psi[t,x,u(t)] \quad , \tag{31.37}$$

where

$$\psi(t,x,u) = -\frac{1}{2}[\varphi'_x(t,x,u)\varphi(t,x,u) + \varphi'_u(t,x,u)] \quad . \tag{31.38}$$

Proof. *First step of the proof.* The vibrosolution $x_*(t)$ of equation (31.37) may be represented (cf., Chapters 12 and 13) in the form

$$x_*(t) = Q[u(t),0,z_*(t),t] \qquad (t_o \leq t \leq t_1) \quad ,$$

where $Q(u,u_o,x_o,t)$ is the solution of the differential equation

$$\frac{dx}{dt} = \varphi[t,x,u]$$

at a fixed t, satisfying the condition $x(u_o) = x_o$, and $z_*(t)$ is the standard solution of the differential equation

$$\frac{dz}{dt} = Q_4'\{0,u(t),Q[u(t),0,z,t],t\} +$$

$$+ \kappa Q_3'\{0,u(t),Q[u(t),0,z,t],t\}\psi\{t,Q[u(t),0,z,t],u(t)\} , \qquad (31.39)$$

satisfying the initial condition

$$z_*(t_o) = Q[0,u(t),y_o,t_o] . \qquad (31.40)$$

To prove Theorem 31.3, it is enough to show that

$$\lim_{n\to\infty} \|z_n(t) - z_*(t)\|_{t_o,t_1} = 0 , \qquad (31.41)$$

where

$$z_n(t) = Q[0,Pv_n(t),PAv_n(t),t] \qquad (31.42)$$

and P is the operator which to each function $v(t) \in \Sigma$ assigns the continuous function $u(t) = Pv(t)$, identical with $v(t)$ in the discontinuity points of $v(t)$ and linear on each interval between the adjacent discontinuity points.

Second step of the proof. This step consists in constructing a special representation of the function (31.42).

Suppose that

$$t_o = \sigma_o^n < \sigma_1^n < \ldots < \sigma_{k_n}^n < \sigma_{k_n+1}^n = t_1$$

are the discontinuity points of the function $v_n(t)$. The function $z_n(t)$ is assumed to be differentiable everywhere except at the points σ_j^n. Take $u_n(t) = Pv_n(t)$ and $x_n(t) = PAv_n(t)$. By definition,

$$\frac{dz_n}{dt} = Q_2'[0,u_n(t),x_n(t),t]u_n'(t) +$$

$$+ Q_3'[0,u_n(t),x_n(t),t]x_n'(t) + Q_4'[0,u_n(t),x_n(t),t] \quad .$$

Hence, in view of the equality

$$Q_2'(0,u,x,t) = - Q_3'(0,u,x,t)\varphi(t,x,u) \quad ,$$

we can write

$$\frac{dz_n}{dt} = Q_4'[0,u_n(t),x_n(t),t] +$$

$$+ Q_3'[0,u_n(t),x_n(t),t] \{x_n'(t) - \varphi[t,x_n(t),u_n(t)]u_n'(t)\} \quad .$$

At the same time, for $\sigma_i^n < t < \sigma_{i+1}^n$,

$$u_n'(t) = \frac{u(\sigma_{i+1}^n) - u(\sigma_i^n)}{\sigma_{i+1}^n - \sigma_i^n}$$

and

$$x_n'(t) = \varphi[\sigma_i^n,x_n(\sigma_i^n),u_n(\sigma_i^n)] \frac{u(\sigma_{i+1}^n) - u(\sigma_i^n)}{\sigma_{i+1}^n - \sigma_i^n} \quad .$$

Therefore, for $\sigma_i^n < t < \sigma_{i+1}^n$,

$$\frac{dz_n}{dt} = Q_4'[0,u_n(t),x_n(t),t] + Q_3'[0,u_n(t),x_n(t),t]\{\varphi[\sigma_i^n,x_n(\sigma_i^n),u_n(\sigma_i^n)] -$$

$$- \varphi[t,x_n(t),u_n(t)]\} \frac{u(\sigma_{i+1}^n) - u(\sigma_i^n)}{\sigma_{i+1}^n - \sigma_i^n} \quad .$$

Because of the equality

$$\varphi(t,x,u) - \varphi(s,y,v) = \varphi_x'(s,y,v)(x-y) +$$

$$+ \varphi_u'(s,y,v)(u-v) + \omega(t,x;x,y;u,v) \quad ,$$

where

$$|\omega(t,s;x,y;u,v)| \leq c(|t-s| + |x-y|^2 + |u-v|^2)$$

and

$$x_n(t) - x_n(\sigma_i^n) = \varphi[\sigma_i^n, x_n(\sigma_i^n), u_n(\sigma_i^n)] \frac{u_n(\sigma_{i+1}^n) - u_n(\sigma_i^n)}{\sigma_{i+1}^n - \sigma_i^n} (t - \sigma_i^n) ,$$

for $\sigma_i^n < t < \sigma_{i+1}^n$, the equality

$$\frac{dz_n}{dt} = - Q_3'[0,u_n(t),x_n(t),t] \, (t-\sigma_i^n) \left[\frac{u(\sigma_{i+1}^n) - u(\sigma_i^n)}{\sigma_{i+1}^n - \sigma_i^n} \right]^2 \times$$

$$\times \{\varphi_x'[t,x_n(t),u_n(t)]\varphi[\sigma_i^n,x_n(\sigma_i^n),u_n(\sigma_i^n)] + \varphi_u'[t,x_n(t),u_n(t)]\} +$$

$$+ Q_4'[0,u_n(t),x_n(t),t] + \omega_1[t,\sigma_i^n;x_n(t),x_n(\sigma_i^n);u_n(t),u_n(\sigma_i^n)] \qquad (31.43)$$

is satisfied with

$$\omega_1(t,s;x,y;u,v) =$$

$$= \omega(t,s;x,y;u,v) Q_3'[0,u_n(t),x_n(t),t] \frac{u(\sigma_{i+1}^n) - u(\sigma_i^n)}{\sigma_{i+1}^n - \sigma_i^n} .$$

Now, by the estimate

$$|\varphi[\sigma_i^n,x_n(\sigma_i^n),u_n(\sigma_i^n)] - \varphi[t,x_n(t),u_n(t)]| \leq$$

$$\leq c_1[|\sigma_i^n - t| + |x_n(\sigma_i^n) - x_n(t)| + |u_n(\sigma_i^n) - u_n(t)|] \quad (\sigma_i^n < t < \sigma_{i+1}^n),$$

and due to (31.43), we end up with the following equation to be studied,

$$\frac{dz_n}{dt} = 2Q_3'[0,u_n(t),x_n(t),t]\, \psi\,[t,x_n(t),u_n(t)] \left[\frac{u(\sigma_{i+1}^n) - u(\sigma_i^n)}{\sigma_{i+1}^n - \sigma_i^n} \right]^2 \times$$

$$\times (t-\sigma_i^n) + Q_4'[0,u_n(t),x_n(t),t] + \Omega_n(t) \quad (\sigma_i^n < t < \sigma_{i+1}^n) . \qquad (31.44)$$

In this equation,

$$|\Omega_n(t)| \le |Q_3'[0,u_n(t),x_n(t),t]|\cdot|\varphi_x'[t,x_n(t),u_n(t)]| \frac{[u(\sigma_{i+1}^n)-u(\sigma_i^n)]^2}{\sigma_{i+1}^n-\sigma_i^n}\times$$

$$\times c_1[|t-\sigma_i^n| + |x_n(t)-x_n(\sigma_i^n)| + |u_n(t)-u_n(\sigma_i^n)|] +$$

$$+ |Q_3'[0,u_n(t),x_n(t),t]| \frac{|u_n(\sigma_{i+1}^n) - u_n(\sigma_i^n)|}{\sigma_{i+1}^n - \sigma_i^n}\times$$

$$\times c[|t-\sigma_i^n| + |x_n(t)-x_n(\sigma_i^n)| + |u_n(t)-u_n(\sigma_i^n)|] .$$

We can easily see that the functions $Q_3'[0,u_n(t),x_n(t),t]$ and $\varphi_x'[t,x_n(t),u_n(t)]$ are uniformly bounded, and

$$|x_n(t) - x_n(\sigma_i^n)| \le c_2|u_n(t) - u_n(\sigma_i^n)| .$$

Thus, *the last term in* (31.44) *admits the bound*

$$|\Omega_n(t)| \le M\left[\frac{|u_n(\sigma_{i+1}^n) - u_n(\sigma_i^n)|^3}{\sigma_{i+1}^n - \sigma_i^n} + |u_n(\sigma_{i+1}^n) - u_n(\sigma_i^n)|\right] . \quad (31.45)$$

Third step of the proof. By (31.42), we can conclude that

$$x_n(t) = Q[u_n(t),0,z_n(t),t]$$

and, due to (31.44), the function $z_n(t)$ is a solution of the differential equation

$$\frac{dz}{dt} = Q_3'\{0,u_n(t),Q[u_n(t),0,z,t],t\}\psi\{t,Q[u_n(t),0,z,t],u_n(t)\}\alpha_n(t) +$$

$$+ Q_4'\{0,u_n(t),Q[u_n(t),0,z,t],t\} + \Omega_n(t) . \quad (31.46)$$

In this step, we shall examine what are the properties of the functions that enter the right-hand side of equation (31.46).

The coefficient at $\alpha_n(t)$ and the second term on the right-hand side respectively converge on every finite interval to

$$Q_3'\{0,u(t),Q[u(t),0,z,t],t\}\,\psi\,\{t,Q[u(t),0,z,t],u(t)\}$$

and to

$$Q_4'\{0,u(t),Q[u(t),0,z,t],t\} \quad ,$$

uniformly with respect to z .

Since the sequence $v_n(t)$ has asymptotic 2- intensity κ , functions $\alpha_n(t)$ satisfy the equalities

$$\lim_{n\to\infty} \int_{\tau_1}^{\tau_2} \alpha_n(t)dt = \kappa(\tau_2 - \tau_1) \qquad (t_o \leq \tau_1 < \tau_2 \leq t_1) \quad .$$

By (31.45) it follows that

$$\lim_{n\to\infty} \int_{t_o}^{t_1} |\Omega_n(t)|dt = 0 \quad .$$

In order to complete the proof, we only need to note that by the above properties of the functions entering the right-hand side of equation (31.46), and due to the convergence of the initial values $z_n(t_o)$ to the number (31.40), the uniform on $[t_o,t_1]$ convergence of the solutions $z_n(t)$ of equations (31.46) to the solutions $z_*(t)$ of the limit equation, i.e., correctness of (31.41), follows.

The above passage to the limit may be performed by using standard constructions that justify the Bogolyubov-Krylov averaging principle (cf., [7,87] for instance).

∎

A passage from equations (31.31) to (31.37) is similar to the transfer from the Stratonovich solution of stochastic equations to the Ito solutions. This analogy is given a more detailed treatment in Chapter 32. The links of the equations (31.13) and equations (31.23), (31.25) and (31.30) exhibit an analogous nature. Theorems close to assertions of this chapter may be proved for other, quite different classes of admissible inputs, at various input-output correspondences and for various notions of intensity. Once more, let us point out a high interest in a detailed analysis of the resulting boundary layers.

31.5. Construction of κ- outputs for hysterons

Constructions of the preceding section are still applicable if equation (31.31) is studied subject to constraints $\gamma_-(t,u)$ and $\gamma_+(t,u)$, provided that the resulting constrained system represents a standard hysteron (constant or variable). In such a case, the following natural analogue of Theorem 31.3 is true: *each continuous input* $u(t)$ $(t_0 \leq \leq t \leq t_1)$ *produces a* κ*- output* $x(t)$ *which is a vibro-solution of equation* (31.37) *subject to the constraints* $\gamma_-(t,u)$, $\gamma_+(t,u)$.

By virtue of this property, the techniques developed in Chapters 11- 13 for an analysis of vibro-solutions turn out applicable to the study of κ- outputs .

At the end of this chapter, let us observe that the notion of κ- -vibrosolution can also be introduced for the differential equations

$$\frac{dx}{dt} = \varphi[t,x,u(t)]\,u'(t) + \psi[t,x,u(t)]$$

which are more general than (31.31). Conversely, one can recover the differential equations that correspond to those κ-vibrosolutions.

Similar constructions are possible also in the case of vector inputs $u(t)$, provided that the appropriate Frobenius condition (cf., Chapter 20) is satisfied. A detailed treatment of such a situation would deserve an interest. Another interesting question is how to introduce κ-vibrosolutions for equations with vector inputs, considered without assuming the Frobenius condition.

To conclude, let us note that the transfer from the standard outputs to κ-outputs offers a comfortable way to encountering effects of self-magnetization type in hysterons.

32. Closure modulo sets of Wiener measure zero

32.1. A general scheme

Constructions we develop in this section represent an extension of techniques of Chapter 31.

Let a measure μ be defined on a Banach space E so that all open balls in E are measurable and have positive measure in that space. Consider an operator A acting in E , whose domain $\mathcal{D}$ has *full measure*. For any element $u \in E$ define a certain set $\mathcal{D}(u) \subset \mathcal{D}$, also having full measure, such that $u_n \in \mathcal{D}(u)$ and $u_n \to u$ imply the convergence of Au_n to some point x . The equality

$$x = W(u) \qquad (u \in E) \tag{32.1}$$

defines a *closure of the operator* A *modulo sets of measure* μ *zero*, or shorter a μ*-closure*.

If the operator A describes input-output correspondences of some transducer, smallness of the norm in E corresponds to smallness of an non-controllable perturbation of the input and, at last, the measure μ reflects "probability" of a realization of the perturbations, then it is more natural to describe the input-output correspondences by the μ-closure (32.1) rather than by the operator A . The operator (32.1) certainly may be different from A also on elements of $\mathcal{D}$.

32.2. Main theorem

In the sequel, we shall take $C = C(t_o,t_1)$ as the space E and the Wiener measure as μ (see, for instance, [24,77]).

Let us consider an arbitrary sequence $L = \{\Lambda_n\}$ of the refining partitionings of the interval $[t_o,t_1]$ by the points

$$t_o = \sigma_o^n < \sigma_1^n < \sigma_2^n < \ldots < \sigma_{k(n)}^n = t_1 \quad (n = 1, 2, \ldots) \ . \tag{32.2}$$

The limit

$$\Phi_2[u(t);\tau_1,\tau_2;L] = \lim_{n\to\infty} \sum_{\tau_1<\sigma_i^n<\tau_2} |u(\sigma_i^n) - u(\sigma_{i-1}^n)|^2 \tag{32.3}$$

will be called a *quadratic variation of the function* $u(t) \in C$ *on the interval* $[\tau_1,\tau_2] \subset [t_o,t_1]$, provided this limit does exist. The functions $u(t) \in C$ will be called L*-regular* if the limit (32.3) exists for all τ_1, $\tau_2 \in [t_o,t_1]$ and

$$\Phi_2[u(t);\tau_1,\tau_2;\ L] = \tau_2 - \tau_1 \quad . \tag{32.4}$$

The set $\mathcal{D} = \mathcal{D}_L$ of all L-regular functions has full Wiener measure (in fact, only this property of the Wiener measure is used in the present section). Clearly, for every L-regular function $u(t) \in C$,

$$\lim_{n\to\infty} \sum_{\tau_1 < \sigma_i^n < \tau_2} |u(\sigma_i^n) - u(\sigma_{i-1}^n)|^3 = 0 \quad (t_0 \leq \tau_1 \leq \tau_2 \leq t_1) \ . \tag{32.5}$$

Neither smooth functions nor functions of bounded variation are L-regular. For any function $u(t) \in C$, we can construct a sequence of partitions L such that $\Phi_2[u(t);t_0,t_1;L] = 0$, i.e., there are no functions which would preserve the L-regularity property for all sequences of partitions L.

Let us consider the differential equation

$$\frac{dx}{dt} = \varphi[t,x,u(t)]u'(t) + \psi[t,x,u(t)] \quad , \tag{32.6}$$

with the initial condition

$$x(t_0) = x_0 \quad . \tag{32.7}$$

For an input $u(t) \in C$ and a partitioning Λ_n of the interval $[t_0,t_1]$ with points (32.2), define a piecewise linear continuous function $x_n(t)$ which satisfies condition (32.7) and has the derivative

$$x'(t) = \varphi[\sigma_{i-1}^n, x_n(\sigma_{i-1}^n), u(\sigma_{i-1}^n)] \frac{u(\sigma_i^n) - u(\sigma_{i-1}^n)}{\sigma_i^n - \sigma_{i-1}^n} +$$

$$+ \psi[\sigma_{i-1}^n, x_n(\sigma_{i-1}^n), u(\sigma_{i-1}^n)] \qquad (\sigma_{i-1}^n < t < \sigma_i^n) \tag{32.8}$$

(the function $x_n(t)$ is reminiscent of Euler's broken line). If the sequence $x_n(t)$ is uniformly convergent to some limit $y(t) = Au(t)$, then this limit will be called an *L-solution of the problem* (32.6) - (32.7). The arguments employed in the proof of Theorem 31.3 imply the following.

<u>Lemma 32.1.</u> Let $\varphi(t,u)$ be a twice continuously differentiable

function, continuous together with its first derivatives, and $\psi(t,x,u)$ be a continuously differentiable, bounded function. Then any L-regular input $u(t)$ determines an L-solution $Au(t)$ which coincides with the vibro-solution of the equation

$$\frac{dx}{dt} = \varphi[t,x,u(t)]u'(t) + \psi[t,x,u(t)] -$$

$$- \frac{1}{2}\{\varphi'_u[t,x,u(t)] + \varphi'_x[t,x,u(t)]\varphi[t,x,u(t)]\} \quad , \qquad (32.9)$$

corresponding to the initial condition (32.7). ∎

As already mentioned, the set of L-regular inputs $u(t)$ has full Wiener measure. Thus, Lemma 32.1 gives rise to the following corollary, quite surprising and interesting.

Theorem 32.1. Suppose that the right-hand side of equation (32.6) satisfies all hypotheses of Lemma 32.1 . Then the operator A, which determines the L-solutions of problem (32.6) - (32.7) that correspond to L-regular inputs, admits the closure

$$x(t) = W[t_0,x_0]u(t) \qquad (t_0 \leq t \leq t_1) \qquad (32.10)$$

modulo sets of Wiener measure zero, defined on the whole space C. Values of the operator (32.10) are vibro-solutions of equation (32.9) that satisfy the initial condition (32.7). ∎

By Theorem 32.1, values of the operator (32.10) do not depend on a choice of the sequence of partitions Λ_n. Let us emphasize that for a smooth input $u(t)$ the function (32.10) is a normal solution to the equation (32.9) rather than to the original equation (32.6)!

32.3. Passage to integral equations

Let us assume that a sequence of partitions $L = \{\Lambda_n\}$ has been chosen. The limit

$$(I) \int_{\tau_1}^{\tau_2} \xi(t)\,du(t) =$$

$$= \lim_{n\to\infty} \sum_{\tau_1 < \sigma_i^n < \tau_2} \xi(\sigma_i^n)\,[u(\sigma_i^n) - u(\sigma_{i-1}^n)] \tag{32.11}$$

will be called the *Ito L-integral* of the continuous function $\xi(t)$ $(\tau_1 \le t \le \tau_2)$ with respect to the continuous function $u(t)$ $(\tau_1 \le t \le \tau_2)$, provided this integral exists. Further we shall use the name of an *Ito-L-solution of the equation* (32.6) *with initial condition* (32.7) for a continuous function $x(t)$ that satisfies the integral equation

$$x(t) = x_0 + (I)\int_{t_0}^{t} \varphi[s,x(s),u(s)]du(s) + \int_{t_0}^{t} \psi[s,x(s),u(s)]ds$$

$$(t_0 \le t \le t_1)\ . \tag{32.12}$$

Theorem 32.2. Suppose that the right-hand side of equation (32.6) fulfils the hypotheses of Lemma 32.1. Then the vibro-solution of equation (32.9), that satisfies initial condition (32.7), is an Ito L-solution of the equation (32.6) for any L-regular input $u(t)$.

Proof. The proof proceeds according to a simple scheme. Let $x_*(t)$ be a vibro-solution of equation (32.9) corresponding to an L-regular input $u_*(t)$. By Theorem 13.1,

$$x_*(t) = Q[u_*(t),0,z_*(t),t] \qquad (t_0 \le t \le t_1)\ , \tag{32.13}$$

where $z_*(t)$ is the normal solution of the differential equation

$$\frac{dz}{dt} = Q_4'\{0,u_*(t),Q[u_*(t),0,z,t],t\} + $$
$$+ Q_3'\{0,u_*(t),Q[u_*(t),0,z,t],t\}\,\Psi\{t,Q[u_*(t),0,z,t],u_*(t)\} \tag{32.14}$$

and

$$\Psi(t,x,u) = \psi(t,x,u) - \frac{1}{2}\,[\varphi_u'(t,x,u) + \varphi_x'(t,x,u)\,\varphi(t,x,u)]\ .$$

Then, for each $j \le k(n)$, the equality

$$x_*(\sigma_j^n) = x_o + \sum_{i=0}^{j} \{Q[u_*(\sigma_i^n),0,z_*(\sigma_i^n),\sigma_i^n] -$$

$$- Q[u_*(\sigma_{i-1}^n),0,z_*(\sigma_{i-1}^n),\sigma_{i-1}^n]\}$$

holds, or with an accuracy up to higher order terms

$$x_*(t) = x_o + \sum_{t_o < \sigma_i^n < t} \{Q_4'[u_*(\sigma_{i-1}^n),0,z_*(\sigma_{i-1}^n),\sigma_{i-1}^n](\sigma_i^n - \sigma_{i-1}^n) +$$

$$+ Q_3'[u_*(\sigma_{i-1}^n),0,z_*(\sigma_{i-1}^n),\sigma_{i-1}^n][z_*(\sigma_i^n) - z_*(\sigma_{i-1}^n)] +$$

$$+ \varphi[\sigma_{i-1}^n,x_*(\sigma_{i-1}^n),u_*(\sigma_{i-1}^n)][u_*(\sigma_i^n) - u_*(\sigma_{i-1}^n)] +$$

$$+ \Psi[\sigma_{i-1}^n,x_*(\sigma_{i-1}^n),u_*(\sigma_{i-1}^n)][u_*(\sigma_i^n) - u_*(\sigma_{i-1}^n)]^2 -$$

$$- \psi[\sigma_{i-1}^n,x_*(\sigma_{i-1}^n),u_*(\sigma_{i-1}^n)][u_*(\sigma_i^n) - u_*(\sigma_{i-1}^n)]^2\} \quad . \qquad (32.15)$$

By virtue of equality (32.14) and due to properties of the operator $Q(u,u_o,x_o,t)$, after passing with n to the limit in equation (32.15), we come to (32.12).

∎

An answer to the question concerning uniqueness of the Ito L-solutions for L-regular inputs is not clear. We only can prove that the operator defined in terms of the vibro-solutions to the equation (32.9) (which are uniquely defined) is the unique, continuous in C and causal selector of the multi-valued operator assigning the set of all Ito L-solutions of the integral equation (32.12) to any L-regular input $u(t)$.

Similarly to the Ito L-integral, we can define the *Stratonovich L-integral*

$$(S)\int_{\tau_1}^{\tau_2} \xi(t)du(t) = \lim_{n\to\infty} \sum_{\tau_1 < \sigma_i^n < \tau_2} \xi\left(\frac{\sigma_{i-1}^n + \sigma_i^n}{2}\right)[u(\sigma_i^n) - u(\sigma_{i-1}^n)] \qquad (32.16)$$

and, accordingly, the *Stratonovich L-solution of problem* (32.6) -

(32.7) as a continuous solution of the integral equation

$$x(t) = x_o + (S) \int_{t_o}^{t} \varphi[s,x(s),u(s)]du(s) + \int_{t_o}^{t} \psi[s,x(s),u(s)]dx \quad (t_o \leq t \leq t_1).$$

Passing from the Ito integral to Stratonovich integral requires only substitution of the term $\xi(\sigma_{i-1}^n)$ in the integral sum by $\xi(\frac{1}{2}\sigma_{i-1}^n + \frac{1}{2}\sigma_i^n)$. This substitution yields quite different values of the integrals and different solutions of the corresponding integral equations.

Theorem 32.3. Suppose that all hypotheses of Theorem 32.2 are satisfied. Then the vibro-solutions of equation (32.6), corresponding to the initial condition (32.7), are Stratonovich L-solutions of the equation (32.6) at any L-regular input $u(t)$.

Proof. The proof is analogous to that of Theorem 32.2. ∎

32.4. Equations with constraints

All considerations of this section apply also to equations with some constraints. The suitable statements can be then treated as theorems on the properties of the transducer that represents a certain hysteron. Upon passing from the Stratonovich solutions to Ito solutions, effects of self-magnetization type can be taken into account.

32.5. Implications for stochastic equations

In this section, the reader is supposed to be familiar with basic concepts of the theory of stochastic processes.

Let us consider the scalar stochastic differential equation

$$dx = \varphi(t,x,w)dw + \psi(t,x,w)dt \quad . \tag{32.17}$$

Here w is a standard Wiener process defined on the whole probabilistic space Ω . By the Wiener process we mean an appropriately constructed function $w(t;\omega)$ $(t_o \leq t \leq t_1,\ \omega \in \Omega)$, continuous in t for almost all $\omega \in \Omega$, and measurable in ω for every t .

Let $x = x(t;\omega)$ be an Ito solution of the equation (32.17), satisfying the initial condition (32.7). By definition, for almost all $\omega \in \Omega$, the function $x(t;\omega)$ is an Ito L-solution of equation (32.6) (here L denotes a fixed sequence of the partitions), corresponding to the input $w(t;\omega)$. Therefore, as a consequence of Theorem 32.2 we can conclude

Theorem 32.4. Suppose that the functions $\varphi(t,x,u)$ and $\psi(t,x,u)$ are sufficiently smooth and (32.10) is the operator constructed on the basis of the vibro-solutions to equation (32.9). Then the stochastic process defined by

$$x(t;\omega) = W[t_0,x_0]w(t;\omega) \qquad (t_0 \leq t \leq t_1,\ \omega \in \Omega) \tag{32.18}$$

is an Ito solution of the stochastic differential equation (32.17), satisfying the initial condition (32.7). ∎

In order to get an analogous statement for the Stratonovich solutions of the stochastic equation (32.17), one only has to replace the vibro-solutions of equation (32.9) by those of (32.6).

Theorem 32.4 offers a method of constructing solutions to the stochastic equation (32.17) on the basis of the vibro-solutions of some differential equation (32.9). The method is reversible: each vibro-solution $x(t) = W[t_0,x_0]u(t)$ of equation (32.9) may be constructed on the basis of the stochastic process $x(t;\omega)$ which is a solution of the equation (32.17). To this end, for each $\varepsilon > 0$ we must construct a minimal closed set $X[u(t);\varepsilon] \subset C = C(t_0,t_1)$, including almost all functions $x(t;\omega)$ for $\omega \in \Omega_\varepsilon$, where

$$\Omega_\varepsilon = \{\omega \in \Omega: \|w(t;\omega) - u(t)\|_{t_0,t_1} < \varepsilon\} .$$

The sets $X[u(t);\varepsilon]$ "shrink" to the function $W[t_0,x_0\ u(t)]$ in C with respect to the Hausdorff metric, i.e.,

$$\lim_{\varepsilon\to 0}\ \sup_{\omega\in\Omega_\varepsilon}\ \max_{t_0\leq t\leq t_1} |x(t;\omega) - W[t_0,x_0]u(t)| = 0$$

(provided, for example, that the hypotheses of Lemma 32.1 hold).

Proof. It is sufficient to note that, by virtue of the classical Ito formula, the stochastic process

$$y(t;\omega) = Q[0,w(t;\omega),x(t;\omega),t]$$

is a solution of the stochastic differential equation

$$dx = \{Q_4'[w,0,Q(0,w,x,t),t] + Q_3'[w,0,Q(0,w,x,t),t]\ \psi(t,x,w)\}dt \quad ,$$

and then make use of Theorem 13.1.

∎

Part 7. Complex hysteresis nonlinearities

Our program will be completed
with a few short items ...
Those who sit it out will find a thread
which ties our story together ...
O'Henry

33. Parallel connections and bundles of hysterons

33.1. Complex nonlinearities

In the last part of our book, we shall consider transducers that admit representation in the form of systems of hysterons and other elementary components (static elements, relays, etc.), convenient for purposes of an analysis. Such system can be treated in a similar way to spectral decompositions, multiple integrals and other typical structures in linear analysis.

Chapters 33 and 34 are mainly devoted to a study of the transducers W that reduce to simple open systems without feedback loops, composed of a finite number of hysterons $W^1, \ldots, W^N$ and some static elements. In general, such transducers are not deterministic. As an appropriate state, instead of the pair input-output, one takes the set $\{u,z_1,\ldots,z_N\} \in R^{N+1}$, where u is the input of W and z_j denote the outputs of single hysterons W^j within the transducer W. It is also useful to introduce the notion of an *associate transducer* Q characterized by the same block-diagram as W and having output vectors z with components z_j, but deterministic, in contrast to W. Operators that assign the variable state $q(t) = \{u(t),Q[t_0,z(t_0)]u(t)\}$ to an input $u(t)$ $(t \geq t_0)$ at the initial state $\{u(t_0),z(t_0)\}$ are referred to as *input-state operators* (or correspondences) *of the transducer* W .

For determining the output $x(t) = W[t_o,z(t_o)]u(t)$ of the transducer W at a given output of the associate transducer Q , an additional rule must be introduced. Operators $W[t_o,z(t_o)]$ are called *input-output operators of the transducer* W .

Complex nonlinearities of hysteresis type are usually non-controllable. We are thus interested in distinguishing controllability domains from the whole domain $\Omega(W) = \Omega(Q)$ of the feasible states. Complex transducers which can be represented by sequential and parallel connections (without feedback loops) of a finite number of static and vibro-correct components, are themselves static and vibro-correct; this important property is evident.

33.2. Parallel connections

Let hysterons $W^1, \dots, W^N$ with domains $\Omega(W^1), \dots, \Omega(W^N)$ of feasible states and with the input-output correspondences

$$z_j(t) = W^j[t_o,z_j(t_o)]u(t) \quad (j = 1, \dots, N) \quad . \tag{33.1}$$

be given. Define

$$\Omega(W) = \{\{u,z_1,\dots,z_N\} : \{u,z_j\} \in \Omega(W^j),\ u \in R^1\} \quad . \tag{33.2}$$

By a *parallel connection of the hysterons* W^j *with weights* ξ_j we shall mean a transducer W with the domain (33.2) of feasible states, such that for each initial state

$$q(t_o) = \{u_o,z_o\} = \{u(t_o,z_1(t_o),\dots,z_N(t_o)\} \in \Omega(W) \subset R^{N+1},$$

all continuous scalar inputs $u(t)$ $(t \geq t_o)$, satisfying the condition $u(t_o) = u_o$, are admissible and the output corresponding to a prescribed input is given by the equality

$$x(t) = W[t_o,z_o]u(t) = \sum_{j=1}^{N} \xi_j\, W^j[t_o,z_j(t_o)]u(t) \quad (t \geq t_o) \quad . \tag{33.3}$$

An example of the parallel connection of hysterons is shown in Fig. 33.1,a.

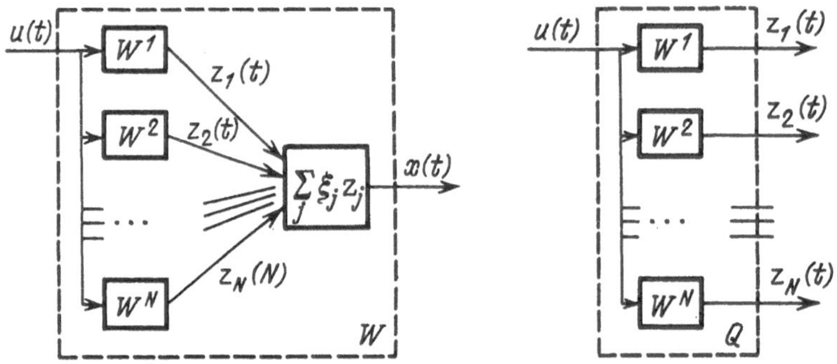

Fig. 33.1

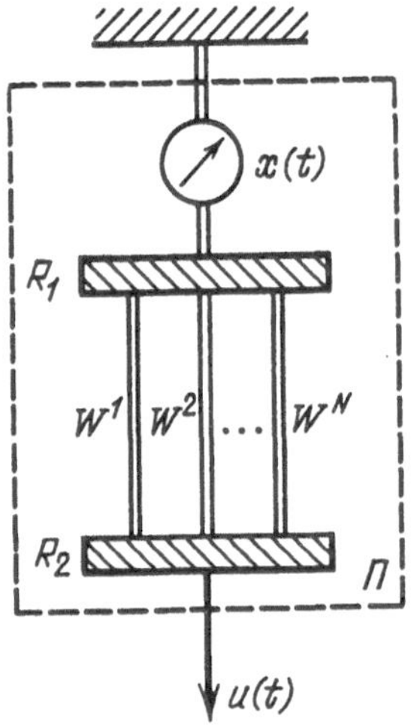

Fig. 33.2

The transducer Q that corresponds to a parallel connection of hysterons forms a *bundle of hysterons* (see Figure 33.1,b). Input--output correspondences of the bundle Q have the form

$$z(t) = Q[t_0,z_0]u(t) = \{W^1[t_0,z_1(t_0)]u(t), \ \ldots \ , W^N[t_0,z_N(t_0)]u(t)\}. \tag{33.4}$$

Hence, the law which describes the state $q(t)$ $(t \geq t_0)$ for a parallel connection of hysterons and for the corresponding bundle of hysterons is given by

$$q(t) = \{u(t), Q[t_0, z_0]u(t)\} \quad . \tag{33.5}$$

Consider the system Π (see Fig. 33.2), composed of parallel elasto-plastic threads $W^1, \ldots, W^N$ and two bars R_1, R_2 which fasten them. The bar R_1 is stationary while R_2 can move along the u-axis. The state of each thread W^j at the time moment t is then defined by the pair $\{u(t), z_j(t)\}$, where $u(t)$ represents a deformation of the thread, equal to the coordinate of the bar R_2, and $z_j(t)$ is the corresponding stress. If every thread represents a hysteron and the dependence of the stress (output $z_j(t)$) upon deformation (input $u(t)$) is characterized by the operator (33.1), then the suitable dependence between the deformation $u(t)$ and the total stress $x(t) = z_1(t) + \ldots + z_N(t)$ is described by the operator (33.3) with the weights $\xi_j = 1$. Therefore, the system Π can be treated as a parallel connection of hysterons $W^1, \ldots, W^N$. Similar systems are often used as phenomenological models of elasto-plastic bars.

33.3. Completely controllable restrictions

By a *controllable restriction of the transducer* V we shall mean a controllable transducer V_c such that the domain $\Omega(V_c)$ of its feasible states is contained in $\Omega(V)$, and the input-output and input-state correspondences (cf., Section 33.1) coincide with those of the transducer V. Furthermore, the class of admissible inputs for the restriction V_c coincides at every initial state from $\Omega(V_c)$ with the class of inputs admissible for V. Let us note that there exist transducers which do not admit any controllable restriction; on the other hand, other transducers have several controllable restrictions.

By a *completely controllable restriction of the transducer* V we shall understand a controllable restriction V_{cc} of V, such that for any initial state from $\Omega(V)$ there exists an admissible input transforming this initial state into a state from $\Omega(V_{cc})$. If a completely controllable restriction does exist, then it is uniquely defined.

In the general case, bundles of hysterons are non-controllable. In this section we shall construct and then consider completely controllable restrictions for bundles of hysterons, hence also such restrictions of the parallel connections of hysterons. Further, the concepts and notations of Section 3.2 will be used. They will refer to different hysterons W^j which are indexed by j .

Theorem 33.1. Each bundle Q of the hysterons $W^1, \ldots, W^N$ admits a completely controllable restriction.

Proof. For a bundle Q, consider the domain $\Omega(Q) \subset R^{N+1}$ of its feasible states. Take the set Ω_ℓ which comprises the states

$$q_\ell(u) = \{u, \Phi_\ell^1(u), \ldots, \Phi_\ell^N(u)\} \quad (-\infty < u < \min_j a_\ell^j) \tag{33.6}$$

in that domain. By $\Omega(Q_{cc})$ we shall denote the set of all states (33.5) corresponding to $\{u(t_0), z_0\} \in \Omega_\ell$ at any admissible input $u(t)$ and any $t \geq t_0$. By taking the restriction of the feasible states domain $\Omega(Q)$ of the bundle Q to $\Omega(Q_{cc})$, we obtain a completely controllable restriction Q_{cc} of Q . For the proof it is enough to observe the following.

First, if $\{u(t_0), z_0\} \in \Omega_\ell$ then for any non-increasing admissible input $u(t)$ the states (33.5) belong to Ω_ℓ for all $t \geq t_0$. Secondly, for any $u_1 < \min a_\ell^j$ there exists $u_0 > u_1$ such that if $\{u_0, z_0\} \in \Omega(W)$ then

$$\{u_1, Q[t_0, z_0]u(t_1)\} \in \Omega(Q_{cc}) \quad , \tag{33.7}$$

where $u(t) = u_0 + t_0 - t$, $t_1 = u_0 + t_0 - u_1$.

∎

So far, no direct characterization has been given to completely controllable restrictions of finite bundles of arbitrary hysterons. A description applicable in the most important cases is offered by the following

Theorem 33.2. Suppose that for any hysteron W^j from the bundle Q each point of the curve Φ_ℓ^j (respectively, of the curve Φ_r^j) either belongs to the set $\Phi_\ell^j \cap \Phi_r^j$ or is the left (respectively, the right) end-point of the graph $\varphi_j(M)$ of one of the functions $\varphi_j(u,M)$ $(u_\ell^j(M) < u < u_r^j(M))$, defining the hysteron W^j . Then the point

$$q_* = \{u_*, z_1^*, \ldots, z_N^*\} \tag{33.8}$$

belongs to $\Omega(Q_{cc})$ if and only if for any pair W^i, W^j of hysterons from the bundle Q the inequality

$$[u_\ell^i(u_*,z_i^*) - u_\ell^j(u_*,z_j^*)][u_r^i(u_*,z_i^*) - u_r^j(u_*,z_j^*)] \leq 0 \tag{33.9}$$

holds.

Proof. Suppose that the inequalities (33.9) hold. Then for all i,j one of the intervals $[v_\ell^i,v_r^i]$, $[v_\ell^j,v_r^j]$ (we use here the notations $v_\ell^i = u_\ell^i(u_*,z_i^*)$, $v_r^i = u_r^i(u_*,z_i^*)$) is contained in the other (see Fig. 33.3). Accordingly, without loss of generality we can assume that the relations

$$v_\ell^1 \leq v_\ell^2 \leq \ldots \leq v_\ell^N \leq u_* \leq v_r^N \leq \ldots \leq v_r^2 \leq v_r^1 \tag{33.10}$$

are satisfied.

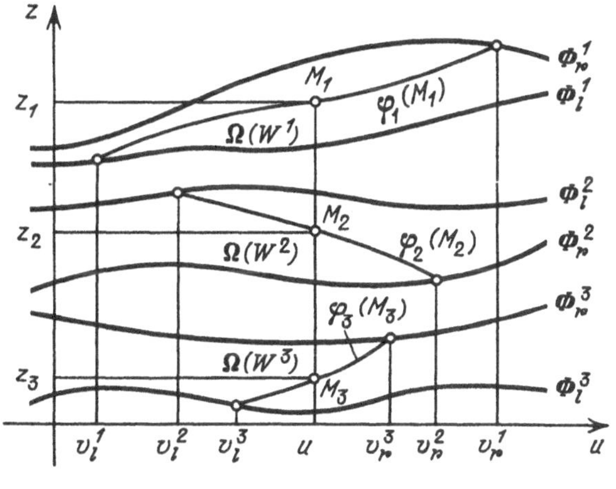

Fig. 33.3

Let us construct a continuous input $u(t)$ $(0 \leq t \leq N)$ which is linear on each of the intervals $[i-1,i]$ $(i = 1,\ldots,N)$, while for integer t and even N (the case of odd N can be treated analogously) assumes the values

$$u(0) = v_\ell^1,\ \ u(1) = v_r^2,\ \ u(2) = v_\ell^3,\ \ldots,\ u(N-1) = v_r^N,\ \ u(N) = u_* \ .$$

Then, in view of (33.5) and (33.10), the state

$$q_0 = \{v_\ell^1, \Phi_\ell^1(v_\ell^1), \ldots, \Phi_\ell^N(v_\ell^N)\} \in \Omega_\ell \quad , \tag{33.11}$$

which corresponds to control $u(t)$, , at the time moment $t = n$ will pass to a state q_* . Since $q_0 \in \Omega(W_{cc})$ (see the proof of Theorem 33.1), also $q_* \in \Omega(W_{cc})$.

To complete the proof, it is enough to take any initial state satisfying the conditions (33.9) and show that these conditions are fulfilled by all states (33.5) at any continuous input . Due to the vibro-correctness of the bundle, we can confine ourselves to the case of piecewise monotone inputs; in turn, by the semigroup property - to the case of monotone inputs. Only two hysterons enter each of the inequalities (33.9); hence we can restrict the considerations to the bundles which comprise only two hysterons. But in the case of a bundle composed of two hysterons which satisfy hypotheses of the theorem and for a monotone input, the proof is obvious.

∎

33.4. Periodic inputs

Every hysteron is monocyclic (see Section 28.1), thus the variable state (33.5), corresponding to a T-periodic input $u(t)$ $(t \geq t_0)$, is T-periodic for $t \geq t_0 + T$. Consequently, also parallel connections and bundles of hysterons $W^1, \ldots, W^N$ are monocyclic. The closed curve

$$u = u(t), \quad x = W[t_0, z_0]u(t) \quad (t_0 + T \leq t \leq t_0 + 2T) \tag{33.12}$$

is a *hysteresis loop*.

For a hysteron, the hysteresis loop corresponding to a fixed T-periodic input either had a non-zero width - then it was completely determined by the input without being dependent on an initial state of the hysteron - or it degenerated to a curve without loops - then, as a rule, it was dependent not only on the input but also on the initial state.

In the case of a parallel connection of hysterons, the hysteresis loop is uniquely defined if all the hysteresis loops

$$u = u(t), \quad x = W^j[t_0, z_j(t_0)]u(t) \quad (t_0 + T \leq t \leq t_0 + 2T; j = 1, \ldots, N) \tag{33.13}$$

are uniquely characterized by the input. In the general situation, even the non-degenerate loops (33.12) may depend on the initial states of the bundle of hysterons.

Suppose now that a T-periodic input $u(t)$ has only one local minimum and one local maximum on any interval of the length T (for example, $u(t)$ can be assumed in the harmonic form $\alpha + \beta \sin 2\pi t/T$). Then each curve (33.12) separates a part of the plane which will be called a hysteresis loop, as well.

Theorem 33.3. Suppose that there are two hysteresis loops corresponding to a T-periodic input $u(t)$ $(t \geq t_0)$ in the case of a parallel connection W of hysterons $W^1, \ldots, W^N$. Then the areas of both those loops are equal and for any u_*, intersections of the line $u = u_*$ with both loops are segments of the same length.

Proof. The second assertion of the theorem follows by virtue of (33.3) and properties of hysterons. In order to prove the first assertion, it suffices to refer to Cavalieri's principle.

∎

In Figure 33.4, two loops that correspond to the inputs $u(t)$, as considered in Theorem 33.3, are shown. The segments of equal length, representing intersections of the loops with the line $u = u^*$, are indicated by the thickened lines.

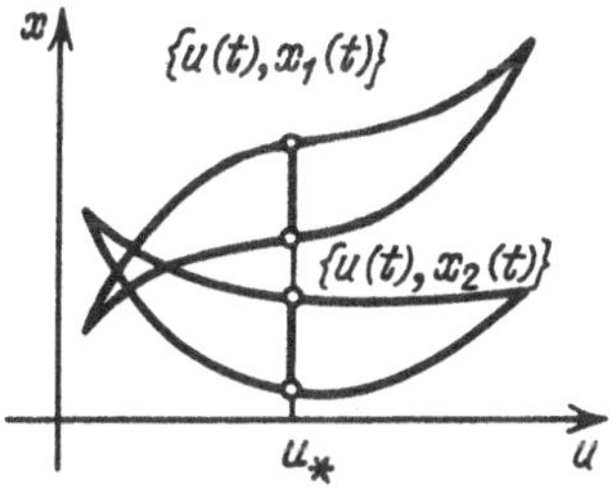

Fig. 33.4

By definition, the parallel connection of hysterons has *positive spin*, if every hysteresis loop (33.12) that corresponds to the input $u(t)$ which has only one local maximum and one local minimum within one period, can be run around by points $\{u(t), x(t)\}$ $(t_0 + T \leq t \leq t_0 + 2T)$ in the positive direction (counter-clockwise). Analogously, one can introduce a *negative spin* of the parallel connection. For a simple hysteron, these definitions coincide with those introduced earlier. Clearly, any parallel connection of hysterons with positive (negative) spins and positive coefficients ξ_j has also positive (negative) spin. If the parallel connection W of hysterons W^j has positive spin and the input $u(t)$ has, as previously, exactly one local maximum and one local minimum within one period, then the area σ of the hysteresis loop is given by the equality

$$\sigma = -\int_{t_0+T}^{t_0+2T} W[t_0,z_0]u(t)du(t) \quad . \tag{33.14}$$

If, however, the transducer W has negative spin, then for the area σ we have the equality

$$\sigma = \int_{t_0+T}^{t_0+2T} W[t_0,z_0]u(t)du(t) \quad . \tag{33.15}$$

By an *oriented area of the hysteresis loop* we shall mean the functional

$$S[u(t);T] = \int_{t_0+T}^{t_0+2T} W[t_0,z_0]u(t)du(t) \quad , \tag{33.16}$$

entering the equalities (33.14), (33.15). The oriented area of the loop is defined, independently of the initial state $\{u_0,z_0\}$, for any continuous and piecewise monotone, T-periodic input $u(t)$, for any smooth periodic input, etc. $S[u(t);T]$ admits the representation

$$S[u(t);T] = S^1[u(t);T] + \ldots + S^N[u(t);T] \quad , \tag{33.17}$$

where $S^j[u(t);T]$ denote the oriented areas of the hysteresis loops corresponding to hysterons W^j.

We can define the oriented area of the hysteresis loop also for any continuous, T-periodic input $u(t)$, by setting

$$S[u(t);T] = \lim_{n\to\infty} S[u_n(t);T] \quad , \tag{33.18}$$

where the functions $u_n(t)$ are continuously differentiable, T-periodic and uniformly convergent to $u(t)$. The limit (33.18) exists and does not depend upon the choice of a sequence $u_n(t)$.

There exists no hysteresis loop which would correspond to any smooth, almost periodic (asymptotically almost periodic) input. As an analogue to the value $T^{-1} S[u(t);T]$ one can take the limit

$$S_1[u(t)] = \lim_{\tau\to\infty} \frac{1}{\tau} \int_{t_0}^{t_0+\tau} W[t_0,z_0]u(t)du(t) \quad , \tag{33.19}$$

which is independent of z_0. The functional (33.19) can be extended by continuity to the set of all almost periodic (asymptotically almost periodic) inputs.

Let us denote by $X[T]$ the operator which transforms any T-periodic input $u(t)$ $(t \geq t_0)$ into the set $X[T]u(t)$ of all T-periodic (for $t \geq t_0 + T$) outputs of the parallel connection W of hysterons W^1, ..., W^N. As it has been already mentioned, the operator $X[T]$ is not necessarily single-valued, nevertheless it admits continuous single-valued selectors. In particular, the following evident property takes place.

Theorem 33.4. Suppose that functions $\Phi^1_\ell(u),\dots, \Phi^N_\ell(u)$ are defined for all $u \in (-\infty,\infty)$. Then the continuous single-valued operator which maps any T-periodic input into the corresponding T-periodic output $x(t)$ with values on the interval $[t_0 + T, t_0 + 2T]$, defined by the equality

$$x(t) = \sum_{j=1}^{N} \xi_j W^j[t_0, \Phi^j_\ell[u(t_0)]] \quad , \tag{33.20}$$

is a selector of the operator $X[T]$.

∎

Instead of (33.20), one can construct selectors of the form

$$x(t) = \sum_{j=1}^{N} \xi_j W^j[t_0,h_j[u(t_0)]]u(t) \quad (t_0 + T \leq t \leq t_0 + 2T) \quad , \tag{33.21}$$

with some appropriately adjusted functions $h_j(u)$. One can comfortably work with the selectors (33.21) (in particular, chosen in the form (33.20)), since many important properties of those selectors follow quite simply from the relevant properties of the hysterons W^j (see Parts 1 - 3).

33.5. An important example

Of special interest are parallel connections W and bundles Q of stops $U^1, \ldots, U^N$ which have the elasticity modulus equal to 1, the yield limits $\pm h_j$ with $h_1 > h_2 > \ldots > h_N > 0$, and the weight coefficients $\xi_1, \ldots, \xi_N$. The domain of feasible states of these transducers comprises the points $\{u, z_1, \ldots, z_N\}$ with any arbitrary first component and $|z_j| \leq h_j$. The operators (33.3) and (33.4), constructed for the parallel connection and bundle of the stops U^j, fulfil the Lipschitz conditions

$$\|W[t_0,z_0]u(t) - W[t_0,z_0]v(t)\|_{t_0,t_1} \leq$$
$$\leq 2(|\xi_1| + \ldots + |\xi_N|)\ \|u(t) - v(t)\|_{t_0,t_1} \tag{33.22}$$

and

$$\|Q[t_0,z_0]u(t) - Q[t_0,z_0]v(t)\|_{t_0,t_1} \leq 2\sqrt{N}\ \|u(t) - v(t)\|_{t_0,t_1} . \tag{33.23}$$

Let us now consider the case of periodic inputs. Since the spin of every stop is negative, for positive ξ_j, the parallel connection of stops has also negative spin. Different hysteresis loops, corresponding to the same T-periodic input $u(t)$, are congruent. If the input $u(t)$ has exactly one point of local maximum and one point of local minimum within one period, where $\min u(t) = m$ and $\max u(t) = M$, then the oriented area $S[u(t);T]$ of the corresponding hysteresis loops is determined by the equality

$$S[u(t);T] = \sum_{h_j < \frac{M-m}{2}} 2\xi_j h_j (M - m - 2h_j) . \tag{33.24}$$

Consider now completely controllable restrictions of the bundle of stops. As a consequence of Theorem 33.2 we get the following.

Theorem 33.5. Let Q_{cc} be a completely controllable restriction of the bundle Q of stops $U^1,\ldots,U^N$ which have the elasticity modulus 1 and the yield limits $\pm h_j$ where $h_1 \geq h_2 \geq \ldots \geq h_N$. Then the corresponding domain $\Omega(Q_{cc})$ of all feasible states consists of the points $\{u,z_1,\ldots,z_N\}$ with any arbitrary u , such that

$$h_1 - z_1 \geq h_2 - z_2 \geq \ldots \geq h_N - z_N \tag{33.25}$$

and

$$h_1 + z_1 \geq h_2 + z_2 \geq \ldots \geq h_N + z_N \ . \tag{33.26}$$

∎

By this theorem, the set $\Omega(Q_{cc})$ is convex, closed and contains all states of the form $\{u,0,\ldots,0\}$.

33.6. Remarks

a. In a similar way, also more general transducers W can be treated. For this, the last term $\Sigma\, \xi_j z_j$ in the block-diagram depicted in Figure 33.1,a is to be replaced by an element F with the output

$$x(t) = F[z_1(t),\ldots,z_N(t)] \ . \tag{33.27}$$

To characterize the resulting transducer $W = F(Q)$, obtained in this way, one can use the same bundle Q of hysterons $W^1, \ldots , W^N$.

b. The constructions of this and the subsequent sections remain valid if instead of the standard hysterons one takes variable hysterons (cf., Part 3), multidimensional hysterons (Part 4), admits phenomena of the self-magnetization type (cf., Part 6), etc.

c. Fig. 33.5 comprehends the block-diagram of a transducer W , more general than the simple parallel connection of hysterons. The input is now chosen as a vector-function $\bar{u}(t)$ with values in some R^k, and the output is a vector-function $\bar{x}(t)$ with values in some R^m . The block-diagram contains not only the hysterons $W^1,\ldots,W^N$ but also elements $(\cdot,\bar{e}_j(t))$ which represent scalar products of the input $u(t)$ and given vector-functions $\bar{e}_j(t)$ having values in R^k ; their outputs are scalar-

-valued (similarly as the outputs $z_j(t)$ of the hysterons). The coefficients $\bar{\xi}_j(t)$ are vector-valued, with values in R^m, hence also the functions $\bar{y}_j(t)$ are vector-valued. For such transducer, the states can be taken in the form $\{v_1,z_1;\dots;v_N,z_N\} \in R^{2N}$. One can easily express the corresponding relations input-state and input-output in an explicit form. For example,

$$\bar{x}(t) = \sum_{j=1}^{N} \bar{\xi}_j(t)\, W^j[t_0,(\bar{u}(t_0),\bar{e}_j(t_0))](\bar{u}(t),\bar{e}_j(t)) \quad . \tag{33.28}$$

If $k = m = N$, $\bar{e}_j(t) \equiv \bar{\xi}_j(t) \equiv \bar{e}_j$, vectors $\bar{e}_j$ form an orthonormal basis in R^N, all the hysterons W^j are chosen in the form of stops with elasticity modulus 1 and arbitrary threshold values (yield limits), then the transducer W represents a multidimensional stop with characteristic in the form of a rectangular parallelepiped in R^N (cf., Part 5).

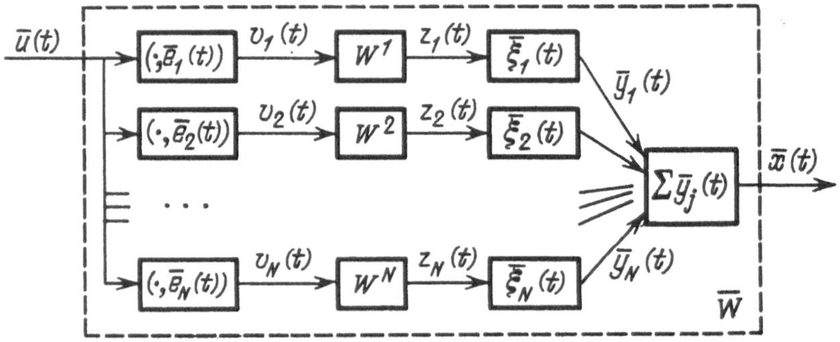

Fig. 33.5

34. Sequential connections of hysterons

34.1. Sequential connections and cascades

The block-diagram of a sequential connection of N hysterons $W^1, \dots, W^N$ is shown in Fig. 34.1,a . That diagram represents a transducer W with scalar input u(t) and scalar output x(t). The corresponding transducer Q is shown in Fig. 34.1,b. It has scalar-valued input u(t) and vector-valued output $\{v_1(t),v_2(t),\dots,v_N\}$, with values in R^N. The transducer Q is referred to as a *cascade of hysterons*.

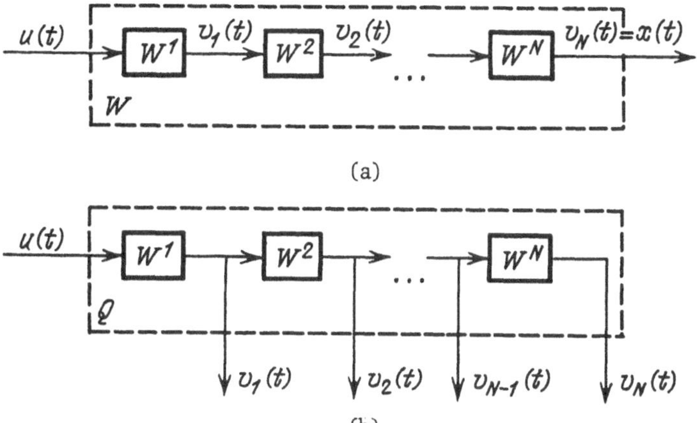

Fig. 34.1

For the sequential connection W and the cascade Q of hysterons, the states are specified by vectors $\{u,v_1,\dots,v_N\} \in R^{N+1}$, such that $\{u,v_1\} \in \Omega(W^1)$, $\{v_1,v_2\} \in \Omega(W^2)$, ..., $\{v_{N-1},v_N\} \in \Omega(W^N)$. For an initial state $\{u_o,v_1^o,\dots,v_N^o\} \in \Omega(W)$ (at $t = t_o$), as an admissible input we shall take any continuous function $u(t)$ $(t \geq t_o)$ which satisfies the condition $u(t_o) = u_o$. The input-output correspondences of the sequential connection W and cascade Q are given by the recursive relations

$$\begin{aligned} v_1(t) &= W^1[t_o,v_1^o]u(t) && (t \geq t_o), \\ v_2(t) &= W^2[t_o,v_2^o]v_1(t) && (t \geq t_o), \\ &\dots\dots\dots\dots\dots\dots\dots\dots && \\ x(t) = v_N(t) &= W^N[t_o,v_N^o]v_{N-1}(t) && (t \geq t_o). \end{aligned} \tag{34.1}$$

Clearly, the transducers W and Q are static and vibro-correct. The cascade Q is deterministic, whereas this property does not hold for the sequential connection W.

The authors do not know any simple description of the controllable or completely controllable restriction of a cascade of hysterons which have a sufficiently general form (similar to those introduced in Chapter 33 for the bundles).

Let the input $u(t)$ $(t \geq t_o)$ of the hysteron W^1 be T-periodic. Then its output $v_1(t)$ is also T-periodic for $t \geq T + t_o$. Consequently, the output $v_2(t)$ of the hysteron W^2 will be T-periodic for $t \geq 2T+t_o$, etc. Thus, the scalar output $x(t)$ of the sequential connection W and the vector output $v(t) = \{v_1(t), \ldots, v_N(t)\}$ of the cascade Q, corresponding to a T-periodic input $u(t)$ $(t \geq t_o)$, are T-periodic for $t \geq NT + t_o$. The closed curve

$$u = u(t), \quad x = x(t), \quad (NT + t_o \leq t \leq NT + T + t_o) \tag{34.2}$$

represents a hysteresis loop. In contrast to the parallel connections of hysterons (cf., Theorem 33.3), for their sequential connection the area of hysteresis loop (34.2) usually depends not only upon T-periodic input but also upon the initial state.

Let us assume that the input $u(t)$ $(t \geq t_o)$ is T-periodic. By a straightforward reasoning we can conclude that the output $x(t)$ of transducer W is then also T-periodic for $t \geq NT + t_o$. By using some more refined arguments, we would conclude the T-periodicity of the output at $t \geq \frac{1}{2} NT + T + t_o$. The last bound does not admit any further sharpening.

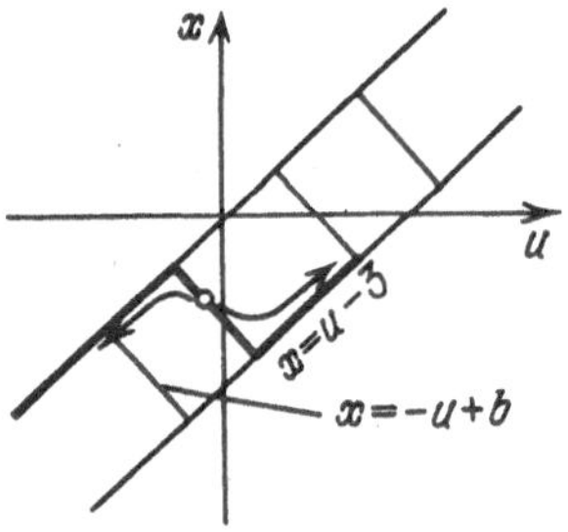

Fig. 34.2

Indeed, let us consider the sequential connection W of N identical hysterons W^o with the defining systems of curves given in Fig. 34.2. For the initial state $q_o = \{0,0,\ldots,0\} \in R^{N+1}$, the input $u(t) = \sin t$ $(t \geq 0)$ is admissible and 2π-periodic, while the output $x(t)$ becomes 2π-periodic only at $t \geq N\pi + \pi/2$. It is to be noted that the initial state q_o belongs to the domain of feasible states for the completely controllable restriction of the transducer W.

34.2. Sequential connections of plays and stops

In a number of important cases one can replace the construction of state spaces for sequential connections W of hysterons (based on introducing the appropriate cascades) by a simpler scheme. Let us consider three representative examples.

a. Suppose that each hysteron W^j in the transducer W (cf., Figure 34.1,a) has the form of a play L^j with characteristics $\Gamma_\ell^j(u)$ and $\Gamma_r^j(u)$ such that

$$\lim_{u\to-\infty} \Gamma_\ell^j(u) = -\infty , \quad \lim_{u\to\infty} \Gamma_r^j(u) = \infty . \tag{34.3}$$

Define the continuous functions

$$\bar{\Gamma}_\ell(u) = \Gamma_\ell^N\{\dots \Gamma_\ell^2[\Gamma_\ell^1(u)] \dots\} , \tag{34.4}$$

$$\bar{\Gamma}_r(u) = \Gamma_r^N\{\dots \Gamma_r^2[\Gamma_r^1(u)] \dots\} \tag{34.5}$$

on their natural domains. Functions (34.4) and (34.5) are non-decreasing, their values fill the whole real axis, $\bar{\Gamma}_r(u) \le \bar{\Gamma}_\ell(u)$ for all u, and

$$\lim_{u\to-\infty} \bar{\Gamma}_\ell(u) = -\infty , \quad \lim_{u\to\infty} \bar{\Gamma}_r(u) = \infty . \tag{34.6}$$

Consequently, the functions (34.4) and (34.5) can be treated as the characteristics of some generalized play $\bar{L}$.

For the sequential connection W of plays, relations (34.1) assume the form

$$v_1(t) = L^1[t_o,v_1^o]u(t),\dots, x(t) = v_N(t) = L^N[t_o,v_N^o]v_{N-1}(t) . \tag{34.7}$$

The last of functions (34.7) coincides with the output of the play $\bar{L}$ that corresponds to the same input, i.e.,

$$x(t) = \bar{L}[t_o,x(t_o)]u(t) \quad (t \ge t_o) , \tag{34.8}$$

where $x(t_o) = v_N^o$. However, $\{u(t_o),v_1^o,\dots,v_N^o\} \in \Omega(W)$ implies that

$\{u(t_0),v_N^0\} \in \Omega(\bar{L})$ and, conversely, it follows from $\{u(t_0),x_0\} \in \Omega(\bar{L})$ that there exist $v_1^0,\dots,v_{N-1}^0$ for which $\{u(t_0),v_1^0,\dots,v_{N-1}^0,x_0\} \in \Omega(W)$. Therefore the transducer W is in a natural sense equivalent to the single play $\bar{L}$. This implies, in particular, that a transducer which corresponds to a sequential connection of plays is deterministic.

Let us consider the cascade Q comprising two plays L^1 and L^2. Construct the play $\bar{L}$ and consider the bundle Q_1 of the plays L^1 and $\bar{L}$. The state space of the cascade is contained in the state space of the bundle. One can easily see that the cascade of the plays L^1 and $\bar{L}$ is in a natural sense equivalent to the completely controllable restriction of the bundle Q_1. Hence, in particular, it follows that the cascade Q is controllable.

b. Let us now consider the sequential connection W of two stops U^1, U^2 with the elasticity moduli E_i and the yield limits $\pm h_i$ $(i=1,2)$, respectively. Denote by Q the corresponding cascade comprising these two stops. Suppose that the inequality

$$h_2 \leq h_1E_2 \tag{34.9}$$

is fulfilled.

By $\bar{U}$ we shall denote the stop with elasticity modulus E_1E_2 and the yield limits $\pm h_2$. For the transducer W, system (34.1) reduces to the form

$$v_1(t) = U^1[u_0,v_1^0]u(t), \quad x(t) = v_2(t) = U^2[t_0,v_2^0]v_1(t) \quad . \tag{34.10}$$

By virtue of (34.10), there is a link between the operators $W[t_0;v_1^0,v_2^0]$ which characterize input-output relations of the sequential connection W and the operators $\bar{U}[t_0,x_0]$ of the input-output correspondences for the stop $\bar{U}$. This relation can be described by a simple rule whose form depends upon the value of v_2^0. We shall distinguish the following three cases:

$$v_2^0 < -h_1E_2 + v_1^0E_2 + h_2 \quad , \tag{34.11}$$

$$-h_1E_2 + v_1^0E_2 + h_2 \leq v_2^0 \leq h_1E_2 + v_1^0E_2 - h_2 \quad , \tag{34.12}$$

$$h_1E_2 + v_1^0E_2 - h_2 < v_2^0 \quad . \tag{34.13}$$

For any admissible input $u(t)$ $(t \geq t_o)$ *, the equality*

$$W[t_o;v_1^o,v_2^o]u(t) = \bar{U}[t_o,v_2^o]u(t) \tag{34.14}$$

holds in the case (34.11) *for all* t *such that*

$$\min_{t_o \leq s \leq t} u(s) \leq u(t_o) - \frac{2h_2}{E_1E_2} \quad ;$$

in the case (34.12) *for all* $t \geq t_o$ *, and in the case* (34.13) *for all* t *such that*

$$\max_{t_o \leq s \leq t} u(s) \geq u(t_o) - \frac{2h_2}{E_1E_2} \quad .$$

Clearly, this statement does not mean that the transducer W is deterministic.

If the condition (34.9) is fulfilled, then the cascade composed of two stops U^1 and U^2 admits a completely controllable restriction, equivalent to a completely controllable restriction of the bundle of stops U^1 and $\bar{U}$.

c. If, instead of (34.9), we have

$$h_2 > h_1E_2 \quad , \tag{34.15}$$

then the parallel connection W of stops U^1 and U^2 should be compared with the stop $\bar{\bar{U}}$, characterized by the elasticity modulus E_1E_2 and the yield limits $\pm h_1E_2$, rather than with the stop $\bar{U}$. Also now it is useful to distinguish three cases:

$$v_2^o < h_1E_2 + v_1^oE_2 - h_2 \quad , \tag{34.16}$$

$$h_1E_2 + v_1^oE_2 - h_2 \leq v_2^o \leq -h_1E_2 + v_1^oE_2 + h_2 \quad , \tag{34.17}$$

$$-h_1E_2 + v_1^oE_2 + h_2 < v_2^o \quad . \tag{34.18}$$

In the case (34.16), *for any admissible input* $u(t)$ $(t \geq t_o)$ *and for all* t *such that*

$$\min_{t_0 \leq s \leq t} u(s) \leq u(t_0) - \frac{h_1 + v_1^0}{E_1} ,$$

the equality

$$W[t_0;v_1^0,v_2^0]u(t) = \bar{\bar{U}}[t_0,v_1^0E_2]u(t) - h_2 + h_1E_2$$

holds. In the case (34.17), *the equality*

$$W[t_0;v_1^0,v_2^0]u(t) = \bar{\bar{U}}[t_0,v_1^0E_2]u(t) + v_2^0 - v_1^0E_2 ,$$

analogous to (34.14), *is satisfied. In the case* (34.18), *for all* t *such that*

$$\max_{t_0 \leq s \leq t} u(s) \geq u(t_0) + \frac{h_1 - v_1^0}{E_2} ,$$

the equality

$$W[t_0;v_1^0,v_2^0]u(t) = \bar{\bar{U}}[t_0,v_1^0E_2]u(t) + h_2 - h_1E_2$$

takes place.

In the case (34.15), the cascade composed of the stops U^1 and U^2 is non-controllable, it admits controllable restrictions but none of them is completely controllable.

All statements of this section can be directly deduced from the suitable definitions, at least in the case of monotone inputs. To admit also piecewise monotone inputs, one should use the semigroup property. Further passage to all continuous inputs requires employing the vibro-correctness of the involved transducers.

34.3. Compensators

Let us consider a transducer W with a given domain $\Omega(W)$ of the feasible states, a class of inputs $u(t)$ $(t \geq t_0)$ admissible at an initial state $q_0 = \{u_0,z_0\} \in \Omega(W)$, operators

$$x(t) = W[t_0,z_0]u(t) \qquad (t \geq t_0) \tag{34.19}$$

of the input-output relations and operators

$$z(t) = Q[t_0,z_0]u(t) \qquad (t \geq t_0) \tag{34.20}$$

describing the input-state correspondences. The transducer W^{-1} will be called a *left compensator of the transducer* W or simply *compensator of the transducer* W, if for any state $q_0 = \{u_0,z_0\} \in \Omega(W)$ there is a state $\{z_0,r_0\} \in \Omega(W^{-1})$ such that every output (34.19) is admissible for W^{-1}, provided that

$$W^{-1}[t_0,r_0]W[t_0,z_0]u(t) \equiv u(t) \qquad (t \geq t_0) \ . \tag{34.21}$$

Equality (34.21) means first that $\{u_0,z_0,r_0\}$ belongs to the domain $\Omega(\bar{W})$ of the feasibles states of the sequential connection $\bar{W} = W^{-1}W$ (see Fig. 34.3) and, secondly, that for the initial state $\{u_0,z_0,r_0\}$ and any admissible input $u(t)$, the output $x(t)$ coincides with $u(t)$.

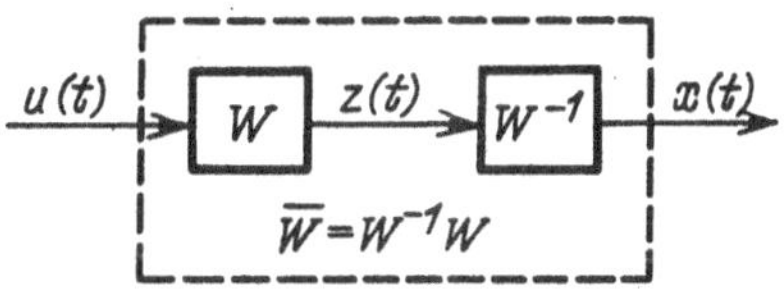

Fig. 34.3

Analogously we can define the *right and bilateral compensators*. The compensators are also referred to as *inverse transducers*.

For an arbitrary static element F with strongly increasing (or strongly decreasing) characteristic $f(u)$, the static element with the characteristic $f^{-1}(u)$ ($f^{-1}(u)$ denotes the function inverse to $f(u)$) will play the role of a (bilateral) compensator.

In this section, we are interested in properties of *compensators to hysterons*. An underlying hypothesis that yields the existence of a compensator to the hysteron W consists in assuming strong monotonicity of all curves $x = T(u;M)$ defining for W. For definiteness, we shall assume that all the functions $T(u;M)$ are strongly increasing.

For simplicity, let the functions $\Phi_\ell(u) = \Phi_\ell(u;W)$ and $\Phi_r(u) = \Phi_r(u;W)$ be defined for all $u \in (-\infty,\infty)$, and

$$\lim_{u\to-\infty} [\Phi_\ell(u) + \Phi_r(u)] = -\infty, \qquad \lim_{u\to\infty} [\Phi_\ell(u) + \Phi_r(u)] = \infty \ . \tag{34.22}$$

Let P be the symmetry mapping with respect to the line $x = u$ which transforms the plane $\{u,x\}$ into itself, with $P(u,x) = \{u,x\}$. Let us define a hysteron W^* (cf., Figure 34.4) such that $\Omega(W^*) = P\Omega(W)$; $\Phi_\ell(u;W^*)$ and $\Phi_r(u;W^*)$ are the inverse functions to $\Phi_\ell(u)$ and $\Phi_r(u)$, respectively; every defining curve $T^*(M)$ of the hysteron W^* coincides with the curve $PT(PM)$. By the identity

$$W^*[t_0,u(t_0)]W[t_0,x_0]u(t) \equiv u(t) \qquad (t \geq t_0) \quad , \tag{34.23}$$

which holds for all the admissible inputs $u(t)$ $(t \geq t_0)$, we can conclude the following simple

<u>Theorem 34.1.</u> For any hysteron W, the hysteron W^* represents its compensator. ∎

A hysteron and its compensator have the same spin. Given a compensator to a hysteron W, one can easily construct compensators to the hysterons $-W$, ξW for $\xi \neq 0$, $W + \text{const.}$, etc.

For plays $L[t_0,x_0]u(t)$ and for stops $U[t_0,x_0]u(t)$ there are no compensators, since the suitable defining curves contain horizontal segments. However, there exist compensators to the hysterons

$$W[t_0,x_0]u(t) = L[t_0,x_0]u(t) + \varepsilon\,[u(t) - u(t_0)] \quad ,$$

$$W_1[t_0,x_0]u(t) = U[t_0,x_0]u(t) + \varepsilon\,[u(t) - u(t_0)] \quad ,$$

"close" to the above plays and stops, respectively.

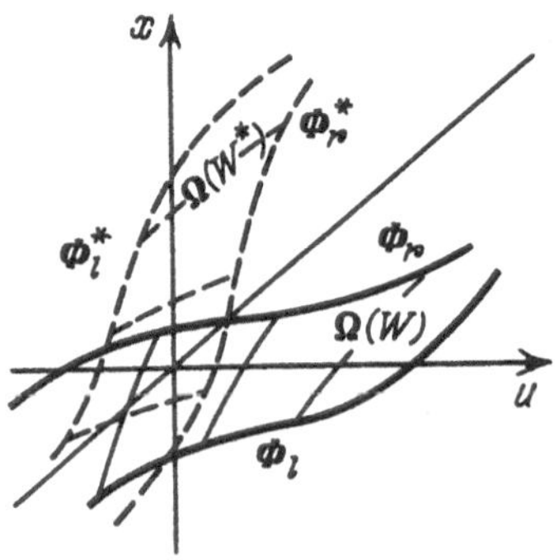

Fig. 34.4

34.4. Complex connections

Let us consider composed of hysterons, static elements and summators, without feedback loops but with a structure more complex than simple parallel or sequential connections. By *chains* we shall mean the maximal in sense of inclusion sequential connections comprehended in such systems. System W shown in Figure 34.5 contains two chains: the first of them comprises hysterons W^1, W^2, W^3, summator Σ and hysteron W^5, the other is composed of hysterons W^1, W^4, summator Σ and hysteron W^5. By $N_0(W)$ we shall denote the maximal number of hysterons within a single chain of the system W. For instance, $N_0(W) = 4$ in the system depicted in Figure 34.5. The corresponding output of system W becomes T-periodic for $t \geqq N_0(W)\,T$. The bound by $N_0(W)T$ is rather rough, as it has been shown in Section 34.1, but no sharper bounds are so far available for general complex systems.

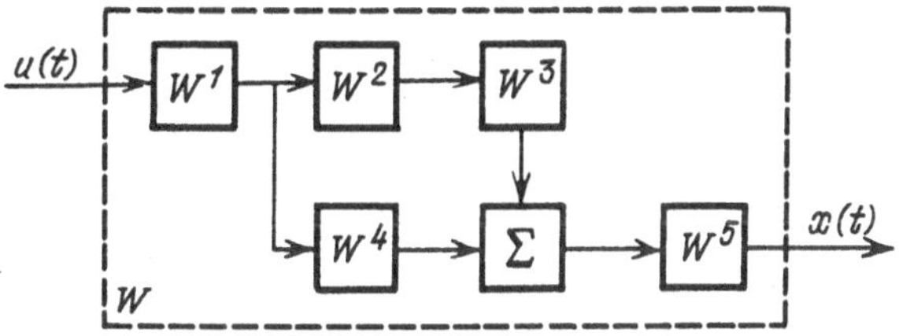

Fig. 34.5

Arbitrary systems composed of stops can be reduced to parallel connections of some other stops, with many basic properties preserved. Consider, for instance, a complex system W without feedback loops, built only of stops, summators and proportional (positive) elements. Any controllable restriction W_c of such system is in a natural sense equivalent to the transducer $W_* + c$, where W_* denotes the completely controllable restriction of a certain parallel connection of stops and c is a constant that depends only upon the choice of restriction W_c. Analogous statements remain valid if the stops are replaced by plays (but not by arbitrary hysterons !).

35. Ishlinskii's material

35.1. Continual systems of hysterons

Some important classes of complex hysteresis nonlinearities may be described by introducing the continual analogues of systems comprising a finite number of hysterons. The continual analogues of parallel connections play a fundamental role in this case.

Let us consider a family $W(\gamma)$ of hysterons. Assume that a parameter γ runs through a certain set Γ. For the case of a finite set Γ, the notions of the parallel connection W and the bundle Q of hysterons $W(\gamma)$ $(\gamma \in \Gamma)$ have been introduced in Chapter 33.

Henceforth, we shall use a more convenient form of the basic relations characterizing the transducers W and Q. The domain Ω of their feasible states comprises the pairs $\{u,z(\gamma)\}$ where $z(\gamma)$ is a function defined on Γ, such that

$$\{u,z(\gamma)\} \in \Omega[W(\gamma)] \qquad (\gamma \in \Gamma) . \tag{35.1}$$

The output $x(t)$ of the parallel connection W is scalar-valued and, at an input $u(t)$ $(t \geq t_0)$, it admits the representation

$$x(t) = \sum_{\gamma} \xi(\gamma)\, W[t_0,z_0(\gamma);\gamma]u(t) \qquad (t \geq t_0). \tag{35.2}$$

At any fixed t, the function

$$z(\gamma;t) = Q[t_0,z_0(\gamma)]u(t) = W[t_0,z_0(\gamma);\gamma]u(t) \quad , \tag{35.3}$$

defined on Γ, represents the output of bundle Q.

At a first glance, the passage to continuum Γ seems to be quite easy: one should introduce some measure on Γ and replace finite sum in the right-hand side of (35.2) by the integral

$$x(t) = \int_{\Gamma} \xi(\gamma)W[t_0,z_0(\gamma);\gamma]u(t)d\mu(\gamma) \quad . \tag{35.4}$$

However, this simplicity is fictitious. First, it may be necessary to use in (35.4) various notions of the integral (integrals in the Riemann, Lebesgue, Stjeltjes or Bochner sense; strong, weak or singular integrals, etc.). Secondly, various classes of functions $\{u, z(\gamma)\}$ (continuous, integrable up to some power, etc.) may be taken as the sets of feasible states. All these difficulties did not occur for finite sets Γ.

Integral representations analogous to (35.4) can easily be constructed for vector-valued nonlinearities of hysteresis type. So far, no detailed analysis has been developed for such nonlinearities. In the subsequent sections, we shall consider systems that have the form of the continual parallel connections of stops, or equivalently, analogous connections of ideally plastic fibres.

A study of the continual analogues of sequential connections and cascades of hysterons remains still an open problem. In that case, an immediate reminiscence of the multiple integrals arises.

35.2. Ishlinskii's transducer

Let us consider a one-parameter family $U(h)$ $(0 < h < \infty)$ of stops with the elasticity modulus 1 and with the threshold values $\pm h$, the same for all elements of the system.

Suppose that a non-decreasing left-continuous function $\Xi = \Xi(h)$ $(h \geq 0)$, satisfying the conditions

$$\lim_{h \to \infty} \Xi(h) = 0 \tag{35.5}$$

and

$$\int_0^\infty |\Xi(h)| dh < \infty \tag{35.6}$$

is given. By (35.5) it follows that the function $\Xi(h)$ is non-positive everywhere. Condition (35.6) is equivalent to the inequality

$$\int_0^\infty h \, d\Xi(h) < \infty \quad . \tag{35.7}$$

Denote by Z the family of continuous functions $z(h)$ $(h \geq 0)$ such that

$$|z(h)| \leq h \qquad (0 \leq h < \infty) \quad . \tag{35.8}$$

We shall treat the pairs $\{u,z(h)\}$, where $u \in (-\infty,\infty)$ and $z(h) \in Z$, as the elements of the set $\Omega(W)$ of feasible states of the transducer W whose input-output relations are described by the equality

$$x(t) = W[t_0,z_0(h);\Xi]u(t) = \int_0^\infty U[t_0,z_0(h);h]u(t)\,d\Xi(h) \quad (t \geq t_0) \tag{35.9}$$

and the variable state $\{u(t),z(h;t)\}$ is given by

$$z(h;t) = Q[t_0,z_0(h);\Xi]u(t) = U[t_0,z_0(h);h]u(t) \quad (t \geq t_0) \ . \tag{35.10}$$

The integral in the right-hand side of (35.9) is to be understood as improper.

$$\int_0^\infty U[t_0,z_0(h);h]u(t)d\Xi(h) = \lim_{A\to\infty} \int_0^A U[t_0,z_0(h);h]u(t)d\Xi(h) \ ,$$

with the limit taken in the standard Riemann-Stjeltjes sense.

The transducer W we have just defined is usually referred to as *Ishlinskii's transducer.*

Directly by definition (and due to some properties of stops), any Ishlinskii's transducer is static and vibro-correct; furthermore, it is non-deterministic and non-controllable. If the weight function $\Xi(h)$ is differentiable and $\xi(h)$ denotes its derivative, then (35.9) can be given the form

$$x(t) = \int_0^\infty \xi(h)\ U[t_0,z_0(h);h]u(t)dt \ . \tag{35.11}$$

For a complete characterization of Ishlinskii's transducer, we have to show that at any initial state $\{u(t_0),z_0(h)\} \in \Omega(W)$ and any continuous input $u(t)$ $(t \geq t_0)$, the integral in the right-hand side of the equality (35.9) is convergent, and the output $x(t)$ defined by this equality is continuous. Both these properties are evident; the joint continuity of the function $U[t_0,z_0(h);h]u(t)$ in h and t follows by the same property of stops, while the uniform in t convergence of the integrals is ensured by the estimate (35.7).

The notion of Ishlinskii's transducer admits various easy generalizations. For this, it is enough to pass to other classes of stops or to other notions of integrals in the defining equality (35.9). Such modification does not create any serious difficulties, though a resulting exposition becomes quite tedious due to many non-essential technicalities. Here we shall only discuss two special cases.

a. Let non-decreasing functions $\Xi(h)$, defined on the open half-line $(0,\infty)$ and satisfying the conditions (35.5), (35.6), be taken as the weight functions. These functions can diverge to $-\infty$ as $h \to 0$. The use of the weight functions unbounded frow below is convenient, for instance, at an analysis of the behaviour of rigid materials subject to small deformations. One can also introduce Ishlinskii's transducer with any non-decreasing weight function $\Xi(h)$ $(0 < h < \infty)$ which satisfies (35.5) and the condition

$$\int_0^A |\Xi(h)|dh < \infty \qquad (A > 0) \quad , \tag{35.12}$$

weaker than (35.6). To this end, one only has to change the class of feasible states. In particular, feasible states can be taken in the form of pairs $\{u,z(h)\}$, with a function $z(h)$ not only satisfying the estimate (35.8) but also bounded on the half-line. By this boundedness, the integrals in the definition of Ishlinskii's transducer are convergent uniformly with respect to t from any finite time interval.

b. Let us now assume that the weight function $\Xi(h)$ can increase exclusively in N points $h_1, \dots ,h_N$. Then, the corresponding Ishlinskii's transducer W is reduced to a finite system of ideally plastic fibres

$$x(t) = \sum_{j=1}^{N} W[t_0,z_j;E_j,H_j]u(t) \tag{35.13}$$

in a parallel connection, where $W(E_j,H_j)$ is the fibre with the elasticity modulus E_j and yields limits $\pm H_j$, such that

$$E_j = \Xi(h_j + 0) - \Xi(h_j - 0), \quad H_j = E_j h_j \quad (j = 1,\dots,N) \quad . \tag{35.14}$$

Therefore, the class of transducers W (characterized by the equality (35.9)) contains parallel connections of "almost arbitrary" finite families of hysterons $W(E_j,H_j)$. Only those families will remain out of the scope which simultaneously contain pairs $W(E_i,H_i)$ and $W(E_j,H_j)$ such that $E_iH_j = E_jH_i$ (e.g., representing two fibres with the same elasticity modulus and the same yield limits). Such an exclusion is meaningless; it might be avoided by passing in (35.9) from the integrals over the half-line to integrals over some more complex domains (this corresponds to a parametrization of the elasto-plastic fibre not only by its elasticity modulus and yield limits but also by some additional characteristics).

35.3. Loading and unloading functions

Equality (35.9) is important from several view-points: for an analysis of the properties of Ishlinskii's transducer , for understanding the possibility of using arbitrary continuous functions (not necessarily piecewise monotone) as the inputs, for an analysis of the dynamics of closed-loop systems with Ishlinskii's transducers as elementary components, etc. However, for determining the output corresponding to a given input one should rather look for some other analysis techniques.

As it has been already mentioned, the transducer $\xi U(h)$ describes an ideally plastic fibre $W(E,H)$ with the elasticity modulus $E = \xi$ and with the yield limits $\pm H$ where $H = \xi h$. The function

$$X_+(u;\xi,h) = \begin{cases} -\xi h, & \text{for } u \leq -h, \\ \xi u, & \text{for } -h \leq u \leq h, \\ \xi h, & \text{for } u \geq h \end{cases} \tag{35.15}$$

is called a *loading function* of the transducer $\xi U(h)$. The relation

$$X_-(u;\xi,h) = 2X_+(u/2;\xi,h) \quad (-\infty < u < \infty) \tag{35.16}$$

defines then the *unloading function*. The functions (35.15) and (35.16) admit a clear physical interpretation: if an input $u(t)$ $(t \geq t_0)$ is monotone, then

$$\xi U[t_0,0;h]u(t) = X_+[u(t) - u(t_0);\xi,h] \quad ; \tag{35.17}$$

if the input $u(t)$ is monotone on each of the intervals $[t_0,\tau]$ and $[\tau,t_1]$ $(t_0 < \tau < t_1)$, and

$$u(t_1) \in [u(t_0),u(\tau)] \quad , \tag{35.18}$$

then for any initial state $\{u(t_0),z(t_0)\} \in \Omega[U(h)]$, the equality

$$\xi U[t_0,z(t_0);h]u(t) - \xi U[t_0,z(t_0);h]u(\tau) =$$
$$= X_-[u(t) - u(\tau);\xi,h] \qquad (\tau \leq t \leq t_1) \tag{35.19}$$

holds.

Let us continue the study of Ishlinskii's transducer W. The continuous odd functions

$$X_+(u;\Xi) = \int_0^u |\Xi(|h|)|dh \quad (-\infty < u < \infty) \tag{35.20}$$

and

$$X_-(u;\Xi) = \int_0^u |\Xi(|h|/2)|dh = 2X_+(u/2;\Xi) \qquad (-\infty < u < \infty) \tag{35.21}$$

will be called a *loading function* and an *unloading function* for *Ishlinskii's transducer with the weight function* $\Xi(h)$, respectively. Both functions (35.20) and (35.21) are non-decrasing; each of them is concave on the half-line $u \geq 0$. Therefore,

$$X_+(u;\Xi) \leq X_-(u;\Xi) \qquad (0 \leq u < \infty) \quad . \tag{35.22}$$

In Figure 35.1,a-b the graphs of functions (35.15), (35.16) and (35.20) are depicted, respectively. The graphs of loading functions are represented by solid lines, those of unloading functions by dashed lines.

Let us now consider Ishlinskii's transducer as a mathematical model of plastic body treated as composed of infinitely many ideally plastic fibres. In a natural way, $\{u,0\}$ is referred to as the *null-* or *unloaded state*. By the equalities (35.17) and (35.19) it follows that for a monotone input $u(t)$ $(t \geq t_0)$ at an unloaded initial state, the variable stress $x(t)$ admits a simple representation by the loading function

$$x(t) = W[t_0,0;\Xi]u(t) = X_+[u(t) - u(t_0);\Xi] \ . \tag{35.23}$$

If an input $u(t)$ $(t_0 \leq t \leq t_1)$ is monotone on each of the intervals $[t_0,\tau]$ and $[\tau,t_1]$ $(t_0 < \tau < t_1)$, with condition (35.18) fulfilled, then the equality

$$W[t_0,z_0(h);\Xi]u(t) - W[t_0,z_0(h);\Xi]u(\tau) =$$
$$= X_-[u(t) - u(\tau);\Xi] \quad (\tau \leq t \leq t_1) \tag{35.24}$$

holds for any initial state $\{u(t_0),z(t_0)\} \in \Omega(W)$. Equalities (35.23) and (35.24) reflect a physical meaning of the loading and unloading functions, respectively.

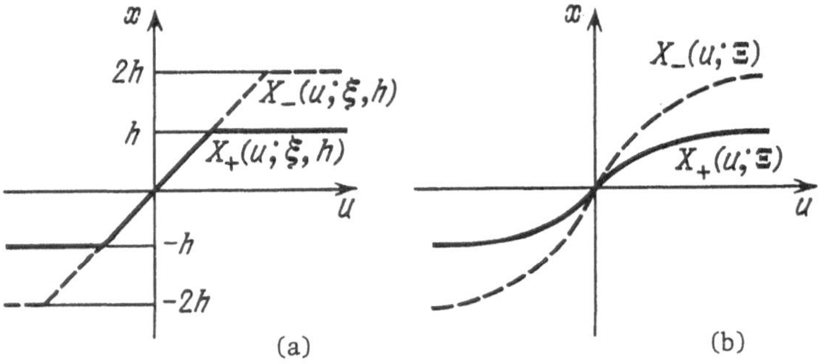

Fig. 35.1

The left-hand side of equality (35.24) can be determined experimentally. Therefore, we can experimentally determine the unloading function and, consequently, the weight function of Ishlinskii's transducer. If the difference $Vu(t) - Vu(\tau)$ has been found for a certain transducer V over a class of inputs $u(t)$ but the knowledge of the difference $u(t) - u(\tau)$ is still not sufficient for determining the value $Vu(t) - Vu(\tau)$ within accuracy of the experiment, then by virtue of (35.24) V cannot be treated as Ishlinskii's transducer.

Relations (35.23) and (35.24) admit a stronger version. Before going into details, we need some definitions.

Take an arbitrary continuous input $u_0(t)$ $(t \geq t_0)$. Denote by $T_1 = T_1[u_0(t)]$ the set of those $t_* \geq t_0$ for which (see Fig. 35.2)

$$|u_o(t) - u_o(t_o)| \leq |u_o(t_*) - u_o(t_o)| \qquad (t_o \leq t \leq t_*) \quad .$$

By $T_2 = T_2[u_o(t)]$ we shall denote the set of all $t_{**} \notin T_1$ such that $u_o(t_{**})$ is one of the end-points of the interval which represents the range of function $u_o(t)$ $(t_o \leq t \leq t_{**})$. The third class $T_3 = T_3[u_o(t)]$ will comprise those $t_{***} \notin T_1 \cup T_2$ for which there exist τ_1 and τ_2 $(t_o \leq \tau_1 < \tau_2 < t_{***})$ such that $u_o(\tau_1) = u_o(t_{***})$ and the range of function $u_o(t)$ on $[\tau_1, t_{***}]$ coincides with interval $[u_o(\tau_2), u_o(t_{***})]$.

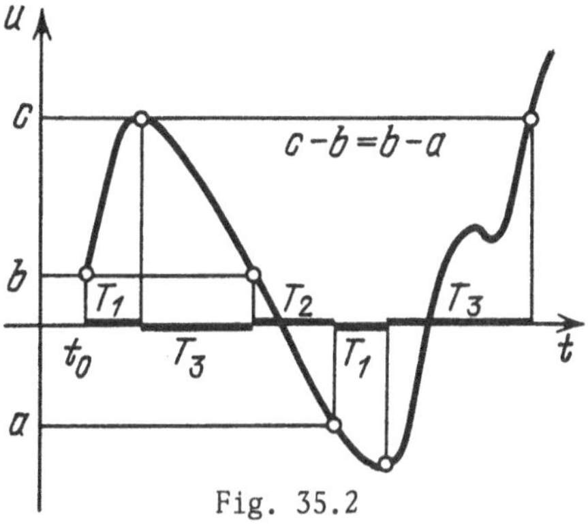

Fig. 35.2

In general, the union $T_1[u_o(t)] \cup T_2[u_o(t)] \cup T_3[u_o(t)]$ does not fill the whole half-line $[t_o, \infty)$ (although it is always dense in that half-line). In case of continuous piecewise monotone inputs $u_o(t)$, the union $T_1[u_o(t)] \cup T_2[u_o(t)] \cup T_3[u_o(t)]$ coincides with $[t_o, \infty)$.

Let us consider an operator V, defined on the set of continuous functions $u(t)$ $(t \geq t_o)$, with values

$$x(t) = V u(t) \qquad (t \geq t_o) \quad , \tag{35.25}$$

such that for $t = 0$, $x(0) = 0$. Let $Y_-(u)$ $(-\infty < u < \infty)$ be a continuous non-decreasing odd function which is concave for $u \geq 0$. Set

$$Y_+(u) = \frac{1}{2} Y_-(2u) \qquad (-\infty < u < \infty) \quad . \tag{35.26}$$

We shall say that the operator V has the *first defining property with respect to the function* $Y_-(u)$ *at the input* $u_o(t)$ $(t \geq t_o)$, if

$$Vu(t_*) = Y_+[u_o(t_*) - u_o(t_o)] \qquad (t_* \in T_1[u_o(t)]) \quad . \tag{35.27}$$

Operator V has the *second defining property*, if $t_{**} \in T_2[u_0(t)]$ implies that

$$Vu(t_{**}) = Y_+[u_{**} - u_0(t_0)] + Y_-[u_0(t_{**}) - u_{**}] \quad , \tag{35.28}$$

where u_{**} is a number such that the range of function $u_0(t)$ defined on $[t_0,t_{**}]$ coincides with $[u_{**},u_0(t_{**})]$. At last, the operator V has the *third defining property with respect to* $Y_-(u)$ *at the input* $u_0(t)$, if from $t_{***} \in T_3[u_0(t)]$ it follows that

$$Vu_0(t_{***}) = Vu_0(\tau_2) + Y_-[u_0(t_{***}) - u_0(\tau_2)] \quad , \tag{35.29}$$

where $\tau_2 \in [t_0,t_{***})$ and the range of function $u_0(t)$ on some interval $[\tau_1,t_{***}] \supset [\tau_2,t_{***}]$ coincides with $[u(\tau_2),u(t_{***})]$, at $u(\tau_1) = u(t_{***})$.

The above definitions give rise to a quite simple construction of functions (35.25) on the class of piecewise monotone inputs. To be more specific: given graph of function $u(t)$, we can construct the sets T_1, T_2 and T_3 , and then take advantage of equalities (35.27) - - (35.29) . If an operator V is vibro-correct, then having constructed the functions (35.25) for piecewise monotone inputs $u(t)$, we can extend those functions to all continuous $u(t)$. Therefore, a vibro-correct operator V that has all defining properties admits complete characterization by single function $Y_-(u)$.

Theorem 35.1. Let W be Ishlinskii's transducer . Then every operator $W[t_0,0;\Xi]$ has all defining properties with respect to the unloading function (35.21) at any continuous input $u(t)$ $(t \geq t_0)$.

Proof. The assertion follows immediately in the case of the transducer $U(h)$ which represents a single stop. Thus, it obviously holds also for the parallel connection of a finite number of stops . To complete the proof, it remains to pass to a suitable limit (taking advantage of the correctness property of Ishlinskii's transducer, considered in Chapter 36).

∎

Theorem 35.1 admits a converse formulation which can be treated as a kind of identification theorem.

35.4. Normal states of Ishlinskii's transducer

The constructions of output signals for Ishlinskii's transducer that use its loading and unloading functions, as described in the preceding section, are no longer applicable if the initial state of transducer is different from zero. In this section, we shall characterize a class of the initial states for which similar constructions remain still applicable.

Often it is convenient to use the name "state" for the second component $z(h)$ of the actual state $\{u, z(h)\}$ of Ishlinskii's transducer W (provided that it does not lead to any misunderstanding).

The state $z(h)$ (which satisfies (35.8)) is called *normal*, if

$$|z(h_1) - z(h_2)| \leqq |h_1 - h_2| \tag{35.30}$$

at all $0 \leqq h_1 < h_2 < \infty$.

In many problems, it is natural to consider only normal states as the only physically feasible. The first reason for this is that, at any initial state $z_o(h)$, function $z(h;t_1)$ fulfils (35.30) for

$$0 \leqq h_1 < h_2 \leqq \max_{t_o \leqq \tau, \sigma \leqq t_1} \frac{1}{2}|u(\tau) - u(\sigma)| \quad .$$

Secondly, if the inequality (35.30) is satisfied at $0 \leqq h_1 < h_2 \leqq a$ by some function $z(h;t_o)$, then it holds with the same h_1, h_2 for $z(h;t)$ at all $t > t_o$. Furthermore, the restriction W_{norm} of a transducer W to the set of normal states contains all its controllable restrictions. At last, restriction W_{norm} is "almost completely controllable" in the following sense: for any initial state $z_o(h)$ and every normal initial state $z_1(h)$, there exists an input $u(t)$ such that the function (35.10) satisfies the relation

$$\lim_{n \to \infty} \sup_{0 \leqq h \leqq n} |z(h;t_o + n) - z_1(h)| = 0 \quad . \tag{35.31}$$

Let $Y_-(u)$ $(-\infty < u < \infty)$ be, as usually, an odd non-decreasing function, concave at $u \geqq 0$. Suppose that another continuous, non-decreasing concave function $Z(u)$ is defined on the half-line $u \geqq 0$ (only on this half-line!), so that the differential inequality

$$\frac{dZ(u)}{du} \leq \frac{dY_-(u)}{du} \tag{35.32}$$

is almost everywhere satisfied. The function $Z(u)$ can equally assume positive and negative values. We define an extension of $Z(u)$ onto the negative half-line by the equality (see Fig. 35.3)

$$Z(u) = \sup_{v \geq 0} [Z(v) + Y_-(u-v)] \quad (u \leq 0) . \tag{35.33}$$

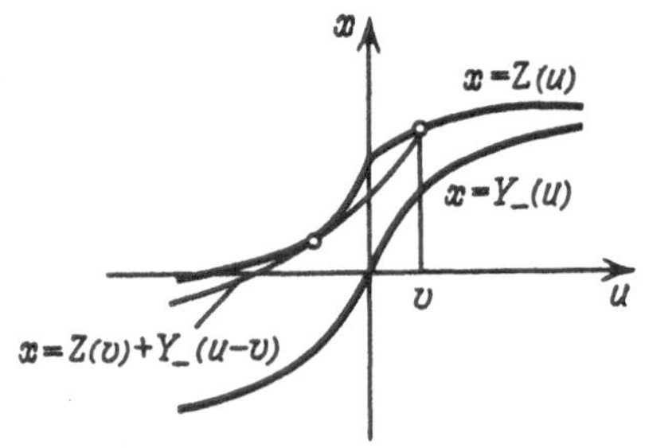

Fig. 35.3

Let us remark that by $Z(u) = Y_+(u)$ at $u \geq 0$ it follows that $Z(u)$ coincides with $Y_+(u)$ also at $u < 0$. If $Z(u) = Y_-(u) + \alpha$ at $u \geq 0$, then $Z(u) = \alpha$ at $u < 0$; in turn, if $Z(u) = \alpha$ at $u \geq 0$, then $Z(u) = Y_-(u) + \alpha$ at $u < 0$. (35.33) implies the following dual equality

$$Z(u) = \inf_{v \leq 0} [Z(v) + Y_-(u-v)] \quad (u \geq 0) . \tag{35.34}$$

Fix an arbitrary input $u_0(t)$ $(t \geq t_0)$. Let t_* be a given real number and $u_0(t_*)$ be one of the end-points of the interval $[u_0(t_*), u_*]$ which represents the range of input $u_0(t)$ on $[t_0, t_*]$ (two cases are possible: $u(t_*) < u_*$ and $u(t_*) > u_*$). Let us construct (see Fig. 35.4) the curves $x = Z[u - u_0(t_0)]$ and $x = Z[u_* - u_0(t_0)] + Y_-(u - u_*)$, where $u \in [u_0(t_*), u_*)$. If those curves have some common points, we shall consider t_* as an element of the set $T_1 = T_1[u_0(t);Z]$; if there are no common points, we shall treat t_* as an element of the set $T_2 = T_2[u_0(t);Z]$. The set $T_3 = T_3[u_0(t);Z]$ is defined by means of $Y_-(u)$, as in Section 35.3.

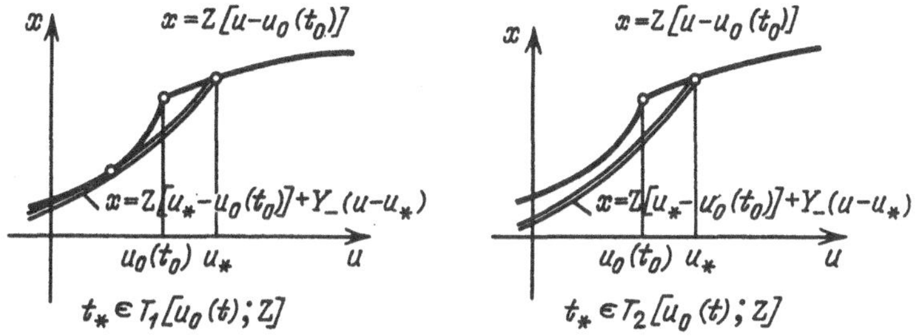

Fig. 35.4

Let us consider an operator V defined on the continuous inputs $u(t)$ $(t \geq t_0)$, with values $x(t)$ satisfying the condition $x(t_0) = Z(0)$.

We shall say that V has the *first defining property with respect to the pair* $\{Y_-(u), Z(u)\}$ *at an input* $u_0(t)$ $(t \geq t_0)$, if from $t_* \in T_1[u_0(t);Z]$ it follows that

$$Vu_0(t_*) = Z[u_0(t_*) - u_0(t_0)] , \qquad (35.35)$$

as in (35.27). In turn, an operator V has the *second defining property*, if $t_{**} \in T_2[u_0(t);Z]$ implies the equality

$$Vu_0(t_{**}) = Z[u_{**} - u_0(t_0)] + Y_-[u_0(t_{**}) - u_{**}] , \qquad (35.36)$$

analogous to (35.27), where the interval $[u_{**}, u_0(t_{**})]$ represents the range of function $u_0(t)$ on $[t_0, t_{**}]$. At last, the operator V has the *third defining property with respect to the pair* $\{Y_-(u), Z(u)\}$, if it has the third defining property with respect to the function $Y_-(u)$ (see (35.29)).

Now V have all the defining properties with respect to the pair $\{Y_-(u), Z(u)\}$ at a piecewise monotone input $u_0(t)$. Then we can construct the function $x(t) = Vu(t)$ by means of the same procedure which has been employed in the preceding section in the case of operators having the defining properties with respect to the pair $\{Y_-(u), Y_+(u)\}$.

Theorem 35.2. Let W be Ishlinskii's transducer with the weight function Ξ. Then for a normal initial state $z_0(h)$ any operator $W[t_0, z_0(h); \Xi]$ has at any continuous input all the three defining properties with respect to the pair $\{Y_-(u), Z(u)\}$, where $Y_-(u)$ is the loading function (35.21) and

$$Z[u(t)] = W[0,z_0(h);\Xi]u_*(t) \quad , \quad u_*(t) = t \quad (t \geq 0) \quad , \tag{35.37}$$

$$Z[u(t)] = W[0,z_0(h);\Xi]u_{**}(h) \quad , \quad u_{**}(t) = -t \quad (t \geq 0) \quad . \tag{35.38}$$

The proof is analogous to that of Theorem 35.1. Technical details are rather tedious, therefore we omit them.

∎

By virtue of Theorem 35.2, we can reconstruct the operator $W[t_0,z_0(h);\Xi]$ from results of an experiment at unknown weight function $\Xi(h)$ and, what is of practical importance, at unknown initial state $z_0(h)$ (provided that the initial state is normal).

Let V be a controllable transducer which has the third defining property with respect to a bounded function $Y_-(u)$ with finite derivative at $u = 0$. Besides, suppose that for any monotone input $u(t)$ all the produced outputs $x(t)$ satisfy the inequality

$$|x(t)| \leq \frac{1}{2} \sup_{u \geq 0} Y_-(u) \quad (t \geq t_0) \quad . \tag{35.39}$$

Then V is in a natural sense equivalent to one of the controllable restrictions of Ishlinskii's transducer with unloading function $Y_-(u)$. The above statement represents a certain form of the identification theorem.

35.5. Periodic inputs

Since all stops are monocyclic and have negative spin, also any Ishlinskii's transducer is monocyclic and its spin is negative.

Let $u(t)$ be a T-periodic input for $t \geq t_0$ The hysteresis loops of Ishlinskii's transducer $W = W(\Xi)$, corresponding to this input, differ from each other only in shiftings along the ordinate axis. In the case of any piecewise monotone input, it is convenient to use the loading and unloading curves for constructing the suitable loop (cf., Section 35.3).

The form of that loop is specified by the third defining property alone. The loop

$$u = u(t), \; x = x_{low}(t) = W[t_0,-h;\Xi]u(t) \quad (t_0 + T \leq t \leq t_0 + 2T) \tag{35.40}$$

is located below all others. The other loops admit the representation

$$u = u(t), \quad x = W[t_0,-h;\Xi]u(t) + \alpha \qquad (t_0 + T \leq t \leq t_0 + 2T) \quad ,$$

with parameters α such that

$$0 \leq \alpha \leq 2 \int_{\frac{1}{2}[\max u(t)-\min u(t)]}^{\infty} |\Xi(h)|dh \quad .$$

For the multi-valued operator Π which assigns to a T-periodic input $u(t)$ the set of all appropriate functions $W[t_0,z_0(h);\Xi]u(t)$ $(t_0 + T \leq t \leq t_0 + 2T)$, its selectors play an important role. For instance, the selector

$$\Pi_0 u(t) = x_{low}(t) + \int_{\frac{1}{2}[\max u(t)-\min u(t)]}^{\infty} |\Xi(h)|dh \qquad (35.41)$$

fulfils the Lipschitz condition

$$\|\Pi_0 u(t) - \Pi_0 v(t)\|_{t_0+T,t_0+2T} \leq |2\Xi(0) - \Xi(\rho)| \cdot \|u(t) - v(t)\|_{t_0,t_0+T} \qquad (35.42)$$

at

$$\max u(t) - \min u(t) \leq \rho \ , \quad \max v(t) - \min v(t) \leq \rho \quad . \qquad (35.43)$$

Let us consider the 2π-periodic input

$$u_0(t) = a \sin t \qquad (t \geq 0) \quad , \qquad (35.44)$$

in particular. Then the 2π-periodic output $x(t)$ which corresponds to Ishlinskii's transducer with homogeneous initial state is defined by the relation

$$x(t) = \begin{cases} X_+[u_0(t);\Xi] & , \text{ for } 0 \leq t \leq \pi/2 \ , \\ X_+(a;\Xi) + X_-[u_0(t) - a;\Xi] & , \text{ for } \frac{\pi}{2} \leq t \leq 3\pi/2 \ , \\ X_+(-a;\Xi) + X_-[u_0(t) + a;\Xi] & , \text{ for } \frac{3\pi}{2} \leq t \leq 5\pi/2 \ . \end{cases} \qquad (35.45)$$

The corresponding hysteresis loop is represented in Figure 35.5 by the thickened line. Values of selectors (35.41) are given by function (35.45). For the loop, its surface area $S[u_0(t)]$ is given by

$$S[u_0(t)] = \int_0^{2a} h(2a-h)d\Xi(h) \quad , \tag{35.46}$$

or, equivalently, by

$$S[u_0(t)] = 2\int_0^{2a} (a-h)\Xi(h)dh \quad . \tag{35.47}$$

The choice of $u_0(t)$ in special form (35.44) did not play any important role in the considerations developed in this section. Analogous formulae remain valid also for inputs with just one local minimum and one local maximum within single period.

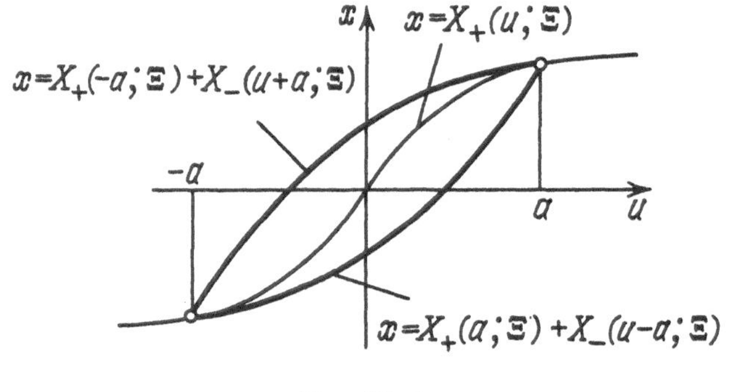

Fig. 35.5

Assume now that a T-periodic input $u_0(t)$ $(t \geq t_0)$ has several different local maxima in the interval $[0,T]$. In such a case, the determination of the area $S[u_0(t);\Xi]$ of a hysteresis loop (clearly, we are concerned with the oriented area) becomes more complicated. Suppose that the function $u_0(t)$ is non-constant on any given interval and $\tau_1 < \tau_2 < \ldots < \tau_k < \tau_{k+1} = \tau_1 + T$ are the points where it achieves locally maximal values. For any τ define

$$\sigma_-(\tau) = \inf\{t\colon u(s) < u(\tau) \quad \text{for} \quad t < s < \tau\}$$

and

$$v_-(\tau) = \min_{\sigma_-(\tau) \leq s \leq \tau} u(s) \quad .$$

Further, set

$$\sigma_+(\tau) = \max\{t\colon v_-(\tau) \leq u(s) \leq u(\tau) \quad \text{for} \quad \tau \leq s \leq t, \quad t \leq \tau + T\}$$

and

$$v_+(\tau) = \min_{\tau \leq s \leq \sigma_+(\tau)} u(s) \quad .$$

Then, the following representation is true:

$$S[u_0(t);\Xi] = \sum_{j=1}^{k} \int_0^{u(\tau_j)-v_+(\tau_j)} [u(\tau_j) - v_+(\tau_j) - 2h]\Xi(h)dh \quad . \tag{35.48}$$

Proof. To give a simple proof of (35.48), we can use an induction argument with respect to k . For $k = 1$, (35.48) represents a natural analogue of (35.47).

∎

35.6. Davidenkov's model

Different weight functions generate various Ishlinskii's transducers. In many applications one assumes that on a certain interval $[0,h_0]$ the weight function is defined by the equality

$$\Xi(h) = -\alpha + \beta h^\gamma \quad , \tag{35.49}$$

where $\alpha, \beta, \gamma > 0$ (moreover, we have $\beta h_0^{\gamma} \leq \alpha$). In this case,

$$X_+(u,\Xi) = \alpha u - \frac{\beta}{1+\gamma} |u|^{\gamma+1} \operatorname{sign} u \qquad (|h| \leq h_0)$$

and

$$X_-(u;\Xi) = \alpha u - \frac{\beta}{(1+\gamma)2^\gamma} |u|^{\gamma+1} \operatorname{sign} u \quad (|h| \le 2h_0) \ .$$

For the input (35.44) (at $a \le h_0$), equality (35.47) reduces to the form

$$S[u_0(t)] = \frac{\beta\gamma}{(1+\gamma)(2+\gamma)} (2a)^{\gamma+2} \ .$$

35.7. Controllable restrictions of Ishlinskii's bundles

Let us denote by $\chi(h_0)$ the class of continuous functions $\alpha(h)$ $(0 \le h \le h_0)$ which are piecewise linear on each interval $[\delta_0, h_0]$, where $\delta_0 > 0$ with $\alpha(0) = 0$ and $|\alpha'(h)| = 1$ almost everywhere. By $\chi(\infty)$ we shall denote the class of functions whose restrictions to any interval $[0,h_0]$ belong to $\chi(h_0)$. With the use of the classes $\chi(h_0)$ and $\chi(\infty)$, it is easy to give characterizations of the controllable and completely controllable restrictions of Ishlinskii's transducer W with bundles Q (cf., (35.10)).

The bundle Q of stops $U(h)$ $(0 \le h < \infty)$ does not admit completely controllable restrictions. One may easily describe its controllable restrictions.

Two functions belonging to $\chi(\infty)$ will be called *equivalent* if they assume the same values at sufficiently large h. Then $\chi(\infty)$ admits a decomposition into *equivalence classes*. Any equivalence class K is a subset of the class Z of functions which satisfy estimate (35.8). The set of all pairs $\{u, z(h)\}$, with $z(h) \in K$, is a subset Ω_K of the domain $\Omega(Q)$ of feasible states for the bundle. By virtue of (35.10), the restriction Q_K of the bundle Q to Ω_K will be controllable. Different equivalence classes define various controllable restrictions of the bundle; every controllable restriction is a restriction of the form Q_K.

The above considerations are independent of any special choice of the weight function. If the weight function $\Xi(h)$ is constant on a certain interval, then the bundle Q which corresponds to Ishlinskii's transducer $W = W(\Xi)$ cannot contain stops with any value h from that interval.

For instance, let $\Xi(h) = 0$ at $h \ge h_0$. In this case, we shall consider the bundle Q of plays $U(h)$ $(0 \le h \le h_0)$. The restriction of bundle Q to the set of states $\{u,z(h)\}$ for which $z(h) \in \chi(h_0)$ is completely controllable.

36. Properties of Ishlinskii's material

36.1. Continuity of Ishlinskii's operator

In this section, we shall study properties of the operator (35.9), i.e.,

$$x(t) = W[t_0,z_0(h);\Xi]u(t) = \int_0^\infty U[t_0,z_0(h);h]u(t)d\Xi(h) \quad (t \geq t_0), \tag{36.1}$$

where $\Xi(h)$ is the weight function of Ishlinskii's transducer $W = W(\Xi)$. Let us recall that the state of Ishlinskii's transducer is defined either by the pairs $\{u,z(h)\}$ or by the functions $z(h)$ which are continuous on $[0,\infty)$ and fulfil the condition

$$|z(h)| \leq h \quad (0 \leq h < \infty) \quad . \tag{36.2}$$

The state evolves according to the relation

$$z(h;t) = Q[t_0,z_0(h)]u(t) = U[t_0,z_0(h);h]u(t) \quad (t \geq t_0,\ 0 \leq h < \infty). \tag{36.3}$$

a. *Properties of the output at a fixed input*

Clearly, the output (36.1) is continuous at any fixed input. *If an input* $u(t)$ *has the modulus of continuity* $\lambda(\varepsilon)$ *on the interval* $[t_0,t_1]$ *where* $t_0 \leq t_1 < t_2 < \infty$ *, then for each state* $z_0(h)$ *the corresponding output* (36.1) *has the modulus of continuity*

$$\lambda_1(\varepsilon) = X\_[\lambda(\varepsilon);\Xi] \quad , \tag{36.4}$$

where $X\_(u;\Xi)$ *is the unloading function* (35.21). In other words, for $t_1 \leq \tau < \sigma \leq t_2$, the estimate

$$|W[t_0,z_0(h);\Xi]u(\tau) - W[t_0,z_0(h);\Xi]u(\sigma)| \leq$$

$$\leq \int_0^{\lambda[\max_{\tau \leq t, s \leq \sigma} |u(t)-u(s)|]} \left|\Xi\left(\frac{h}{2}\right)\right| dh \tag{36.5}$$

is true.

For the proof it is convenient to establish an analogous estimate in the case of a single stop and integrate that estimate with respect to $d\Xi(h)$.

∎

Since the unloading function $X_-(u;\Xi)$ is concave for $u \geq 0$, we get the bound

$$|X_-(u;\Xi)| \leq |\Xi(0)| \; |u| \; . \tag{36.6}$$

Thus, Ishlinskii's transducer (at arbitrarily fixed initial state) maps each space H_α of the inputs $u(t)$ $(t_0 \leq t \leq t_1)$ into the same H_α . In particular, outputs $x(t)$ are Lipschitz continuous if the appropriate inputs $u(t)$ have the same property. The Lipschitz constants for the inputs and outputs differ by the factor $|\Xi(0)|$ (a-analogous relation is true for the suitable Hölder conditions).

For any input $u(t)$ $(t \geq t_0)$, the corresponding output (36.1) is uniformly bounded:

$$|x(t) - x(t_0)| \leq X_-[\|u(\tau) - u(t_0)\|_{t_0,t};\Xi] \leq 2 \int_0 |\Xi(h)|dh \quad (t \geq t_0) \; .$$

b. *Vibro-correctness modulus*

Let us treat $W[t_0,z_0(h);\Xi]$ as an operator in the space $C(t_0,t_1)$ of continuous functions. In view of the obvious properties of stops, it follows that *for every initial state* $z_0(h)$ *and any* $u(t), v(t) \in C(t_0,t_1)$, *the estimate*

$$\|W[t_0,z_0(h);\Xi]u(t) - W[t_0,z_0(h);\Xi]v(t)\|_{t_0,t_1} \leq$$

$$\leq 2X_+[\|u(t) - v(t)\|_{t_0,t_1};\Xi] \tag{36.7}$$

is satisfied.

By virtue of estimate (36.7), operators (36.1) are continuous in the space $C(t_0,t_1)$. Treated as operators in any Hölder space, the operators (36.1) are continuous, too.

(36.7) yields the following more rough estimate

$$\|W[t_0,z_0(h);\Xi]u(t) - W[t_0,z_0(h);\Xi]v(t)\|_{t_0,t_1} \leq$$
$$\leq 2\,|\Xi(0)|\;\|u(t) - v(t)\|_{t_0,t_1} \tag{36.8}$$

which implies the Lipschitz continuity of operators (36.1).

In general, the Lipschitz constant in condition (36.8) has optimal value. Nevertheless, in some special cases condition (36.8) may still be refined. For example,

$$\|W[t_0,0;\Xi]u(t) - W[t_0,0;\Xi]v(t)\|_{t_0,t_1} \leq$$
$$\leq |2\Xi(0) - \Xi(\rho)|\;\|u(t) - v(t)\|_{t_0,t_1} \qquad (u(t),\ v(t) \in B(\rho))\ , \tag{36.9}$$

where $B(\rho)$ is the class of inputs $u(t)$ which satisfy the condition $\|u(t) - u(t_0)\|_{t_0,t_1} \leq \rho$.

Let $W[t_0;\Xi]$ be now the operator which assigns to an input $u(t)$ $(t_0 \leq t \leq t_1)$ the set of all outputs (36.1) at any given initial state $z_0(h)$. Referring to estimate (36.9), let us remark that the multi-valued operator $W[t_0;\Xi]$ does not admit any selector which would satisfy on $B(\rho)$ Lipschitz condition with coefficient smaller than $|2\Xi(0) - \Xi(\rho)|$.

Estimate (36.8) admits refinement for all normal states $z_0(h)$ (cf., Section 35.4). We shall denote by $B(\alpha,\beta)$ $(\alpha \leq 0, \beta \geq 0)$ the set of functions $u(t) \in C(t_0,t_1)$ such that

$$\alpha \leq u(t) - u(t_0) \leq \beta \qquad (t_0 \leq t \leq t_1)\ . \tag{36.10}$$

Then, for every normal initial state $z_0(h)$ and any $u(t),\ v(t) \in B(\alpha,\beta)$, the estimate

$$\|W[t_0,z_0(h);\Xi]u(t) - W[t_0,z_0(h);\Xi]v(t)\|_{t_0,t_1} \leq$$
$$\leq |2\Xi(0) - \Xi(h_0)|\;\|u(t) - v(t)\|_{t_0,t_1}$$

is valid, with

$$h_0 = \min\{h: -h - \alpha \leq z_0(h) \leq h - \beta\} \tag{36.11}$$

(of course, provided that the above minimum exists).

c. *Dependence upon initial states*

Every stop is Lipschitz continuous with respect to its initial state, with Lipschitz constant 1 . Thus, for any input $u(t)$ $(t \geq t_0)$, by the definition of operator (36.1) it follows that

$$|W[t_0,z_0(h);\Xi]u(t) - W[t_0,z_1(h);\Xi]u(t)| \leq$$
$$\leq \int_0^\infty |z_0(h) - z_1(h)|d\Xi(h) \quad (t \geq t_0) \quad . \tag{36.12}$$

Upon slight modification, estimate (36.12) can be given a sharper form. To this end, let us define

$$\eta_+(h) = \begin{cases} z_0(h) - z_1(h) \ , & \text{if } z_0(h) \geq z_1(h) \ , \\ 0 \ , & \text{if } z_0(h) \leq z_1(h) \end{cases}$$

and

$$\eta_-(h) = |z_0(h) - z_1(h)| - \eta_+(h) \quad .$$

Then

$$|W[t_0,z_0(h);\Xi]u(t) - W[t_0,z_1(h);\Xi]u(t)| \leq$$
$$\leq \max\left\{\int_0^\infty \eta_+(h)d\Xi(h), \int_0^\infty \eta_-(h)d\Xi(h)\right\} \quad (t \geq t_0) \quad .$$

For the normal states $z_0(h)$ and $z_1(h)$, the bound (36.12) can be strongly sharpened. Then,

$$|W[t_0,z_0(h);\Xi]u(t) - W[t_0,z_1(h);\Xi]u(t)| \leq$$

$$\leq \sup_{h_0 > 0} \left| \int_{h_0}^{\infty} [z_0(h) - z_1(h)]d\Xi(h) \right| \qquad (t \geq t_0) \ .$$

36.2. Correctness with respect to weight functions

Let us consider two Ishlinskii's transducers $W_1 = W(\Xi_1)$ and $W_2 = W(\Xi_2)$. In this section, we shall discuss the possibility of constructing an estimate on the "closeness measure" for the pair W_1 , W_2 by using weight functions $\Xi_1(h)$ and $\Xi_2(h)$. This question turns out especially important in identification problems. Further, it determines the possibility of constructing approximations to Ishlinskii's transducer either by the parallel connection of a finite number of plays or by some transducers with smooth weight functions.

The sets Z of all feasible states $z(h)$ of transducers W_1 and W_2 coincide. By Z_{norm} we shall denote the corresponding set of normal states (cf., Section 35.4) .

Define

$$\theta(W_1,W_2) = \sup_{z_1(h)\in Z} \ \inf_{z_2(h)\in Z} \ \sup_{u(t)\in C(t_0,t_1)} \|W_1[t_0,z_1(h)]u(t) - W_2[t_0,z_2(h)]u(t)\|_{t_0,t_1} \tag{36.13}$$

and

$$\rho(W_1,W_2) = \max\{\theta(W_1,W_2),\ \theta(W_2,W_1)\} \ . \tag{36.14}$$

The quantities (36.13) and (36.14) are analogous to (10.1) and (10.2), respectively. Further, let

$$\rho_{norm}(W_1,W_2) = \sup_{z_0(h)\in Z_{norm}} \ \sup_{u(t)\in C(t_0,t_1)} \|W_1[t_0,z_0(h)]u(t) - W_2[t_0,z_0(h)]u(t)\|_{t_0,t_1} \ . \tag{36.15}$$

For arbitrary weight functions $\Xi_1(h)$ *and* $\Xi_2(h)$, *the estimate*

$$\rho_{norm}(W_1,W_2) \leq \int_0^{\infty} |\Xi(h) - \Xi_2(h)|dh \tag{36.16}$$

is satisfied. If the weight functions are continuous, then

$$\rho(W_1,W_2) \le \int_0^\infty |\Xi_1(h) - \Xi_2(h)|dh \quad . \tag{36.17}$$

Proof. Estimates (36.16) and (36.17) follow directly from properties of stops.

∎

Let us remark the following:

a. In view of the estimate (36.17), any Ishlinskii's transducer $W = W(\Xi)$ may be approximated with an arbitrary accuracy by the parallel connection of a finite number of stops. It would be interesting to find effective estimates on the number of the stops, sufficient for an ε - approximation of the considered Ishlinskii's transducer .

b. Let $z_0(h)$ be a fixed normal state. It is convenient to characterize the distance between operators $W_1[t_0,z_0(h);\Xi_1]$ and $W_2[t_0,z_0(h);\Xi_2]$ by the value

$$\rho[W_1,W_2;z_0(h);\alpha,\beta] = \sup_{u(t)\in B(\alpha,\beta)} ||R_1u(t) - R_2u(t)||_{t_0,t_1} \quad ,$$

where

$$R_iu(t) = W_i[t_0,z_0(h);\Xi_i]u(t) - W_i[t_0,z_0(h);\Xi_i]u(t_0)$$

and $B(\alpha,\beta)$ is defined by the inequalities (36.10). Then for arbitrary weight functions $\Xi_1(h)$ and $\Xi_2(h)$ the following estimate

$$\rho[W_1,W_2;z_0(h);\alpha,\beta] \le \int_0^{h_0} |\Xi_1(h) - \Xi_2(h)|dh \quad ,$$

analogous to (36.16), holds with h_0 defined by (36.11).

In particular, for $z_0(h) \equiv 0$ we get the estimate

$$\rho[W_1,W_2\ 0;-a,a] \le \int_0^a |\Xi_1(h) - \Xi_2(h)|dh \quad .$$

c. The estimate (36.17) is not true for arbitrary weight functions $\Xi_1(h)$ and $\Xi_2(h)$. If the function $\Xi_2(h)$ is continuous and $\Xi_1(h)$ arbitrary, then we can claim that

$$\theta(W_1,W_2) \leq \int_0^\infty |\Xi_1(h) - \Xi_2(h)|dh \quad .$$

36.3. Unilateral estimates

Let us consider a stop $U(h)$ with the modulus of elasticity 1 and with the yield limits $\pm h$. For an absolutely continuous input $u(t)$ $(t \geq t_o)$, also the output

$$x(t) = U[t_o,x_o;h]u(t) \qquad (t \geq t_o)$$

will be continuous (cf., Chapter 6). Moreover, almost everywhere

$$x'(t) - u'(t) \begin{cases} \geq 0 \ , & \text{if} \quad x(t) = -h \\ = 0 \ , & \text{if} \quad -h < x(t) < h \\ \leq 0 \ , & \text{if} \quad x(t) = h \ . \end{cases} \tag{36.18}$$

By (36.18) and analogous estimates for the function

$$y(t) = U[t_o,y_o;h]v(t)$$

it follows that for almost all $t \geq t_o$ the inequality

$$\{[x'(t) - u'(t)] - [y'(t) - v'(t)]\} \ [x(t) - y(t)] \leq 0$$

holds, or equivalently, re-written in a more convenient form,

$$[x(t) - y(t)] \frac{d}{dt} [u(t) - v(t)] \geq \frac{1}{2} \frac{d}{dt} [x(t) - y(t)]^2 \quad .$$

The last estimate implies that

$$\int_{t_o}^{t} \{U[t_o,x_o;h]u(t) - U[t_o,y_o;h]v(t)\}d[u(t) - v(t)] \geq$$

$$\geq \frac{1}{2}\{U[t_o,x_o;h]u(t) - U[t_o,y_o;h]v(t)\}^2 - \frac{1}{2}(x_o - y_o)^2 \quad . \qquad (36.19)$$

Let us now consider Ishlinskii's transducer $W = W(\Xi)$. As a consequence of (36.19), we get the main result of this section.

Theorem 36.1. Let the inputs $u(t)$, $v(t)$ $(t \geq t_o)$ be absolutely continuous. Then

$$\int_{t_o}^{t} \{W[t_o,z_1(h);\Xi]u(t) - W[t_o,z_2(h);\Xi]v(t)\}d[u(t) - v(t)] \geq$$

$$\geq \frac{1}{2}\int_0^\infty \{U[t_o,z_1(h);h]u(t) - U[t_o,z_2(h);h]v(t)\}^2 d\Xi(h) -$$

$$- \frac{1}{2}\int_0^\infty [z_1(h) - z_2(h)]^2 d\Xi(h) \quad . \qquad (36.20)$$

∎

Let us now formulate some conclusions following from Theorem 36.1. If $z_1(h) = z_2(h) = z(h)$, then (36.20) implies the inequality

$$\int_{t_o}^{t} \{W[t_o,z(h);\Xi]u(t) - W[t_o,z(h);\Xi]v(t)\}d[u(t) - v(t)] \geq 0 \quad (t \geq t_o) \ . \qquad (36.21)$$

This equality allows us to apply monotonicity techniques (for operators that are monotone in Minty's sense) to the study of differential equations which include Ishlinskii's transducers.

Suppose now that the inputs $u(t)$, $v(t)$ $(t \geq t_o)$ are T-periodic Let $x(t), y(t)$ be T-periodic outputs of the transducer $W = W(\Xi)$, corresponding to these inputs. Then the estimate

$$\int_{t_*+T}^{t_*+2T} [x(t) - y(t)]d[u(t) - v(t)] \geq 0 \qquad (36.22)$$

follows as a consequence of (36.20). The relation (36.22) and its analogue referring to almost periodic inputs $u(t)$, $v(t)$ play an important role at a study of forced vibrations in systems including Ishlinskii's transducers .

In addition, we shall postulate that the weight function $\Xi(h)$ fulfils the condition

$$\frac{1}{2}\int_0^\infty h^2 \, d\Xi(h) = \int_0^\infty h|\Xi(h)|dh = A < \infty \quad .$$

Then, by (36.20), we can conclude the estimate

$$\int_{t_0}^{t} \{W[t_0,z_1(h);\Xi]u(t) - W[t_0,z_2(h);\Xi]v(t)\}d[u(t) - v(t)] \geqq -A \quad .(36.23)$$

Estimates of the form (36.23) are useful in the stability and dissipativity analysis of systems including Ishlinskii's transducers .

At the end, let us remark that direct analogues of Theorem 36.1 are true for multidimensional stops and for their parallel connections.

37. Finite systems of relays

37.1. Block-diagrams with relays

Since in the case of a relay all piecewise monotone inputs are admissible (cf., Section 28.7), one can construct transducers with arbitrary relays, actuators and summators, such that the corresponding block-diagrams do not include feedback loops. Such transducers W are static; they admit controllable restrictions but in general none of those restrictions is completely controllable. If all transmission coefficients in a certain block-diagram are non-negative, then the corresponding transducers are monotone with respect to inputs, initial conditions, threshold values, etc. In a more detail, we shall discuss here only the case of parallel connections of non-ideal relays defined on continuous inputs.

37.2. Parallel connections and bundles of relays

Let us consider a transducer W with the block-diagram depicted in

Fig. 33.1,a, and with W^j representing non-ideal **relay** $R^j = R(\alpha_j,\beta_j)$. As usual, the threshold values are subjected to the inequalities $\alpha_j < \beta_j$. The corresponding transducer (see Fig. 33.1,b) will assume the form of the bundle Q of relays R^j. The domains $\Omega(W)$ and $\Omega(Q)$ of the feasible states for the parallel connection W and bundle Q of relays $R^1, \dots, R^N$ coincide. They can be described by the equality

$$\Omega(W) = \{\{u,r_1,\dots,r_N\}:\{u,r_j\} \in \Omega(R^j);\ \ j = 1,\dots,N\} \ . \tag{37.1}$$

Let us recall (cf., Chapter 28) that $\Omega(R^j)$ consists of the pairs $\{u,r_j\}$ with an arbitrary first component and the second component equal to 0 at $u \leq \alpha_j$, equal to 1 at $u \geq \beta_j$ and admitting any of the values 0 or 1 at $\alpha_j < u < \beta_j$.

At the initial state

$$q_o = \{u_o,r_1^o,\dots,r_N^o\} \in \Omega(W) \ , \tag{37.2}$$

we shall admit any continuous input $u(t)$ $(t \geq t_o)$ which satisfies the initial condition $u(t_o) = u_o$. The input-output relations of the parallel connection of relays are described by the equality

$$x(t) = W[t_o,r_1^o,\dots,r_N^o]u(t) = \sum_{j=1}^{N} \xi_j R[t_o,r_j^o:\alpha_j,\beta_j]u(t) \ . \tag{37.3}$$

The dynamics of the bundle (and thus also the input-output relations of the parallel connection of relays) can be described by

$$q(t) = \{u(t),r_1(t),\dots,r_N(t)\} \quad (t \geq t_o) \ , \tag{37.4}$$

where

$$r_j(t) = R[t_o,r_j^o,:\alpha_j,\beta_j]u(t) \qquad (t \geq t_o;\ j = 1,\dots,N) \ . \tag{37.5}$$

As a consequence of the properties of relays given in Chapter 28, analogous properties are true for the parallel connections and bundles of relays. The bundles are deterministic, static and monocyclic; after passing to parallel connections, one looses the first of these properties (except for the trivial case of the non-intersecting intervals

$(\alpha_1,\beta_1), \ldots, (\alpha_N,\beta_N)$). At fixed positive $\xi_1, \ldots, \xi_N$, the bundles and parallel connections of relays are monotone with respect to the initial states, inputs, threshold values, etc.

The bundle Q of relays $R(\alpha_j,\beta_j)$ is controllable if and only if

$$(\alpha_i - \alpha_j)(\beta_i - \beta_j) < 0 \quad (i \neq j) \ . \tag{37.6}$$

In the general case, the bundle Q admits a completely controllable restriction Q_{cc} for which the domain $\Omega(Q_{cc})$ of the feasible states is defined by the equality

$$\Omega(Q_{cc}) = \{\{u,r_1,\ldots,r_N\} \in \Omega(Q): r_i \leq r_j \ \text{ for } \ \alpha_i \leq \alpha_j \ \text{ and } \ \beta_i \leq \beta_j\}. \tag{37.7}$$

It deserves noticing that the corresponding transducer W_{cc} is deterministic if and only if the inequalities

$$(\alpha_i - \alpha_j)(\beta_i - \beta_j) \geq 0 \quad (i,j = 1,\ldots,N) \tag{37.8}$$

are fulfilled.

The number of various feasible outputs for the bundle Q_{cc} of relays $R^1, \ldots, R^N$ is finite. For controllable bundles, this number is equal 2^N . If, however, the bundle Q is non-controllable, then that number is essentially smaller, equal to $1 + N + N_1$, where N_1 denotes the number of controllable bundles (comprising two, three or more relays) which can be selected from the bundle Q . In particular, if the inequalities (37.8) are satisfied, then the transducer Q_{cc} admits $1 + N$ different outputs.

37.3. Independent perturbations of inputs

Parallel connections of relays represent discontinuous transducers . We are going to discuss briefly the closure and convexification procedures for the input-output relations (37.3). The inputs $u_0(t)$ and outputs $x(t)$ will be considered on a finite interval $[t_0,T]$.

In this section, we shall assume that there are no relations between the perturbations $h_1(t),\ldots,h_N(t)$ of the input $u(t)$, corresponding to different relays $R^1,\dots,R^N$. Subject to those perturbations, the output $x(t)$ of the parallel connection of relays can be characterized by the relation

$$x(t) = \sum_{j=1}^{N} \xi_j \; R[t_o,r_j^o;\alpha_j,\beta_j][u(t) + h_j(t)] \quad . \tag{37.9}$$

Let the input-output relations (37.3) be treated as operators from the space $C(t_o,T)$ into $L_q(t_o,T)$ with $1 \le q < \infty$. Then, by virtue of (37.9) and due to Theorem 28.2,

$$\bar{W}[t_o,r_1^o,\dots,r_N^o]u(t) = \sum_{j=1}^{N} \xi_j \bar{R}[t_o,r_j^o;\alpha_j,\beta_j]u(t) \quad . \tag{37.10}$$

The multi-valued operators (37.10) may be regarded as the input-output correspondences of a transducer with the block-diagram given by Fig. 33.1,a where W^j are represented by the closures $\bar{R}(\alpha_j,\beta_j)$ of relays $R(\alpha_j,\beta_j)$.

In Section 28.5, we have introduced notions of the convexification for relay (those operations represented a natural idealization of some generalized play). If parameters $\xi_1, \dots, \xi_N$ are non-negative, then the formula

$$\dot{W}^{\square}[t_o,r_1^o,\dots,r_N^o]u(t) = \sum_{j=1}^{N} \xi_j R^{\square}[t_o,r_j^o;\alpha_j,\beta_j]u(t), \tag{37.11}$$

analogous to (37.10), describes the convexification of the parallel connection of relays.

37.4. Arbitrary perturbation of the input

Suppose the same perturbation acts on the input $u(t)$ of all relays $R(\alpha_1,\beta_1), \dots , R(\alpha_N,\beta_N)$ in the parallel connection W . If (37.3) are treated as operators from $C(t_o,t_1)$ into $L_q(t_o,T)$, where $1 \le q < \infty$, then the values

$$x(t) = \bar{\bar{W}}[t_o,r_1^o,\dots,r_N^o]u(t) \tag{37.12}$$

of the closure $\bar{\bar{W}}$ are to be defined as the sets of limits of all the sequences of the form

$$y_n(t) = \sum_{j=1}^{N} \xi_j R[t_o, r_j^o; \alpha_j, \beta_j] u_n(t) \quad ,$$

convergent in L_q , such that

$$\|u_n(t) - u(t)\|_{t_o,T} \to 0 \quad .$$

In general, characterizations of the operator $\bar{\bar{W}}$ are quite cumbersome. If all numbers $\alpha_1, \ldots, \alpha_N$ are different and the same is true for $\beta_1, \ldots, \beta_N$, then $\bar{\bar{W}}$ coincides with $\bar{W}$ and the multi--valued input-output operators are described for $\bar{\bar{W}}$ by the right-hand side of (37.10).

38. Continual systems of relays

38.1. Bundles of relays and CRS - transducers

By a *bundle of the family* $R[\alpha(\gamma), \beta(\gamma)]$ $(\gamma \in \Gamma)$ *of relays* , we shall understand the transducer Q which assigns the output function $z(\gamma;t)$, defined by

$$z(\gamma;\ t) = Q[t_o, z(\gamma;\ t_o)]u(t) = R[t_o, z(\gamma;\ t_o);\ \alpha(\gamma),\ \beta(\gamma)]u(t) \quad (38.1)$$

for $\gamma \in \Gamma$ and $t \geq t_o$, to any continuous input $u(t)$ $(t \geq t_o)$. The pairs $\{u, z(\gamma)\}$, where u is any number and $z(\gamma)$ denotes a function defined on Γ , will represent the state. For any fixed $\gamma = \gamma_*$, the pair of numbers $\{u, z(\gamma_*)\}$ will characterize the feasible states of the relay $R[\alpha(\gamma_*), \beta(\gamma_*)]$. The functions $z(\gamma)$ in the pairs $\{u, z(\gamma)\}$ of the bundle are the characteristic functions for some subsets of the set Γ .

Suppose that a finite measure μ is defined on the set Γ . Then by a *continual system of relays* (CRS) we shall mean the transducer $R = R(\mu)$ whose input-output relations are defined by the equality

$$R[t_o, z_o(\gamma);\ \mu]u(t) = \int_\Gamma R[t_o, z_o(\gamma);\ \alpha(\gamma), \beta(\gamma)]u(t)d\mu(\gamma) =$$

$$= \mu\{\gamma \in \Gamma:\ R[t_0,z_0(\gamma);\ \alpha(\gamma),\ \beta(\gamma)]u(t) = 1\} \quad . \tag{38.2}$$

States of the transducer $R(\mu)$ will be represented only by those pairs $\{u,z_0(\gamma)\}$ in which the function $z_0(\gamma) = z(\gamma;t_0)$ is measurable and, moreover, all functions (38.1) (corresponding to various $t \geq t_0$) are measurable. For $R(\mu)$, the suitable transducer will be represented by restriction $Q(u)$ of the bundle Q to this set. The transducers Q and $Q(u)$ are deterministic, whereas the transducer $R(\mu)$ is non--deterministic.

We shall now confine ourselves to the case of Γ being the half-plane $\alpha < \beta$, i.e., to the case of the set of all non-ideal relays $R(\alpha,\beta)$. The measure μ will be assumed Borelian. Then the formulae (38.2) assume the form

$$R[t_0,z_0(\alpha,\beta);\ \mu]u(t) = \int_{\alpha<\beta} R[t_0,z_0(\alpha,\beta);\ \alpha,\beta]u(t)\ d\mu(\alpha,\beta) =$$

$$= \mu\{\{\alpha,\beta\}:\ \alpha < \beta,\ R[t_0,z_0(\alpha,\beta);\ \alpha,\beta]u(t) = 1\} \quad . \tag{38.3}$$

The domain $\Omega[R(\mu)] = \Omega[Q(\mu)]$ of the feasible states comprises all pairs $\{u,z(\alpha,\beta)\}$ where u is an arbitrary number and $z(\alpha,\beta)$ is the characteristic function of a certain μ-measurable subset of the half-plane $\alpha < \beta$. In all further constructions of this section, it will be convenient to consider the states $\{u,z_1(\alpha,\beta)\}$ and $\{u,z_2(\alpha,\beta)\}$ as indistinguishable if

$$\mu\{\{\alpha,\beta\}:\ z_1(\alpha,\beta) \neq z_2(\alpha,\beta)\} = 0 \quad . \tag{38.4}$$

The passage from an arbitrary set Γ to the half-plane $\alpha > \beta$ slightly restricts the class of the considered systems of relays. Nevertheless, the transducers having the general form (38.1) can be still sufficiently well approximated by transducers (38.3).

Let us consider the possibility of passing from the transducer (38.2) to more general transducers of the form

$$R[t_0,z_0(\gamma);\ \mu,\xi]u(t) = \int_{\Gamma} \xi(\gamma)R[t_0,z_0(\gamma);\ \alpha(\gamma),\beta(\gamma)]u(t)\ d\mu(\Gamma), \tag{38.5}$$

with an arbitrary function $\xi(\gamma)$ and measure μ possibly unbounded.

38.2. Monotonicity of CRS-transducers

Directly by the properties of non-ideal relays (cf., Chapter 28) we can conclude the following.

Theorem 38.1. Suppose that

$$u(t) \leq v(t) \qquad (t \geq t_o) \tag{38.6}$$

and

$$z_1(\alpha,\beta) \leq z_2(\alpha,\beta) \quad (\{u(t_o),z_1(\alpha,\beta)\},\{v(t_o),z_2(\alpha,\beta)\} \in \Omega[R(\mu)]\,). \tag{38.7}$$

Then

$$R[t_o,z_1(\alpha,\beta);\ \mu]u(t) \leq R[t_o,z_o(\alpha,\beta);\ \mu]v(t) \qquad (t \geq t_o)\ . \tag{38.8}$$

■

38.3. Demagnetization function

By a *demagnetization function* we shall mean

$$F(\xi,\eta) = \begin{cases} \mu\{\{\alpha\ \beta\}:\ \xi \leq \alpha < \beta \leq \eta\}, & \text{for } \xi < \eta\ , \\ 0 & \text{for } \xi = \eta\ , \\ -\mu\{\{\alpha\ \beta\}:\ \eta \leq \alpha < \beta \leq \xi\}, & \text{for } \xi > \eta\ . \end{cases} \tag{38.9}$$

This function is monotone with respect to each of its variables ξ and η : non-increasing in ξ and non-decreasing in η ; at $\xi \leq \eta$ it is left-semicontinuous in ξ and right-semicontinuous in η ; the equality

$$\lim_{\xi\to-\infty,\ \eta\to\infty} F(\xi,\eta) = \mu\{\{\alpha,\beta\}:\ \alpha < \beta\} < \infty$$

holds. At last,

$$F(\xi,\eta) + F(\xi_1,\eta_1) \geq F(\xi,\eta_1) + F(\xi_1,\eta) \qquad (\xi \leq \xi_1 < \eta_1 \leq \eta)\ . \tag{38.10}$$

The difference between the left-hand side and the right-hand side of inequality (38.10) is the measure of the rectangle $\xi \leq a < \xi_1$, $\eta_1 < \beta \leq \eta$.

Let us note that *any antisymmetric function* $F(\xi,\eta)$ *having the above properties is the demagnetization function of some continual system of relays.*

In the theory of CRS, the demagnetization function plays a role analogous to the role of the unloading function in the theory of Ishlinskii's materials. Values of the function $F(\xi,\eta)$ can be determined by experiments, since for the input

$$u_0(t) = \begin{cases} \xi + (\eta - \xi)t \ , & \text{if } 0 \leq t \leq 1 \ , \\ \eta + (\xi - \eta)(t-1), & \text{if } 1 \leq t \leq 2 \end{cases}$$

and any initial state $\{\xi, z_0(\alpha,\beta)\}$ (unknown to us!), the following equality holds:

$$R[0,z_0(\alpha,\beta);\ \mu]u_0(1) - R[0,z_0(\alpha,\beta);\ \mu]u_0(2) = F(\xi,\eta) \ . \qquad (38.11)$$

An equality of the form (38.11) remains true under quite weak conditions. Consider a continuous input $u(t)$ $(t \geq t_0)$ and any arbitrary initial state $\{u(t_0), z_0(\alpha,\beta)\}$. Let $t_0 \leq \tau_1 < \tau_2 < \tau_3 < \infty$. Suppose that the values of the input $u(t)$ on $[\tau_1,\tau_2]$ cover the interval $[u(\tau_1),u(\tau_2)]$, and its values on $[\tau_2,\tau_3]$ the interval $[u(\tau_2),u(\tau_3)]$ where $u(\tau_3) \in [u(\tau_1),u(\tau_2)]$. Then the equality

$$R[t_0,z_0(\alpha,\beta);\ \mu]u(\tau_3) = R[t_0,z_0(\alpha,\beta);\ \mu]u(\tau_2) - F[u(\tau_2),u(\tau_3)] \qquad (38.12)$$

holds. This equality is an analogue of the third defining property for Ishlinskii's material.

38.4. Periodic inputs

Since the output of any relay $R(\alpha,\beta)$, corresponding to a T-periodic input $u(t)$ $(t \geq t_0)$, is T-periodic at $t \geq t_0 + T$, the same property holds also for the output of every transducer $R(\mu)$.

Since the spin of any relay is negative, also the *spin of the transducer* $R(\mu)$ *is negative.*

The hysteresis loop corresponding to a T-periodic input is fully characterized by the initial state. As usual, different loops can be constructed just by translation along the ordinate axis.

Suppose that the input $u_o(t)$ has exactly one local maximum and

$$m = \min u_o(t), \quad M = \max u_o(t) \quad .$$

Then the oriented area of the hysteresis loop that corresponds to input $u_o(t)$ is negative and can be represented by the demagnetization function (38.9) according to the equality

$$S[u_o(t);\ \mu] = \int_m^M [F(m,h) + F(h,M)]dh - (M-m)F(m,M) \quad . \tag{38.13}$$

For the inputs $u_o(t)$ which have several local maxima within one period, a natural analogue of (38.13) is true. It is also of interest to establish analogues of the one-sided estimates (cf., Section 36.3) for the transducers $R(\mu)$ defined on periodic or almost periodic inputs. Such analogues remain true under fairly general hypotheses, but some special notions must be introduced what exceeds the scope of the present book.

38.5. Determination of the output

Let us consider a fixed continuous input $u(t)$ $(t \geq t_o)$. Define

$$m_1(t) = \min_{t_o \leq s \leq t} u(s), \quad M_1(t) = \min_{t_o \leq s \leq t} u(s) \quad . \tag{38.14}$$

For any $\xi \in [m_1(t), m_2(t)]$, by $\tau(\xi;t)$ let us denote the maximal point s of the interval $[t_o,t]$, such that $u(s) = \xi$. Next, we introduce

$$m_2(t) = \min_{\tau[M_1(t);t] \leq s \leq t} u(s), \quad M_2(t) = \max_{\tau[m_2(t);t] \leq s \leq t} u(s) \quad . \tag{38.15}$$

Further, by induction we can define the numbers

$$m_{k+1}(t) = \min_{\tau[M_k(t);t] \leq s \leq t} u(s), \quad M_{k+1}(t) = \max_{\tau[m_{k+1}(t);t] \leq s \leq t} u(s) \quad . \tag{38.16}$$

Obviously,

$$m_1(t) \leq m_2(t) \leq \ldots \leq m_k(t) \leq \ldots \leq u(t) \leq \ldots$$
$$\ldots \leq M_k(t) \leq \ldots \leq M_2(t) \leq M_1(t) \tag{38.17}$$

and

$$\lim_{k\to\infty} m_k(t) = \lim_{k\to\infty} M_k(t) = u(t) \quad . \tag{38.18}$$

In particular, the sequences (38.16) may be finite. In such a case, starting from a certain index, they coincide with $u(t)$.

Let

$$\kappa(\alpha,\beta;\ t) = \max_k \kappa_k(\alpha,\beta;\ t) \qquad (\alpha < \beta,\ t \geq t_0) \quad , \tag{38.19}$$

where

$$\kappa_k(\alpha,\beta;\ t) = \begin{cases} 1\ , & \text{if} \quad \alpha < m_{k+1}(t) \quad \text{and} \quad \beta \leq M_k(t) \quad , \\ 0\ , & \text{otherwise} \end{cases}$$

and $m_k(t)$, $M_k(t)$ are taken in the form (38.16).

Theorem 38.2. For any initial state $\{u(t_0), z_0(\alpha,\beta)\}$, the output of the bundle $Q(\mu)$ is characterized by

$$z(\alpha,\beta;\ t) = \begin{cases} z_0(\alpha,\beta)\ , & \text{if} \quad \alpha < m_1(t) \quad \text{or} \quad \beta > M_1(t) \quad , \\ \kappa(\alpha,\beta;\ t) & \text{otherwise} \quad , \end{cases} \tag{38.20}$$

and the output $x(t)$ of the transducer $Q(\mu)$ is given by

$$x(t) = \mu\{\{\alpha,\beta\}:\ z(\alpha,\beta;\ t) = 1\} \quad . \tag{38.21}$$

Proof. To show the assertion, one has to analyze the dependence of single relays (cf., Chapter 28) upon sequence (38.16).

∎

38.6. Vibro-correctness

Any relay is a discontinuous transducer. Nevertheless, under quite weak assumptions, CRS-transducers remain vibro-correct.

By Ψ we shall denote the class of functions $\psi(v)$ $(v \geq 0)$ which are Lipschitz continuous with Lipschitz constant 1, i.e.,

$$|\psi(v) - \psi(u)| \leq |v - u| .$$

Theorem 38.3. The transducer $R(\mu)$ is uniformly vibro-correct if and only if the measure of any curve $\alpha + \beta = \psi(\beta - \alpha)$ $(\alpha < \beta)$, where $\psi(v) \in \Psi$, equals zero.

Proof. (i) *Necessity.* Suppose that $\psi_0(v) \in \Psi$ and the measure of the curve

$$\beta + \alpha = \psi_0(\beta - \alpha) \quad (0 < \beta - \alpha < \gamma_0) \tag{38.22}$$

is equal to zero. For any fixed $\varepsilon > 0$, we can construct a broken line Γ which comprises a horizontal half-line and a finite set of horizontal and vertical segments such that the curve (38.22) is located between the shifts $\Gamma_1(\varepsilon)$ and $\Gamma_2(\varepsilon)$ of Γ, in direction of the straight-line $\beta = \alpha$, of the magnitude ε and $-\varepsilon$, respectively (see Fig. 38.1). Line Γ may be here constructed in such a way that all its points N_1, located right of the point N_2, are located not over N_2. The measure of the open strip G bounded by curves $\Gamma_1(\varepsilon)$ and $\Gamma_2(\varepsilon)$ is not smaller than the measure of curve (38.22) and, thus, is at least equal $\mu_0 > 0$.

The peaks of the curve $\Gamma_1(\varepsilon)$ will be denoted (as shown in Fig. 38.1) by $\{m_2,M_1\}$, $\{m_2,M_2\}$, ..., $\{m_{n+1},M_n\}$, $\{m_{n+1},M_{n+1}\}$. Let us construct a continuous input $u(t)$, linear on each interval $[k-1,k]$, by defining its values for integer k by

$$u(0) = M_1, \quad u(1) = M_1, \quad u(2) = m_2, \quad u(3) = M_2, \quad u(4) = m_3, \quad \ldots,$$
$$u(2k-1) = M_k, \quad u(2k) = m_{k+1} .$$

We also define

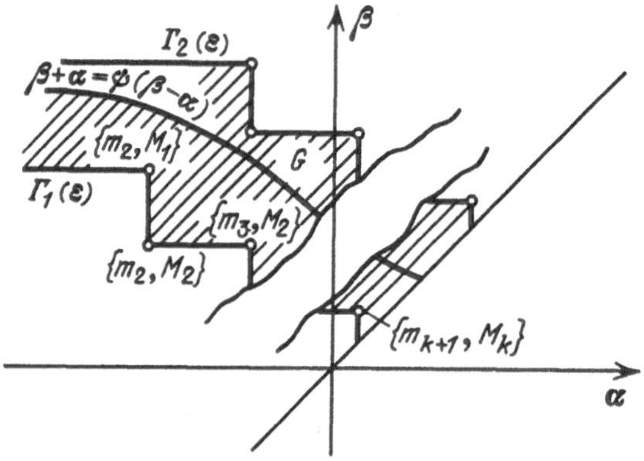

Fig. 38.1

$$v(t) = \begin{cases} u(t) + \sqrt{2}\,\varepsilon t \quad , & \text{if } 0 \leq t \leq 1 \ , \\ u(t) + \sqrt{2}\,\varepsilon \quad , & \text{if } t \geq 1 \ . \end{cases}$$

$z_0(\alpha,\beta)$ will denote the initial state

$$z_0(\alpha,\beta) = \begin{cases} 1 \ , & \text{for } \beta \leq M_1 \ , \\ 0 \ , & \text{for } \beta > M_1 \ . \end{cases}$$

Then the outputs

$$x(t) = R[t_0, z_0(\alpha,\beta);\ \mu]u(t) \quad ,$$

$$y(t) = R[t_0, z_0(\alpha,\beta);\ \mu]v(t) \quad ,$$

by virtue of Theorem 38.2, fulfil the inequality

$$|x(2k) - y(2k)| \geq \mu G \geq \mu_0 \quad .$$

At the same time, the norm of $u(t) - v(t)$ is 2ε. Thus, because ε is arbitrary, $R(\mu)$ cannot represent any uniformly vibro-correct transducer.

(ii) *Sufficiency*. For any function $\psi(u) \in \Psi$ and each $\varepsilon > 0$, we introduce the corresponding strip

$$G(\psi,\varepsilon) = \{\{\alpha,\beta\}: \psi(\beta-\alpha) - 2\varepsilon \leqq \alpha+\beta \leqq \psi(\beta-\alpha)\} \qquad (38.23)$$

and define

$$\lambda(\varepsilon) = \sup_{\psi(v)\in\Psi} \mu G(\psi;\varepsilon) \quad (\varepsilon > 0) \ . \qquad (38.24)$$

If the measure of any curve $\alpha+\beta = \psi(\beta-\alpha)$ $(\alpha < \beta)$, where $\psi(v) \in \Psi$, is equal zero, then the function (38.24) is finite and converges to zero as $\varepsilon \to 0$. To prove the sufficiency, we shall derive the estimate

$$\|R[t_0,z_0(\alpha,\beta);\ \mu]u(t) - R[t_0,z_0(\alpha,\beta);\ \mu]v(t)\|_{t_0,t_1} \leqq$$
$$\leqq 2\lambda[\|u(t)-v(t)\|_{t_0,t_1}] \ . \qquad (38.25)$$

The monotonicity of transducer $R(\mu)$ with respect to initial states and admissible inputs enables us to prove, instead of (38.25), the following simpler estimate

$$R[t_0,z_0(\alpha,\beta);\ \mu]u(t_*) - R[t_0,z_1(\alpha,\beta);\ \mu]v(t_*) \leqq 2\lambda(\varepsilon)$$
$$(t_* \in [t_0,t_1]) \ , \qquad (38.26)$$

where $v(t) = u(t) - \varepsilon$ and

$$z_1(\alpha,\beta) = \begin{cases} 0 & , \text{ if } \alpha \geqq u(t_0) - \varepsilon \ , \\ z_0(\alpha,\beta) & , \text{ otherwise} \ . \end{cases}$$

Let us consider the values $z_0(\alpha,\beta;t)$ and $z_1(\alpha,\beta;t)$ of the outputs for bundle $Q(u)$ that appropriately correspond to the inputs $u(t)$, $v(t)$ and the initial states $z_0(\alpha,\beta)$, $z_1(\alpha,\beta)$. Due to Theorem 38.2, these values coincide outside the union of the strip

$$F(\varepsilon) = \{\{\alpha,\beta\}: m_1(t_*) - \varepsilon \leqq \alpha \leqq m_1(t_*),\ \alpha < \beta\} \ ,$$

where $m_1(t_*)$ denotes the first of relations (38.14), and the set (38.23) in which the function $\psi(v)$ is piecewise linear on each interval $[\delta,\infty)$ $(\delta > 0)$, the equality $\psi(0) = 2u(t_*)$ holds and

$$\psi'(v) = \begin{cases} -1, & \text{for } v > M_1(t_*) - m_2(t_*), \\ 1, & \text{for } M_1(t_*) - m_2(t_*) > v > M_2(t_*) - m_2(t_*), \\ -1, & \text{for } M_2(t_*) - m_2(t_*) > v > M_2(t_*) - m_3(t_*), \\ \cdots\cdots & \\ 1, & \text{for } M_{k-1}(t_*) - m_k(t_*) > v > M_k(t_*) - m_k(t_*), \\ -1, & \text{for } M_k(t_*) - m_k(t_*) > v > M_k(t_*) - m_{k+1}(t_*), \\ \cdots\cdots & \end{cases}$$

where $m_k(t_*)$ and $M_k(t_*)$ are given by (38.16). Hence, the difference in the left-hand side of (38.26) can be bounded from above by the measure of union $F(\varepsilon) \cup G(\psi;\varepsilon)$. This yields (38.26). ∎

The operators $R[t_0,z_0(\alpha,\beta);\mu]$ are continuous in every point $u(t)$ of the space $C(t_0,t_1)$ under more general conditions which do not guarantee the uniform vibro-correctness of the transducer $R(\mu)$. *For such a continuity it is necessary and sufficient that the measure of every horizontal and every vertical half-line equals zero.*

It has been shown in the proof of Theorem 38.3 that the function $2\lambda(\varepsilon)$, with $\lambda(\varepsilon)$ defined by the equality (38.24), represents a modulus of vibro-correctness for operator $R[t_0,z_0(\alpha,\beta);\mu]$. The function $\lambda(\varepsilon)$ represents here a lower bound for all vibro-correctness moduli. This leads to an important consequence: all operators $R[t_0,z_0(\alpha,\beta);\mu]$ are Lipschitz continuous if and only if the function (38.24) admits the estimate $\lambda(\varepsilon) \le \gamma\varepsilon$. In particular, the following property is obviously true.

Theorem 38.4. Let μ be a measure with bounded support and bounded density. Then all operators $R[t_0,z_0(\alpha,\beta);\mu]$ are Lipschitz continuous in the space $C(t_0,t_1)$ (on their domains), with the same Lipschitz coefficient. ∎

In a natural way, the transducer $R(\mu)$ is vibro-correct with respect to the initial states and measure μ. Let us note here that for arbitrary initial states $\{u(t_0),z_0(\alpha,\beta)\}$ and $\{u(t_0),z_1(\alpha,\beta)\} \in \Omega[R(\mu)]$, the estimate

$$|R[t_0,z_0(\alpha,\beta);\mu]u(t) - R[t_0,z_1(\alpha,\beta);\mu]u(t)| \le$$

$$\leq \sup_{\xi,\eta} \left| \int_{\substack{\alpha \leq \xi, \\ \beta \geq \eta}} [z_0(\alpha,\beta) - z_1(\alpha,\beta)] d\mu(\alpha,\beta) \right| \tag{38.27}$$

is valid.

38.7. Controllable restrictions

In this section, we shall consider only uniformly vibro-correct transducers $R(\mu)$ (cf., Theorem 38.3).

From the class Ψ introduced in Section 38.6, we shall select a subclass Ψ_0 of all continuous functions $\psi(v)$, piecewise linear on every interval $[\delta,\infty)$ $(\delta > 0)$ and satisfying the following conditions: $|\psi'(v)| = 1$ for almost all v and

$$\lim_{v \to \infty} v^{-1} \psi(v) = 0 \ .$$

Two functions $\psi_1(v), \psi_2(v) \in \Psi_0$ will be called *equivalent*, if they coincide at sufficiently large v. We are going to decompose Ψ_0 into equivalence classes. Let us choose an equivalence class K. By Ω_K we shall denote the set of states $\{u,z(\alpha,\beta)\} \in \Omega[R(\mu)]$, where for almost all points $\{\alpha,\beta\}$ with respect to the measure μ,

$$z(\alpha,\beta) = \begin{cases} 1 \ , & \text{if } \ \alpha + \beta < \psi(\beta - \alpha) \ , \\ 0 \ , & \text{if } \ \alpha + \beta > \psi(\beta - \alpha) \end{cases} \tag{38.28}$$

at some $\psi(v) \in K$. The functions $z_0(\alpha,\beta)$ and $z_1(\alpha,\beta)$ will be treated as identical, if they coincide almost everywhere in the sense of measure μ.

As a consequence of Theorem 38.3 it follows that the *restriction* $R_K(\mu)$ $(\Omega[R_K(\mu)] = \Omega_K)$ *of the transducer* $R(\mu)$ *is controllable*. Moreover, every controllable restriction of $R(\mu)$ has the form $R_K(\mu)$. $R(\mu)$ admits a completely controllable restriction if and only if the measure μ has bounded support.

Any *restriction* $R_K(\mu)$ *is almost completely controllable* in the following sense: for any initial state $\{u_0,z_0(\alpha,\beta)\} \in \Omega[R(\mu)]$, any state $\{u_1,z_1(\alpha,\beta)\} \in \Omega_K$ and each $\varepsilon > 0$ there exists a continuous input $u(t)$ $(t_0 \leq t \leq t_1)$ which satisfies the conditions $u(t_0) = u_0$,

$u(t_1) = u_1$ and

$$\int_{\alpha<\beta} |R[t_0,z_0(\alpha,\beta);\mu]u(t_1) - z_1(\alpha,\beta)|d\mu(\alpha,\beta) < \varepsilon \ .$$

In general, the transducer $R(\mu)$ admits various almost completely controllable restrictions. Among them we may select the *maximal almost completely controllable restriction* $R_*(\mu)$. The set $\Omega[R_*(\mu)]$ consists of the pairs $\{u,z(\alpha,\beta)\}$ such that $z(\alpha,\beta)$ have the form (38.28), with $\psi(v)$ belonging to Ψ (not necessarily belonging to Ψ_0). The restriction $R_*(\mu)$ is analogous to the normal restrictions of Ishlinskii's transducers (cf., Chapter 36). It exhibits several remarkable properties. Here we shall mention three of them; proofs of those properties are quite simple, they follow as corollaries from Theorem 38.3.

a. Let a function $\lambda(\varepsilon)$ be chosen in the form (38.24). As already mentioned, the function $2\lambda(\varepsilon)$ represents a vibro-correctness modulus of the transducer $R(\mu)$. The same function will be a vibro-correctness modulus of the restriction $R_*(\mu)$; this vibro-correctness modulus has an optimal value. In particular, if the derivative $\lambda'(0)$ is finite, then the operators of the input-output correspondences for $R_*(\mu)$ are Lipschitz continuous, with optimal Lipschitz constant $\lambda'(0)$.

b. For the operators $R_*[t_0,z_0(\alpha,\beta);\mu]$ that correspond to the restriction $R_*(\mu)$, natural analogues of the defining properties of Ishlinskii's transducers are true (cf., Sections 36.4 and 35.5).

Let the functions $F(\xi,\eta)$ and $\Phi(u)$ be defined so that

$$\Phi(u) = \begin{cases} \sup\limits_{v<u_0} [\Phi(v) + F(v,u)] , & \text{for } u > u_0 , \\ \inf\limits_{v>u_0} [\Phi(v) + F(v,u)] , & \text{for } u < u_0 , \end{cases}$$

where u_0 is a fixed number. Define $\Gamma(u_0) = u_0$ and

$$\Gamma(v) = \begin{cases} \sup\{u < u_0 : \Phi(v) = \Phi(u) - F(v,u)\} , & \text{for } v > u_0 , \\ \inf\{u > u_0 : \Phi(v) = \Phi(u) - F(v,u)\} , & \text{for } v < u_0 . \end{cases}$$

Function $\Gamma(v)$ may be assumed equal to $-\infty$ at $v > 0$ and equal to ∞ at $v < 0$. We shall say that the operator V , defined on piecewise monotone inputs $u(t)$ $(t \geq t_0,\ u(t_0) = u_0)$, has the *first* $\{F,\Phi\}$ *-defining property at the input* $u_0(t)$, if from either

$$u_0(\tau) = \max_{t_0 \leq \sigma \leq \tau} u_0(\sigma), \quad u_0(\tau) \geq \Gamma[\max_{t_0 \leq \sigma \leq \tau} u_0(\sigma)]$$

or

$$u_0(\tau) = \min_{t_0 \leq \sigma \leq \tau} u_0(\sigma), \quad u_0(\tau) \leq \Gamma[\max_{t_0 \leq \sigma \leq \tau} u_0(\sigma)]$$

it follows that

$$Vu_0(\tau) = \Phi[u_0(\tau)] \quad .$$

The operator V will be said to have the *second* $\{F,\Phi\}$ *-defining property at the input* $u_0(t)$, if

$$u_0(\tau) = \max_{t_0 \leq \sigma \leq \tau} u_0(\sigma), \quad u_0(\tau) \leq \Gamma[\min_{t_0 \leq \sigma \leq \tau} u_0(\sigma)]$$

implies that

$$Vu_0(\tau) = \Phi[\min_{t_0 \leq \sigma \leq \tau} u_0(\sigma)] + F[\min_{t_0 \leq \sigma \leq \tau} u_0(\sigma), u_0(\tau)]$$

and if

$$u_0(\tau) = \min_{t_0 \leq \sigma \leq \tau} u_0(\sigma), \quad u_0(\tau) \geq \Gamma[\min_{t_0 \leq \sigma \leq \tau} u_0(\sigma)]$$

yields the equality

$$Vu_0(\tau) = \Phi[\max_{t_0 \leq \sigma \leq \tau} u_0(\sigma)] + F[\max_{t_0 \leq \sigma \leq \tau} u_0(\sigma), u_0(\tau)] \quad .$$

At last, the operator V has the *third* $\{F,\Phi\}$ *- defining property at the input* $u_0(t)$, if for every $t_0 \leq \tau_1 < \tau_2 < \tau_3 < \infty$, such that $u_0(t) \in [u_0(\tau_1), u_0(\tau_2)]$ at $\tau_1 \leq t \leq \tau_2$ and $u_0(t) \in [u_0(\tau_2), u_0(\tau_3)]$ at $\tau_2 \leq t \leq \tau_3$, with

$$u_o(\tau_3) \in [u_o(\tau_1), u_o(\tau_2)] \quad ,$$

the following equality (cf., (38.12)) holds:

$$Vu_o(\tau_3) = Vu_o(\tau_2) + F[u_o(\tau_2), u_o(\tau_3)] \quad .$$

Any operator $R_*[t_o, z_o(\alpha,\beta);\mu]$ *has all three* $\{F,\Phi\}$ *-defining properties, with* $F(\xi,\eta)$ *being the demagnetization function* (38.9) *and*

$$\Phi(u) = \begin{cases} \mu\{\{\alpha,\beta\}:\ z_o(\alpha,\beta) = 1 \quad or \quad \beta < u\} \ , \ if \ u \geq u_o \ , \\ \mu\{\{\alpha,\beta\}:\ z_o(\alpha,\beta) = 1 \quad and \quad \alpha < u\} \ , \ if \ u < u_o \ . \end{cases}$$

The last statement admits a natural conversion which represents a variant of an identification theorem .

c. Unexpectedly easily one may formulate conditions which ensure that the restriction $R_*(\mu)$ (but, obviously, not the transducer $R(\mu)$ itself!) is deterministic. To this end, it is enough that the support of measure μ is contained in a certain curve of the form $\beta - \alpha = \varphi(\beta + \alpha)$ where $\varphi(v)$ $(-\infty < v < \infty)$ is Lipschitz continuous with Lipschitz constant 1, i.e.,

$$|\varphi(u) - \varphi(v)| \leq |u - v| \quad .$$

The deterministic transducer $R_*(\mu)$ represents a hysteron. Moreover, it is a generalized play $L(\Gamma_\ell, \Gamma_r)$ (cf., Chapter 1) with the characteristics

$$\Gamma_\ell(u) = \mu\{\{\alpha,\beta\}:\ \beta - \alpha = \varphi(\beta + \alpha),\ \ \alpha < u\} \quad ,$$

$$\Gamma_r(u) = \mu\{\{\alpha,\beta\}:\ \beta - \alpha = \varphi(\beta + \alpha),\ \ \beta < u\} \quad .$$

39. Rheological models

39.1. Construction of the model

Rheological models describe mechanical systems containing elastic, plastic and viscous elements. Such structures are used for describing a large class of processes in plasticity theory, including the models due to Prandtl, Ivan, Ishlinskii, Novozhilov-Kadashevich, etc. In the sequel we shall only consider models comprising a finite number of elastic and plastic elements.

Assume that components E_i $(i = 1,\dots,n)$ of the model are indexed so that $E_1, \dots, E_p$ $(p < n)$ are plastic, whereas the remaining ones are elastic. For any $t \geq t_o$, the state of an element E_i is characterized by the pair $\{u_i(t), x_i(t)\}$ $(i=1,\dots,n)$, including its stress and deformation. Functions $u_i(t)$ and $x_i(t)$ $(t \geq t_o)$ for $i = 1, \dots, p$ are connected by the relations

$$|u_i(t)| \leq \tau_i \quad , \quad \dot{x}_i(t) \begin{cases} \geq 0, & \text{if } t \in \{\xi: u_i(\xi) = \tau_i\} \\ = 0, & \text{if } t \in \{\xi: |u_i(\xi)| < \tau_i\} \\ \leq 0, & \text{if } t \in \{\xi: u_i(\xi) = -\tau_i\} \end{cases} , \tag{39.1}$$

where $\tau_i > 0$ represents the yield limit of the plastic element E_i. In turn, for $i = p + 1, \dots, n$,

$$u_i(t) = k_i \, x_i(t), \tag{39.2}$$

where $k_i > 0$ is the suitable elasticity modulus. Deformations $x_i(t)$ $(i = 1,\dots,p)$ are postulated to be absolutely continuous and satisfy relations (39.1) almost everywhere.

The rheological model takes form of a planar system of horizontal, rigid, weightless bars $S_1, \dots, S_m$ such that any pair S_i, S_j is either disconnected or connected by some weightless, elastic and plastic elements (cf., Fig. 39.1). Each bar can move only in the vertical direction.

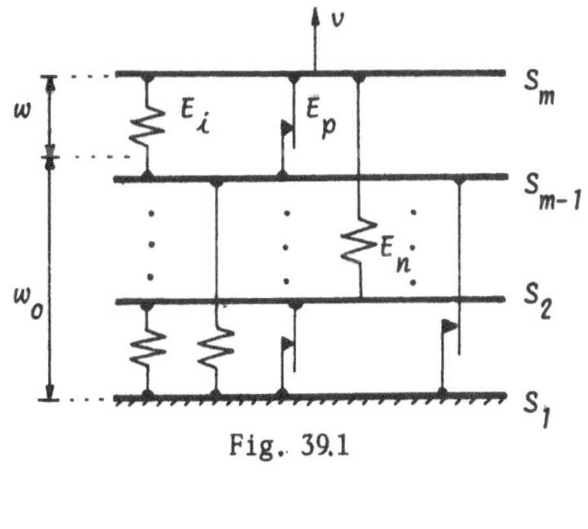

Fig. 39.1

Suppose the bar S_1 is fixed whereas S_m is subject to a vertical force $v(t)$ $(t \geq t_o)$. By w_o denote the distance between the bars S_1 and S_m at rest (i.e., in the state where $v = 0$ and there is no element which would have non-zero deformation or stress), and by $w_1(t)$ $(t \geq t_o)$ the variable distance between these bars. The complete deformation of the rheological model is characterized by the function

$$w(t) = w_1(t) - w_o .$$

To characterize dynamical properties of the rheological model, relations between the variables $w(t)$ and $v(t)$ are to be determined.

39.2. Graphs

In the sequel, some notions and techniques of graph theory are used. We define an oriented graph G_o corresponding to the considered model, such that the bar S_i is represented by vertex s_i and the element E_j by edge e_j. An orientation of any edge of the graph is specified by assuming positive direction of forces acting on the relevant element of the model. Let G be the graph obtained from G_o by adding any oriented edge e_o between the vertices s_1 and s_m. n_c will denote the number of cycles, and $\{\alpha_{iq}\}$, $\{\beta_{iq}\}$ $(i = 1,\dots,m;\ j = 1,\dots,n_k;\ q = 0, 1,\dots,n)$ correspondingly the incidence and cycle matrices of the oriented graph G. We set $u_o(t) = v(t)$, $x_o(t) = w(t)$ $(t \geq t_o)$.

For all $t \geq t_o$, the sum of all forces acting on any bar is zero and the deformations of elements connected with a fixed pair of bars are equal. Therefore,

$$\sum_{q=0}^{n} \alpha_{iq}\, u_q(t) = 0 ,$$
$$\sum_{q=0}^{n} \beta_{jq}\, x_q(t) = 0 \quad (t \geq t_o;\ i = 1,\dots,m;\ j = 1,\dots,n_c) . \tag{39.3}$$

The graph G and vectors $\tau = \{\tau_1,\dots,\tau_p\}$, $k = \{k_{p+1},\dots,k_n\}$ completely define geometric structure and parameters of the model; its dynamic characterization is given by relations (39.1) - (39.3). Thus, instead of dealing with mechanical constructions we can treat any system described by (39.1) ÷ (39.3) as a rheological model. Below we give a precise definition.

Let us take an oriented graph G with m vertices, $(n+1)$ edges and n_c cycles; suppose that $\{\alpha_{iq}\}$ and $\{\beta_{jq}\}$ $(i = 1,\dots,m;\ j = 1,\dots,n_c;$ $q = 0,1,\dots,n)$ are its incidence and cycle matrices, respectively. For each edge e_q $(q = 0,1,\dots,n)$, we define its state as a pair $\{u_q(t), x_q(t)\}$ of functions that are absolutely stable at $t \geq t_0$. Finally, we define vectors $\tau = \{\tau_1,\dots,\tau_p\}$, $k = \{k_{p+1},\dots,k_n\}(p<n)$ with positive components.

By a *rheological model with graph* G *and vectors* τ, k we shall mean a system whose state is defined for each $t \geq t_0$ by the vector

$$Y(t) = \{u_0(t),\ x_0(t),\ u_1(t),\ x_1(t),\ \dots,\ u_n(t)\ x_n(t)\} \in R^{2(n+1)} ,$$

with the state vectors of the edges of G as components; its dynamics is described by relations (39.1) ÷ (39.3).

The edge e_0 of the graph G will be called its *input-output edge*, and in dependence on the considered problem either of the functions $u_0(t)$ or $x_0(t)$ will be treated as an *external action* on the system.

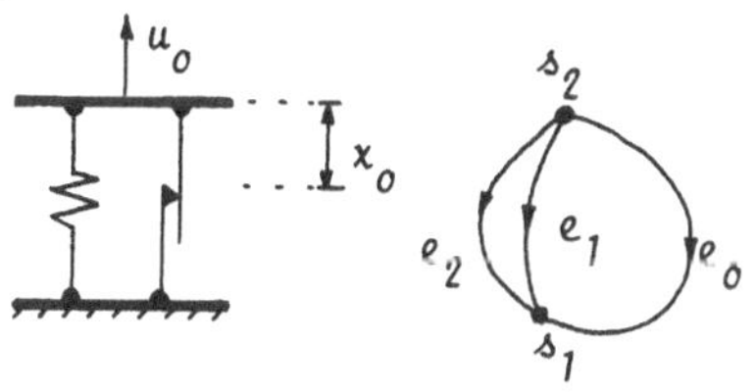

Fig.39.2

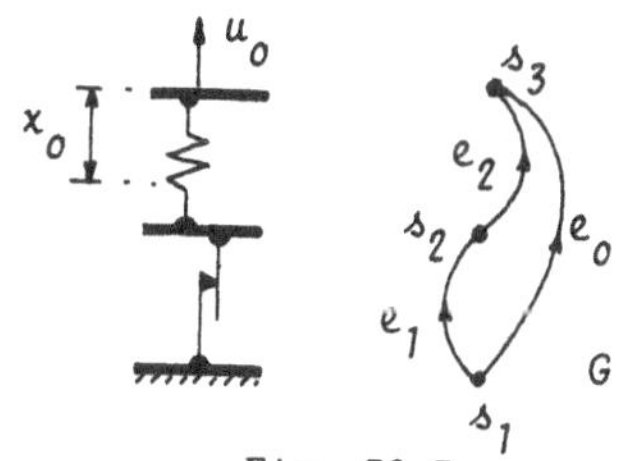

Fig. 39.3

We shall give a few examples. The model shown in Fig. 39.2 describes one-dimensional play with input $u_0(t)$ and output $x_0(t)$. The model shown in Fig. 39.3 refers to a one-dimensional stop. In this case, $x_0(t)$ represents the input, and $u_0(t)$ output of the system. Figure 39.4 shows models of parallel connections of one-dimensional plays and stops; in turn, a simplest version of the Novozhilov-Kadashevich model is depicted in Fig. 39.5.

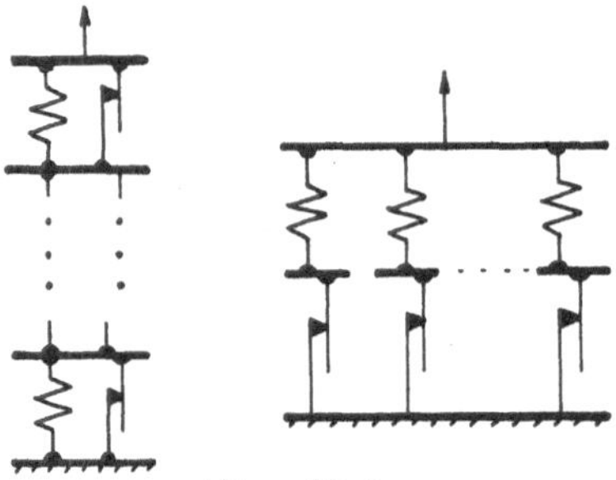

Fig. 39.4

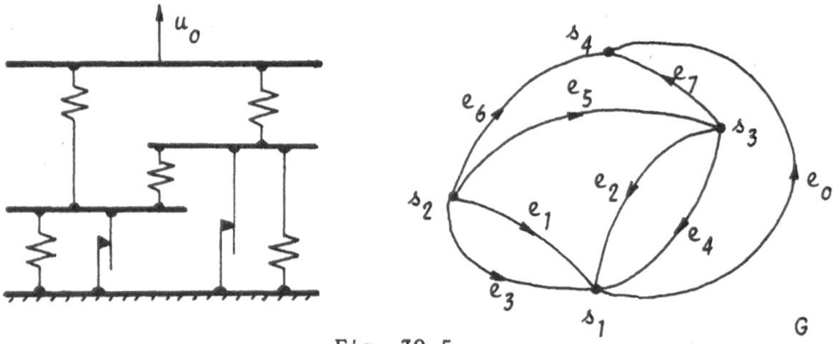

Fig. 39.5

By S_o and S^o we denote vertices which are incident to the input-output edge e_o ; let

$$G_p = G \setminus (e_o \cup e_{p+1} \ldots \cup e_n) \quad .$$

The subgraph G_p of G comprises all the edges which represent plastic elements.

Suppose that the graph G is not identical with its maximal cyclically connected subgraph $\overline{G}$ (i.e., a subgraph of G , such that all pairs of its vertices belong to a certain cycle) containing the vertices S^o and S_o.

Then, in view of (39.3), the states of the edges of $G \setminus \tilde{G}$ remain in no relation to the state of input-output edge e_o . Thus, the graph G can be considered as cyclically connected.

We shall assume moreover that the subgraph G_p contains neither cycles nor paths connecting vertices S_o and S^o , and the subgraph $G \setminus (e_o \cup G_p)$ is connected.

39.3. Transducer M

Provided the postulates of the last section hold, relations (39.2) and (39.3) are equivalent to the equalities

$$[u_{p+1}(t),\dots,u_n(t),x_{p+1},\dots,x_n(t)]^T = B[u_o(t),x_1(t),\dots,x_p(t)]^T, \quad (39.4)$$

$$u_i(t) = c_i\, u_o(t) - \sum_{j=1}^{p} a_{ij}\, x_j(t) \quad (i = 1,\dots,p) , \quad (39.5)$$

$$x_o(t) = \mu\, u_o(t) + \sum_{j=1}^{p} f_j\, x_j(t) \quad , \quad (39.6)$$

where B is a $2(n-p)\times(p+1)$ - matrix, $A = \{a_{ij}\}$ is a positive definite symmetric matrix of order p , $c = \{c_i\}$, $f = \{f_i\}$ are non-zero vectors in R^p and μ is a scalar (in (39.4), ^{T} represents transposition). At a given incidence matrix of the graph G and a vector k, matrices B , A , vectors c , f and scalar μ can be explicitly constructed. Here we omit those constructions for they require an awkward exposition.

As a matter of fact, relations (39.4)$\div$(39.6) imply a disparity of the components of vector Y . Due to (39.5), relations (39.1) comprise only the variables $u_o, x_1, \dots, x_p$, whereas the other components of Y are linear combinations of these variables. Therefore, for a study of the rheological models an auxiliary transducer M will be introduced, with the state defined by the "truncated" vector $\{u_o,x_1,\dots,x_p\} \in R^{p+1}$.

To begin with, let us notice that if the functions

$$u_o(t),\ x_1(t),\ \dots,\ x_p(t) \qquad (t \ge t_o)$$

satisfy (39.1) and (39.5), then the inequality

$$-\tau \le C\, u_o(t) - A\, x(t) \le \tau \quad ,$$

where $x(t) = \{x_1(t),\dots,x_p(t)\} \in R^p$, is fulfilled componentwise at $t \geq t_o$. The set

$$\left\{\{v,x\} = \{v,u_1,\dots,u_p\} \in R^{p+1}: -\tau \leq C\, u_o(t) - A\, x(t) \leq \tau\right\} \tag{39.7}$$

is a prism in R^{p+1} such that, due to non-singularity of the matrix A, its intersections with the planes $v = \text{const.}$ are non-singular p-dimensional parallelepipeds.

Let M be a transducer with the set $\Omega(M)$ of feasible states $\{v,x\} = \{v,x_1,\dots,x_p\} \in R^{p+1}$, defined by (39.7). Suppose that for an absolutely continuous input $v(t)$ $(t \geq t_o)$ the corresponding variable state

$$\{v(t),x(t)\} = \{v(t),\Lambda[t_o,x^o]v(t)\} \quad (t \geq t_o) \tag{39.8}$$

of the transducer M, subject to the initial condition

$$\{v(t_o),x(t_o)\} = \{v_o,x^o\} \in \Omega(M), \tag{39.9}$$

is given by (39.1) with $u_o(t) = v(t)$ and functions $u_i(t)$ $(i = 1,\dots,p)$ defined by (39.5). The output

$$w(t) = M[t_o,x^o]v(t) \quad (t \geq t_o) \tag{39.10}$$

of M is given by (39.6) with $w(t) = x_o(t)$.

Together with the transducer M, we shall consider (cf., Section 33.1) the associate transducer Λ with the output

$$x(t) = \Lambda[t_o,x^o]v(t) \quad (t \geq t_o). \tag{39.11}$$

Operators (39.10) and (39.11) respectively define the input-output relations and input-state relations (39.8) of the transducer M. The above characterization of M is correct in the following sense.

Theorem 39.1. For each absolutely continuous input $v(t)$ $(t \geq t_o)$, relations (39.1), (39.5) define a unique absolutely continuous vector-function $x(t)$ $(t \geq t_o)$, satisfying the initial condition (39.9) and such that

$$\{v(t),x(t)\} \in \Omega(M) \quad (t \geq t_0) \ .$$

A proof of this theorem exploits techniques of the theory of non-linear monotone semigroups (cf., [11,48]). ∎

The input $v(t)$ $(t \geq t_0)$ of M will be called x^0 - *admissible* (or simply admissible as long as x^0 is uniquely specified), if $\{v(t_0),x_0\} \in \Omega(M)$ for some $x^0 \in R^p$.

Single-valued operators (39.10) and (39.11) are defined on the set of absolutely continuous input functions. Their values are also absolutely continuous .

39.4. Properties of the transducer M

In Chapters 33 through 38, we considered transducers which admitted representations in the form of systems composed of relays, one-dimensional hysterons and other simple elements. Analogous representations, containing multidimensional hysterons, turn out useful at a study of the transducer M .

Theorem 39.2. Let $v(t)$ $(t \geq t_0)$ be an admissible absolutely continuous input. Then

$$\Lambda[t_0,x^0]v(t) = A^{-1/2} L\,[t_0,A^{1/2}x^0](A^{-1/2} c\,v(t)) \ , \tag{39.12}$$

where $L = L(Z)$ is a multidimensional play with the characteristic

$$Z = \{x \in R^p\colon -\tau \leq A^{1/2}x \leq \tau\} \ . \tag{39.13}$$

A proof of this theorem is based on a transformation of relations (39.1) and (39.5) to the form (16.19) expressed in terms of

$$\xi = A^{1/2}x \quad , \quad u = A^{1/2}c\,v \ .$$

∎

The set (39.13) is a non-singular polyhedron in R^p . Hence, due to representation (39.12) an extension of many properties of the multidimensional play with polyhedral characteristic (cf., Part 4) onto the transducer M is possible.

In particular, in view of (18.1) and (39.12), operators (39.11) satisfy Lipschitz condition with constant

$$\|A^{-1/2}\| \ \gamma(Z) \quad ,$$

where $\gamma(Z)$ is the constant given by (18.7) for the play with characteristic (39.13).

Therefore, by standard constructions (cf., Section 17.3), operators (39.10) and (39.11) can be extended onto the set of all admissible continuous inputs. We shall preserve the same notations for the extended operators. The transducer Λ is deterministic, static, vibro-correct and convergent on the set of admissible continuous inputs; Theorem 39.2 holds also in this case.

Let us now define a notion of ε_o - convergence **for** transducer M . We shall say that M is ε_o - *convergent* with some positive ε_o , if for every admissible input $v(t)$ $(t \geq t_o)$ and all $x', x'' \in R^p$, such that

$$\{v(t_o),x'\}, \ \{v(t_o),x''\} \in \Omega(M) \quad ,$$

the equality

$$|v(t_1) - v(t_o)| = \varepsilon_o \quad ,$$

fulfilled at some $t_1 > t_o$, implies that

$$\Lambda[t_o,x']v(t) = \Lambda[t_o,x'']v(t) \qquad (t \geq t_1) \quad .$$

Multidimensional plays with polyhedral characteristics are not ε_o-convergent on the set of all admissible vector-functions (they are even not strongly convergent, cf., Section 18.1). Therefore, in view of Theorem 39.2 one could wrongly expect that also any arbitrary transducer M is not ε_o - convergent.

Theorem 39.3. The transducer M is ε_o - convergent for some $\varepsilon_o > 0$ if and only if the vector $A^{-1}c$ has no zero components. ∎

The geometric sense of Theorem 39.3 is clear. Components of the vector $A^{-1}c$ represent coordinates of vector $A^{-1/2}c$ (input for the play (39.12)) in the basis composed of vectors normal to $(p-1)$ - dimensional edges of polyhedron (39.13).

The ε_o -convergence notion is useful at analyzing controllability of a transducer M . Unlike a multidimensional play, the transducer M is not always controllable. In particular, the transducer corresponding to a parallel connection of plays (cf., Chapter 33) was non-controllable.

Theorem 39.4. Let M be an ε_o -convergent transducer at some $\varepsilon_o > 0$. Then it admits a completely controllable contraction.

A proof of this theorem proceeds similarly to that of Theorem 33.1. ∎

A transducer M which is not ε_o -convergent in some cases may admit a completely controllable (unique) restriction, sometimes it even has a continuum of various controllable restrictions.

39.5. Transducer W

In Sections 39.3 and 39.4 we treated the rheological model as a transducer M (deterministic transducer Λ), with the force $v(t)$ playing role of an input signal. Now we are going to discuss that model with deformation $w(t)$ as an input.

As in Section 39.3, relations (39.2) and (39.3) can be transformed to the form

$$[u_{p+1}(t),\ldots,u_n(t),x_{p+1}(t),\ldots,x_n(t)]^T = \tilde{B}[x_o(t),u_1(t),\ldots,u_p(t)]^T , \tag{39.14}$$

$$x_i(t) = \tilde{c}_i\, u_o(t) - \sum_{j=1}^{p} \tilde{a}_{ij}\, u_j(t) \quad (i = 1,\ldots,p) , \tag{39.15}$$

$$u_o(t) = \tilde{\mu}\, u_o(t) + \sum_{j=1}^{p} \tilde{f}_j\, u_j(t) , \tag{39.16}$$

with some matrix $\tilde{B}$, positive definite symmetric matrix $\tilde{A} = \{\tilde{a}_{ij}\}$ of appropriate dimensions, non-zero vectors $\tilde{c} = \{\tilde{c}_j\}$, $\tilde{f} = \{\tilde{f}_i\} \in R^p$ and scalar $\tilde{\mu}$. To construct them, vector k and cycle matrix of the graph G must be given.

Let us consider a transducer W with the set

$$\Omega(W) = \Big\{\{w,u\} = \{w,u_1,\dots,u_p\} \in R^{p+1} \ : \ -\tau \leq u \leq \tau\Big\} \qquad (39.17)$$

of feasible states, such that for an absolutely continuous input $w(t)$ $(t \geq t_o)$ its variable state

$$\{w(t),u(t)\} = \{w(t),V[t_o,u^o]w(t)\} \quad (t \geq t_o) \ , \qquad (39.18)$$

which satisfies the initial condition

$$\{w(t_o),u(t_o)\} = \{w_o,u^o\} \in \Omega(W) \ , \qquad (39.19)$$

is given by relations (39.1), with $x_o(t) = w(t)$ and functions $x_i(t)$ $(i = 1,\dots,p)$ defined by (39.15). The output

$$v(t) = W[t_o,u^o]w(t) \qquad (t \geq t_o) \qquad (39.20)$$

of W is defined by equality (39.16) with $v(t) = u_o(t)$. In turn, the vector-function

$$u(t) = V[t_o,u^o]w(t) \qquad (t \geq t_o) \qquad (39.21)$$

represents the output of the associate transducer V .

In the sequel, our study of rheological models will be based on an analysis of the transducers W and V . The set (39.17) of their feasible states has the form of a right prism in R^{p+1}, with a p-dimensional parallelepiped as base. Any absolutely continuous input $w(t)$ $(t \geq t_o)$ is admissible for W . The definition of transducer W is correct, as it was for transducer M .

<u>Theorem 39.5.</u> For each absolutely continuous input $w(t)$ $(t \geq t_o)$, relations (39.1) and (39.15) define a unique absolutely continuous vector-function $u(t)$ $(t \geq t_o)$ which satisfies the initial condition (39.19) and the inclusion

$$\{w(t),u(t)\} \in \Omega(W) \qquad (t \geq t_o) \ .$$

■

Single-valued operators (39.20) define the input-output relations, and operators (39.21) characterize the input-state relations (39.18) of transducer W . These operators are defined in the space of absolutely continuous functions.

The following statement, analogous to Theorem 39.2, holds for W .

Theorem 39.6. The equality

$$V[t_o,u^o]w(t) = \tilde{A}^{-1/2}\,U[t_o,\tilde{A}^{1/2}u^o](\tilde{A}^{-1/2}\tilde{c}\,w(t))$$

is satisfied on the set of absolutely continuous inputs, with a multidimensional stop (cf., Part 4) having the characteristic

$$\mathcal{Z} = \{x \in R^p : -\tau \le \tilde{A}^{1/2}\,x \le \tau\} \quad .$$

∎

Due to this theorem, several properties of multidimensional stops, studied in Part 4 , carry over to the transducer W. Assertions of Theorems 39.3 and 39.4 remain valid also for W , up to the change of A and c onto $\tilde{A}$ and $\tilde{c}$, respectively.

39.6. Remarks

The hypotheses assumed in Section 39.2 for the graph G are close to the necessary and sufficient conditions for transducers Λ and V to be deterministic. We shall illustrate this analogy. Let us forget about the hypotheses of Section 39.2 and confine ourselves to the class of models such that the graph $G \setminus G_p$ is connected. The last requirement is natural in view of the theory of variable hysterons, developed in Chapter 15. Indeed, without this assumption, as it follows from simple examples, to provide that the rheological models are deterministic, not only the geometric structure of graph G but also relations between components of vectors τ and k must be known. Therefore, if parameters of the considered model vary in time (due to heating, ageing, etc.), the model, initially deterministic, may lose this property at some time instant.

For the models with connected graphs $G \setminus G_p$, the following statement is true.

Theorem 39.7. A transducer Λ (transducer V) is deterministic if and only if the graph G_p does not contain cycles, and the graph $G \setminus (e_o \cup G_p)$ is connected (respectively, the graph G_p contains neither cycles nor paths connecting vertices S_o and S^o). ∎

Throughout Sections 39.3 ÷ 39.5, the same hypotheses as in Section 39.2 were maintained. Under those postulates, both transducers Λ and V were deterministic. In particular, such situation is characteristic for the Ivan and Novozhilov-Kadashevich models. However, there are models for which only one of the considered transducers is deterministic. For instance, in the case of a one-dimensional play only Λ is deterministic, whereas for a one-dimensional stop, only V has this property. As a matter of fact, it was enough to postulate in Sections 39.3 ÷ 39.4 (in Section 39.5) that only Λ was deterministic (respectively, only V was deterministic). Then the scalar μ in (39.6) (respectively, $\tilde{\mu}$ in (39.16)) is equal to zero if and only if Λ is deterministic whereas V is non-deterministic (correspondingly, V is deterministic and Λ non-deterministic).

If simultaneously Λ and V are deterministic in some model, then M and W are mutual compensators (cf., Section 34.3).

Now assume that G is a planar graph. Then there exists the dual graph G^* . A rheological model represented by the graph G^* will be then called *dual* to the model defined by graph G. In this sense, one-dimensional play and stop are dual to each other.

Denote by Λ^* and V^* the transducers corresponding to the dual model.

Theorem 39.8. Let G be a planar graph. Then the transducer Λ (transducer V) is deterministic if and only if the corresponding transducer V^* (respectively, transducer Λ^*) has the same property. ∎

In view of the above results, it seems natural to make use of plays with characteristic different from polyhedrons, as well as infinite-dimensional plays for a study of more general rheological models.

Bibliographic comments

There are hundreds of papers and monographs dealing with various problems related to nonlinearities of hysteresis type. The list of references given in this book by no means pretends to any completeness.

Some information on the results obtained by the authors, as well as on other recent approaches to the study of hysteresis phenomena, together with a comprehensive list of references can be found in [12, 14, 23, 33, 35, 38, 39, 44, 72, 86, 89, 90, 99-101, 120, 125, 128, 143-145, 148-150]. For the Nole-Coleman-Truesdell approach, see [14, 135] . Foundations of the general systems theory are discussed in monographs [41, 87, 150, 165] .

Some of the results exposed in this book are original and have not been published elsewhere.

Nonlinearities which correspond to a play, stop or some more complex hysterons as described in Part 1 are typical for physics, mechanics, control engineering and other branches of science and technology (for instance, cf., [19, 115, 129, 150]). Exposition of Part 1 is based on papers [50, 53, 56] . The results of Chapter 5 have been obtained jointly with V.S. Kozyakin. A part of the results of Chapter 6 is due to N.I. Grachev [26, 27] .

The identification theorem which is the main result of Part 2 has been obtained by the authors jointly with V.S. Kozyakin [50] ; in the book a new proof of that theorem is given. In Chapters 9 and 10 some results originally due to N.P. Panskikh [10] are used; the approximations to the play and stop, considered in Section 10.4, have been introduced in the papers [94, 95] by A.V. Netushil. Their mathematical analysis is due to V.B. Prival'skii [117].

The theory of vibro-correct differential equations, exposed in Part 3, has been developed in [57, 69, 105]. In Part 3 and further, classical theorems on differential inclusions, due to E.E. Viktorovskii [139] and A.F. Filippov [21, 22], are systematically exploited. General theorems on the operator of translation along trajectories of differential equations are given in [51].

The notion of a multidimensional hysteron, introduced in Part 4, dates back to the models developed by von Mises, Saint-Venant, Tresca and others in plasticity theory (cf., [91, 101, 114, 115, 123, 124, 131], in particular). Exposition in Part 4 is based on results of [45, 46, 111, 140, 141]. For the elements of the theory of convex sets, used in that part, we refer to [34, 122]. Some results have their origin in the theory of nonlinear monotone semigroups due to Komura [11, 48] and in variational inequalities techniques [19, 76]. A proof of the last statement in Section 18.5 may be accomplished by means of the techniques developed in [110] for the study of dissipative relay systems. Theorem 19.1 is due to A.A. Vladimirov. For the elements of differential geometry (differentiation with respect to a vector, Frobenius condition, etc.), used in Part 4, we refer to [13, 126]. Some results in Chapter 21 have been obtained jointly with M.I. Kamenskii.

There is an extensive literature on the superposition operators generated by functions f(t,u) continuous in u (cf., [70] for instance). So far, only few results are available for discontinuous superposition operators. The notion of a superpositionally measurable function (SM-function) was introduced in [70]. Exposition in Part 5 is based on results of [60, 65]; recall also the papers [62, 66]. For the classical Michael's theorems we refer to [83, 88]; in turn, the Lusin-Yankov principle was discussed in [1, 34, 79, 160]. Theorems on the measurable selectors and Volterra selectors can be found in [108]. A comprehensive analysis of the multi-valued superposition operator generated by an SM-function was developed by L.N. Lyapin [80, 81], I.V. Shragin [127] and others. A general theory of systems containing ideal and non-ideal relays (arising from numerous applications) was exposed in the monographs by Y.Z. Tsypkin [133], Yu.I. Nejmark [93], in particular. Exposition of Chapter 28 is close to that of the paper [109].

Self-magnetization phenomena were studied in [12, 72, 149]. In

Part 6, results of [63, 106] are used. The model considered in Chapters 29 and 30 is originally due to Madelung and dates back to his paper [82] on the Wiener measure, random processes and stochastic equations; cf., [24, 77], too.

Exposition of Part 7 is based on [58, 60, 61, 68]. Finite systems of stops were used by many authors as models of elasto-plastic bodies A passage to continual systems was accomplished by A.Yu. Ishlinskii. For the related models we refer to [5, 15, 35-38, 101, 113]. In the papers we have just listed nearly no mathematical results are given as far as the involved input-output and input-state operators are concerned. A first model of hysteresis nonlinearities in the form of finite and continual relay systems was proposed and extensively studied by P. Preisach [116]. Unilateral estimates characterizing stability of systems with complex nonlinearities were used by V.A. Yakubovich [159]. Unilateral bounds for relay systems were utilized in [23] at an asymptotic analysis of forced vibrations. The constructions of Section 35.3 are based on the well-known Masing's procedures (cf., [101]).

References

1. Arkin, V. I., Levin, V. L.: Convexity of the values of vector integrals, measurable selection theorems and variational problems. Uspekhi Matem. Nauk *27* (1972) 21-27 [Russian]
2. Becker, R.: Elastische Nachwirkung und Plastizität. Zeitschrift für Physik *33* (1925) 185-212
3. Bennewitz, K.: Über elastische Nachwirkung, elastische Hysteresis und innere Reibung. Zeitschrift für Physik *25* (1924) 417-485
4. Berezovskii, A. A., Nizhnik, L. P.: Mathematical models of hysteresis. In: Proceedings of the 5th International Conference on Nonlinear Vibrations, Vol. 4, Izd. AN USSR, Kiev 1970, 69-71 [Russian]
5. Besseling, J. F.: A theory of elastic, plastic and creep deformations of an initially isotropic material showing anisotropic strain-hardening, creep recovery and secondary creep. J. Appl. Mech. *25* (1958) 529-536
6. Bogolyubov, N. N.: On some statistical methods in mathematical physics. Izd. AN USSR, Kiev 1945 [Russian]
7. Bogolyubov, N. N., Mitropol'skii, A. Yu.: Asymptotic methods in nonlinear vibrations theory. Nauka, Moscow 1974 [Russian]
8. Boltzmann, L.: Zur Theorie der elastischen Nachwirkung. Wien Akad. Sitzber. *70* (1874) 275-306
9. Bouc, R.: Influence du cycle d'hystéresis sur la resonance nonlinéaire d'un circuit série. Coll. Intern. du CNRS (1964) 148-155
10. Bouc, R.: Modêle mathêmatique d'hystêresis, application au circ oscillant a self saturable. Conference on Nonlinear Vibrations, Vol. 4, Izd. AN USSR, Kiev 1970, 100-113
11. Brezis, H.: Operations maximaux monotones et semi groupes de contraction dans les espaces de Hilbert. North-Holland, Amsterdam 1973
12. Bozorth, R. M.: Ferromagnetism. Van Nostrand Reinhold, Toronto 1951
13. Cartan, H.: Calcul diffêrentiel. Dunod, Paris 1967
14. Coleman, B. D., Noll, W.: Foundations of linear viscoelasticity. Reviews Modern Phys. *33* (1961) 239-249
15. Davidenkov, N. N.: On energy scattering at vibrations. Zhurnal Teoreticheskoj Fiziki *8* (1938) [Russian]
16. Della-Torre, E.: Effect of interaction on the magnetisation of single domain particles. IEEE Transaction of Audio and Electroacoustics *AU-14* (1966) 86-92
17. Della-Torre, E.: Measurements of interaction in an assembly of j-iron oxide particles. J. Appl. Phys. *36* (1966) 518-522

18. Drew, J. H.: Periodic solutions of a forced system with hysteresis. Intern. J. Non-Linear Mech. *7* (1972) 93-99
19. Duvaut, G., Lions, J. L.: Inequalities in Mechanics and Physics. Springer-Verlag, Berlin, Heidelberg, New York 1976
20. Emel'yanov, S. V.: Automatic control systems with variable structure. Nauka, Moscow 1967 [Russian]
21. Filippov, A. F.: Classical solutions of differential equations with multivalued right-hand side. Vestnik MGU, ser. matem., No. 3 (1967) 16-26 [Russian]
22. Filippov, A. F.: Differential equations with discontinuous right-hand side. Matem. Sbornik *51* (1968) 99-108 [Russian]
23. Gil'man, T. S., Pokrovskii, A. V.: Forced oscillations of systems with simplest hysteresis nonlinearities. DAN SSSR *262* (1982) 437-450 [Russian]
24. Gikhman, I. I., Skorokhod, A. V.: Stochastic differential equations. Naukova Dumka, Kiev 1968 [Russian]
25. Giltay, J.: On ferromagnetic states. Appl. Sci. Res. *2* (1951) 199-215
26. Grachev, N. I.: Some properties of nonlinear elements with hysteresis. In: Abstracts of the 8th All-Union Conference on Control Problems, Vol. 1, Gosplan ESSR, Tallinn 1980, 40-42 [Russian]
27. Grachev, N. I.: Filtering properties of the generalized nonlinear play . Automatika i Telemekhanika, No. 5 (1982) 47-51 [Russian]
28. Gröger, K.: Zur Theorie des quasistatischen Verhaltens von elastisch--plastischen Körpern. Zeit. Ang. Math. Mech. *58* (H. 1) (1978), 78-85
29. Gröger, K.: Zur Theorie des dynamischen Verhaltens von elastisch-plastischen Körpern. Zeit. Ang. Math. Mech. *58* (H. 11) (1978) 36-41
30. Grothendieck, A.: Sur certain espaces des fonctions holomorphes. Zeitschrift für Reine und Ang. Math. *194* (H. 1, 2) (1953)
31. Hayashi, G. H.: The influence of hysteresis on nonlinear resonance. J. of Franklin Inst. *281* (1966) 379-386
32. Hill, R.: Mathematical theory of plasticity. Gostekhizdat, Moscow 1956 [Russian translation]
33. Il'yushin, A. A.: Plasticity. Foundations of general mathematical theory. Nauka, Moscow 1963 [Russian]
34. Ioffe, A. D., Tikhomirov, V. A.: Theory of extremal problems. North-Holland, Amsterdam 1979
35. Jona, F., Shirane, G.: Ferroelastic crystals. Pergamon Press, New York 1962
36. Ishlinskii, A. Yu.: Some applications of statistical methods to describing deformations of bodies. Izv. AN SSSR, Techn. Ser., No. 9 (1944) 580-590 [Russian]
37. Ishlinskii, A. Yu.: Plasticity. In: 30 years of mechanics in the Soviet Union, Gostekhizdat, Moscow 1950 [Russian]
38. Ishlinskii, A. Yu.: General plasticity theory with linear hardening. Ukrain. Matem. Zhurnal *6* (1954) 430-441 [Russian]
39. Kadashevich, Yu. I., Novozhilov, V. V.: On treatment of microstrains in plasticity theory. Mekhanika Tverdogo Tela, No. 3 (1968) 17-32 [Russian]
40. Kalman, R. E.: On the general theory of control systems. In: Proceedings of the 1st IFAC Congress, Vol. 2, Izd. AN SSSR, Moscow 1961, 521-547
41. Kalman, R. E., Falb, P. L., Arbib, M.: Topics in mathematical system theory. McGraw-Hill, New York 1969
42. Kantorovich, L. V., Akilov, G. P.: Functional analysis. Nauka, Moscow 1977 [Russian]
43. Kartak, K.: A generalisation of the Carathéodory theory of the differential equations. Czech. Math. J. *17* (1967) 482-514

44. Kirenskii, L. V.: Magnetism. Nauka, Moscow 1967 [Russian]
45. Klepcyn, A. F.: Properties of the Mises actuator. In: Analysis of operator equations, Izd. Kuibyshev. Univ., Kuibyshev 1983, 45-52 [Russian]
46. Klepcyn, A. F., Pokrovskii, A. V.: Vibrocorrectness of some elements with hysteresis. In: Proceedings of a Seminar on Dynamics of Nonhomogeneous Systems, All-Union Systems Research Institute, Moscow 1982, 62-69 [Russian]
47. Koiter, W. I.: General theorems for elastic-plastic solids. In: Progress in solid mechanics, North-Holland, Amsterdam 1960, 165-221
48. Komura, Y.: Nonlinear semigroups in Hilbert spaces. J. Math. Soc. Japan *19* (1967) 493-507
49. Kozyakin, V. S.: On the vibrostability of second order differential equations. Uspekhi Mat. Nauk *27* (1972) 241-243 [Russian]
50. Kozyakin, V. S., Krasnosel'skii, M. A., Pokrovskii, A. V.: Vibrostable hysterons. Dokl. AN SSSR *206* (1972) 800-803 [Russian]
51. Krasnosel'skii, M. A.: The operator of translation along the trajectories of differential equations. Translations of Mathematical Monographs, Vol. 19, Amer. Math. Society, Providence 1968
52. Krasnosel'skii, M. A.: Positive solutions of operator equations. P. Noordhoff, Groningen 1964
53. Krasnosel'skii, M. A.: Mathematical description of the vibrations of a material point on an elasto-plastic element. In: Partial differential equations, Nauka, Moscow 1970, 146-149 [Russian]
54. Krasnosel'skii, M. A.: Equations with hysteresis nonlinearities. In: Proceedings of the 7th Int. Conference on Nonlinear Oscillations, Vol. 1.1, Akademie-Verlag, Berlin 1977, 437-458
55. Krasnosel'skii, M. A., Burd, V. Sh., Kolesov, Yu. S.: Nonlinear almost periodic oscillations. Halsted Press, New York 1973
56. Krasnosel'skii, M. A., Darinskii, V. M., Emelin, I. V., Zabreiko, P. P., Lifshitz, E. A., Pokrovskii, A. V.: Operator-hysteron. Dokl. AN SSSR *190* (1970) 29-33 [Russian]
57. Krasnosel'skii, M. A., Pokrovskii, A. V.: Vibrostability of solutions of differential equations. Dokl. AN SSSR *195* (1970) 544-547 [Russian]
58. Krasnosel'skii, M. A., Pokrovskii, A. V.: Systems of hysterons. Dokl. AN SSSR *200* (1971) 733-736 [Russian]
59. Krasnosel'skii, M. A., Pokrovskii, A. V.: Vibrostable differential equations with continuous right-hand side. Proc. of Moscow Math. Soc., Vol. 27 (1972) 93-112 [Russian]
60. Krasnosel'skii, M. A., Pokrovskii, A. V.: Periodic oscillations in systems with switching nonlinearities. Dokl. AN SSSR *216* (1974) 733-736 [Russian]
61. Krasnosel'skii, M. A., Pokrovskii, A. V.: Modelling of transducers with hysteresis by continual systems of relays. Dokl. AN SSSR *227* (1976) 547-550 [Russian]
62. Krasnosel'skii, M. A., Pokrovskii, A. V.: On a discontinuous superposition operator. Uspekhi Mat. Nauk *32* (1977) 169-170 [Russian]
63. Krasnosel'skii, M. A., Pokrovskii, A. V.: Natural solutions of stochastic differential equations. Dokl. AN SSSR *240* (1978) 264-267 [Russian]
64. Krasnosel'skii, M. A., Pokrovskii, A. V.: Identification of some nonlinear transducers. In: Proceedings of the 4th IFAC symposium on identification and system parameter estimation. North-Holland, Amsterdam 1978, 213-221
65. Krasnosel'skii, M. A., Pokrovskii, A. V.: On the continuity points of monotone operators. Comment. Mathem., Tomus specialis in honorem L. Orlicz (1979) 205-216 [Russian]

66. Krasnosel'skii, M. A., Pokrovskii, A. V.: Equations with discontinuous nonlinearities. Dokl. AN SSSR *248* (1979) 1055-1059 [Russian]
67. Krasnosel'skii, M. A., Pokrovskii, A. V.: The method of block-diagrams in mathematical modeling of systems incorporating complex nonlinearities. In: Preprints of 8th IFAC world congress, Kyoto, Japan 1981, v. VI.
68. Krasnosel'skii, M. A., Pokrovskii, A. V.: Oscillations in systems with composite nonlinearities. In: Proc. of the 9th Int. Conf. on Nonlinear Oscillations, Vol. 1, Izd. AN USSR, Kiev 1982 [Russian]
69. Krasnosel'skii, M. A., Rutickii, Ya. B.: Convex functions and Orlicz spaces. P. Noordhoff, Groningen 1961
70. Krasnosel'skii, M. A., Zabreiko, P. P., Pustyl'nik, E. I., Sobolevskii, P. E.: Integral operators in spaces of summable functions. Noordhoff, Leyden 1976
71. Kravchenko, A.A.: On transducers describing one-dimensional rheological models. Dokl. AN SSSR *277* (1984) 525-529 [Russian]
72. Kronmüller, M.: Nachwirkung in Ferromagnetika. Springer-Verlag, Berlin 1968
73. Kuksin, S. B.: Applications of monotone semigroups in theory of ideal elastoplasticity. Uspekhi Matem. Nauk *37* (1982) 189-190 [Russian]
74. Landau, L. D., Lifshitz, E. M.: Fluid mechanics. Pergamon Press, London 1959
75. Levy, M.: Mémoire sur les équations des corps solides dietiles au-delâ de la limite elastique. J. Math. Pures et Appl. *16* (1871) 369-372
76. Lions, J.-L.: Inéquations variationelles d'évolution. In: Proceedings of the International Congress of Mathematicians, Nice 1970
77. Lipcer, R. S., Shiryaev, A. N.: Statistics of random processes. Vols. I, II, Springer-Verlag, New York 1977 - 1978
78. Lur'e, A. I.: Elasticity theory. Nauka, Moscow 1970 [Russian]
79. Luzin, N. N.: Lectures on analytic sets and their applications. Gostekhizdat, Moscow 1953 [Russian]
80. Lyapin, L. N.: Integral operator in the space of multivalued functions. In: Proc. of Tambovsk. Inst. Chem. Eng., Vol. 4 (1970) 34-37 [Russian]
81. Lyapin, L. N.: On the theory of Aumann-Hukuhara integral. In: Proc. of Tambovsk. Inst. Chem. Eng., Vol. 6 (1971) 3-8 [Russian]
82. Madelung, E.: Über Magnetisierung durch schnellverlaufende Ströme und die Wirkungsweise des Rutherford-Marconischen Magnetdetektors. Ann. der Physik *17* (1905) 861-890
83. Massera, J. L., Schäffer, J. J.: Linear differential equations and function spaces. Academic Press, New York 1966
84. Maxwell, I. G.: On the dynamical theory of gases. Philos. Trans. of London *157* (1867) 49-88
85. Mednikov, I. I.: On the stability of a block-diagram. In: Proc. All--Union Inst. Radiation. Techn., Vol. 16 (1968) 74-82 [Russian]
86. Mechanics in the USSR - 50 Years. L. I. Sedov, Ed., Nauka, Moscow 1968 [Russian]
87. Messarović, M., Takahara, Y.: General systems theory: mathematical foundations. Academic Press, New York 1975
88. Michael, E.: Continuous selections. Ann. Math. *63* (1956) 361-382
89. Mitropol'skii, Yu. A.: The averaging method in nonlinear mechanics. Naukova Dumka, Kiev 1971 [Russian]
90. Moskvitin, V. V.: Plasticity under variable loads. Izd. MGU, Moscow 1968 [Russian]
91. Mosolov, P. P., Myasnikov, V. P.: Variational methods in the theory of a visco-plastic medium. Prikl. Matem. Mekhanika *29* (1965) [Russian]
92. Mosolov, P. P., Myasnikov, V. P.: Mechanics of rigid-plastic media. Nauka, Moscow 1972 [Russian]

93. Neimark, Yu. I.: Method of point mappings in nonlinear oscillations theory. Nauka, Moscow 1972 [Russian]
94. Netushil, A. V.: Nonlinear elements of the stop type. Avtom. Telemekh., No. 7 (1968) 175-179 [Russian]
95. Netushil, A. V.: Self-oscillations in systems with negative hysteresis. In: Proc. 5th international conference on nonlinear oscillations, Vol. 4, Izd. AN USSR, Kiev 1970, 393-396 [Russian]
96. Neumark, S.: Concept of complex stiffness applied to problems of oscillations with viscous and hysteresis damping. H. M. Stat. Soc., London 1962
97. Novikov, P. S.: Sur les fonctions implicites measurables. Fund. Math. (1931) 8-25
98. Novikov, P. S.: On consistency of some hypotheses in descriptive set theory. In: Proc. Math. Inst. AN SSSR, Vol. 38 (1951) 279-316 [Russian]
99. Novozhilov, V. V.: On complex loading and perspectives of phenomenological approach to study of microstrains. Prikl. Matem. Mekhanika *28* (1964) 393-400 [Russian]
100. Pal'mov, V. A.: On some model of a composite medium. Prikl. Matem. Mekhanika *33* (1969) 733-768 [Russian]
101. Pal'mov, V. A.: Oscillations of elasto-plastic bodies. Nauka, Moscow 1976 [Russian]
102. Panskikh, N. P.: On correct identification of simplest hysteresis nonlinearities. In: Qualitative and approximate methods for operator equations. Izd. Yaroslawl'. Univ., Yaroslawl' 1979, 103-110 [Russian]
103. Pisarenko, G. S.: Energy scattering in mechanical oscillatons. Izd. AN USSR, Kiev 1962 [Russian]
104. Pisarenko, G. S.: Oscillations of mechanical systems for imperfectly elastic materials. Izd. AN USSR, Kiev 1970 [Russian]
105. Pokrovskii, A. V.: Nonlocal extendibility of solutions of vibrostable differential equations. Dokl. AN SSSR *208* (1973) 1286-1289 [ZM: 297.34044]
106. Pokrovskii, A. V.: On the theory of hysteresis nonlinearities. Dokl. AN SSSR *210* (1973) 896-900 [ZM: 293.45014]
107. Pokrovskii, A. V.: The limiting norm of a linear operator and its applications. Dokl. AN SSSR *249* (1979) 517-520 [ZM: 462.93086]
108. Pokrovskii, A. V.: On general systems with state space. Avtomatika Telemekh., No. 8 (1980) 179-182 [Russian]
109. Pokrovskii, A. V.: On a certain class of complex systems. In: Abstracts of the 8th All-Union conference on control problems, Vol. 1, Gosplan ESSR, Tallin 1980, 55-58 [Russian]
110. Pokrovskii, A. V.: On dissipativity of a certain class of systems. Avtomatika Telemekh., No. 4 (1981) 70-74 [Russian]
111. Pokrovskii, A. V.: On a certain class of discontinuous systems. Avtomatika Telemekh., No. 7 (1981) 47-51 [Russian]
112. Pontryagin, L. S.: Gewöhnliche Differentialgleichungen. Dt. Verl. d. Wiss., Berlin 1965 [Translation from Russian]
113. Prager, W.: A new method of analysing stresses and strains in work-hardening plastic solids. J. Appl. Mech. *23* (1956) 493-496
114. Prager, W.: On ideal locking materials. Soc. Rheology *1* (1957) 169-175
115. Prager, W., Hodge, F. G.: Theory of ideally plastic bodies. Izd. Inostr. Lit., Moscow 1956 [Russian translation]
116. Preisach, P.: Über die magnetische Nachwirkung. Zeitschrift für Physik *94* (1938) 277-302
117. Prival'skii, V. B.: On Netushil's models. In: Qualitative and approximate methods for operator equations, Izd. Yaroslawl'. Univ., Yaroslawl' 1978, 179-189 [Russian]

118. Pugachev, V. S.: Statistical methods in automatic control. In: Optimal systems, statistical methods, Proc. of the 2nd IFAC Congress, Nauka, Moscow 1965 [Russian]
119. Pugachev, V. S., Kazakov, I. E., Evlanov, L. G.: Foundations of statistical theory of automatic systems. Mashinostroenie, Moscow 1974 [Russian]
120. Reiner, M.: Deformation, strain and flow. An elementary introduction to rheology. H.K. Lewis, London 1960.
121. Riesz, F., Szökefalvi-Nagy, B.: Functional analysis. Ungar, New York 1965
122. Rockafellar, R. T.: Convex analysis. Princeton Univ. Press 1970
123. Saint-Venant, M.: Sur l'etablisement des équations des mouvements interieurs opérés dans les corps ductiles au-delà des limites d'elasticite. C. R. Acad. Sci. Paris *70* (1870) 473-480
124. Saint-Venant, M.: Sur les équations du mouvement interieur du solides ductiles. J. Math. Pures et Appl. *16* (1871) 373-382
125. Sedov, L. I.: Continuum mechanics. Nauka, Moscow 1976 [Russian]
126. Shilov, G. E.: Mathematical analysis. Functions of several real variables. Nauka, Moscow 1972 [Russian]
127. Shragin, I. V.: Superpositional measurability. Dokl. AN SSSR *197* (1971) 295-298 [Russian]
128. Sorokin, E. S.: On the theory of internal friction in oscillating elastic system. Gosstroizdat, Moscow 1960 [Russian]
129. Theory of systems with variable structure. Emel'yanov, S. V., Ed., Nauka, Moscow 1970 [Russian]
130. Thompson, J. H. C. Philos. Trans., London, A (1933) 231-257
131. Tresca, H. C. R. Acad. Sci. Paris *59* (1864); *70* (1870)
132. Truesdell, C.: A first course in rational continuum mechanics. Academic Press, New York 1980
133. Tsypkin, Ya. Z.: Impulse automatic systems. Nauka, Moscow 1974 [Russian]
134. Utkin, V. I.: Creep processes in optimization and control problems. Nauka, Moscow 1981 [Russian]
135. Vasil'ev, N. G., Zverev, V. A.: Electronic modelling of the hysteresis characteristics of ferromagnetic materials. Izv. VUZ, Elektromekhanika, No. 9 (1956) 3-17 [Russian]
136. Venets, V. I.: Differential inclusions in convex problems. Avtomatika Telemekh., No. 9 (1979) 5-14 [Russian]
137. Veretennikov, A. Yu.: On the strong solutions and explicit formulas for solving stochastic differential equations. Matem. Sbornik *111* (1980) 434-452 [Russian]
138. Veretennikov, A. Yu.: On the approximation of ordinary differential equations by stochastic ones. Matem. Zametki *31* (1981) 67-81 [Russian]
139. Viktorovski, E. E.: On some generalization of integral curves for the discontinuous directions field. Matem. Sbornik *34* (1954) 213-248 [Russian]
140. Vladimirov, A. A., Klepcyn, A. F.: On some hysteresis elements. Avtomatika Telemekh., No. 7 (1982) [Russian]
141. Vladimirov, A. A., Klepcyn, A. F., Kozyakin, V. S., Krasnosel'skii, M. A., Lifshitz, E. A., Pokrovskii, A. V.: Vector hysteresis nonlinearities of Mises-Tresca type. Dokl. AN SSSR *257* (1981) 506-509 [Russian]
142. Vogel, T.: Théorie des systémes evolutifs. Gauthier-Villars, Paris 1965
143. Volterra, E.: Vibration of elastic systemes having hereditary characteristics. J. Appl. Mech. *17* (1950) 363-371
144. Volterra, V.: Sur la théorie mathématique des phénomènes héreditaires. J. Math. Pures et Appl. *7* (1928) 249-298

145. Volterra, V.: Sulle equazioni integro-differenziali della elascità nel caso della isotropica. Rend. Accad. Lincei *5* (1909) 577-586
146. Volterra, V.: Les problèmes qui ressortent du concept de fonctions de lignes. Sitzungsber. Berlin Math. Ges. *13* (1914) 130-150
147. Volterra, V.: Theory of functionals and of integral and integrodifferential equations. Dover Publ., New York 1959
148. Vonsovskii, S. V.: Modern theory of magnetism. Gostekhizdat, Moscow 1952 [Russian]
149. Vonsovskii, S. V.: Magnetism. Nauka, Moscow 1971 [Russian]
150. Voronov, A. A.: Foundations of automatic control theory. Energiya, Moscow 1980 [Russian]
151. Vroblevskii, N. A.: Physical foundations of magnetic soundrecording. Energiya, Moscow 1970 [Russian]
152. Wang, C.-C.: Stress relaxation and the principle of fading memory. Arch. Rational Mech. Anal. *18* (1965) 117-126
153. Wang, C.-C., Truesdell C.: Introduction to rational elasticity. Noordhoff, Leyden 1973
154. Ward, I. M., Onat, E. T.: Nonlinear mechanical behaviour of oriented polypropylene. J. Mech. Phys. Solids *11* (1963) 217-223
155. Webb, G. F.: Abstract Volterra integro-differential equations and a class of reaction-diffusion equations. Lecture Notes Math., Vol. 737 Berlin 1979, 295-303
156. Wiener, N.: Extrapolation, interpolation, and smoothing of stationary time series. With engineering applications, Cambridge - New York 1949
157. Williams, G.: On internal interaction and concept of thermal isolation. Arch. Rational Mech. Anal. *34* (1969) 245-259
158. Yakubovich, V. A.: Frequency conditions of the absolute stability of controlled systems with hysteresis nonlinearities. Dokl. AN SSSR *149* (1963) 288-291 [Russian]
159. Yakubovich, V. A.: Methods of the theory of absolute stability. In: Methods of the study of nonlinear systems of automatic control, Nelepin, R. A., Ed., Nauka, Moscow 1975 [Russian]
160. Yankov, V.: On the uniformization of A- and B-sets. Dokl. AN SSSR *30* (1941) 591-592 [Russian]
161. Yankovich, B.: On a possibility of approximating hysteresis loop. In: Proceedings of the 5th International Conference on Nonlinear Oscillations, Vol. 4, Izd. AN USSR, Kiev 1970, 503-511 [Russian]
162. Young, L. C.: Lectures on the calculus of variations and optimal control theory. Saunders, Philadelphia 1969
163. Zaanen, A. C.: Linear analysis. North-Holland, Amsterdam 1953
164. Zabreiko, P. P.; Krasnosel'skii, M. A., Lifshitz, E. A.: An elastic--plastic element oscillator. Dokl. AN SSSR *190* (1970) 217-220 [Russian]
165. Zadeh, L. A., Desoer, C. A.: Linear system theory: the state space approach. McGraw-Hill, New York 1963
166. Zadeh, L. A., Polak, E.: System theory. McGraw-Hill, New York 1969
167. Zapas, L. J.: Viscoelastic behaviour under large deformations. J. Res. Nat. Bur. Stand. *70* (1966) 526-543
168. Zeines, B.: Automatic control systems. J. Wiley, Englewood Cliffs, New York 1972
169. Zener, C.: Internal friction in solids. The Physical Review *52* (1937) 230-235
170. Zunde, P.: Interaction in general systems theory. In: Res. 1971 - 1972, Annu. Prog. Rept. Sch. Inform. and Comput. Sci., Atlanta 1972

Subject index

Bundle of hysterons 322

Carathéodory function 213
Cascade of hysterons 260
Chain 341
Characteristic
-, monotone 226
-, of Madelung's hysteron 282
-, of a play 156
-, of a static element 212
-, of a stop 157
-, of type L 221
-, proper 219
-, strictly convex 164
-, superpositionally measurable 241
Classical solution 122
Closure
-, modulo negligence class 247
-, of an element 244
-, of a function 247
-, of an operator 237
-, of a relay 268
-closure of an operator 312
Compensator 339
Cone
-, completely regular 225
-, minihedral 225
-, non-flattened 224
-, normal 155
-, regular 225
-, reproducing 224
-, semi-regular 225
-, solid 224
-, tangent 155
Continual system 371
Convexification 251
-, of an operator 259
-, of a relay 269
-, of a transducer 252

Defining curve 28
Defining property 349
Demagnetization function 373
Density of frame 90
Deviation of hysterons 82
χ-deviation 235
Distance
-, between hysterons 82
-, left & right 41
χ-distance 235

Embedding of spaces

-, compact 48
-, continuous 48
Equation
-, vibro-correct 95
-, vibro-correct on constant inputs 98
-, vibro-correct in a point 102
-, globally vibro-correct 115
-, monotone 118
-approximation of a function 236

Frame of hysterons
δ-frame 89
Frobenius condition 198
Function
-, asymptotically almost periodic 20
-, asymptotically periodic 20
-, Borel measurable 215
-, Borel (mod 0) 241
-, ε-approximation of 236
-, equivalent 358
-, piecewise continuous 233
-, (*)-proper 127
-, superpositionally measurable (SM-function) 214

Hausdorff deviation 234
Hausdorff distance 234
Hausdorff limit 237
Hysteresis loop 24
-, canonical 31
Hysteron 24
-, canonical 31
-, infinitesimal 145
-, monotone 37
-, nondegenerate 39
-, normal 79
-, ordinary 39
-, smooth 138
-, variable with drift 147

Input 1
-, absorbing 187
-, admissible 1
-, with bounded variation 284
Intensity 295
-, asymptotic 296
Ishlinskii's transducer 345
Ito L-integral 315

Loading function 346

Madelung's hysteron 281
-, prehysteron 279
Measurable selection 240
Michael's conditions 235
Modulus of continuity 13
Modulus of uniform convexity 166
Monster 215

Non-flattening 224
Norm, semi-monotone 235

Operator
-, closed 234
-, defining 165
-, lower semicontinuous 234
-, upper semicontinuous 234
-, continuous 234
-, monotone 238
-, completely monotone 238
-, sequential closure of 247
-, weak closure of 247
Ordering, linear 225
-, partial 223

Output 1

Parallel connection 321
Partitioning, mesh size 142
Play
-, generalized 8
-, generalized monocyclic 19
-, generalized monotone 17
-, multidimensional 156
-, ordinary 152
-, variable 145
Point
-, L-regular 50
-, lower singular 66
-, upper singular 65
Prehysteron 62
-, α-close 78
-, locally controllable 80

Regular pair of curves 8
Relay 262
-, threshold value of 262
-, ideal 262
-, non-ideal 264
Restriction of a transducer
-, controllable 323
-, completely controllable 323

Selector of a multi-valued function 233
Sequence
-, extremal 182
-, Hausdorff-convergent 237
-, κ-proper 182
-, weak limit of 249
Semigroup identity 4
Set, infimum of 225
-, invariant 42
-, supremum of 225
Shift operator 105
Solution, of type L 312
-, Ito 315
Spin 20
State, of a transducer 2
-, normal 351
Static element 212
-, L-regular 312
-, stationary 212
-, with multi-valued characteristic 233
Stop 23
-, multidimensional 156
-, variable 156
Stratonovich L-integral 316
Superposition, multi-valued 342

Transducer
-, autonomous 4
-, controllable 4
-, convergent 160
-, ε-convergent 392
-, deterministic 2
-, linear 2
-, monocyclic 19
-, monotone 17
-, physically realizable 3
-, short-memory 2
-, static 4
-, strongly convergent 167
-, vibro-correct 129
-, weakly convergent 208
-, weakly correct 62

Unloading function 346

Variation of function 312
Vibro-correctness 6
-, modulus of 162

-, necessary condition 98
-, sufficient condition 204
Vibro-solution 113
-, of constrained equation 131
-, of type κ 305
Volterra operator 3

Weak closure of an operator 250